2011 IEEE Applied Power Electronics Colloquium

(IAPEC 2011)

Johor Bahru, Malaysia
18 – 19 April 2011

IEEE Catalog Number:	CFP1177N-PRT
ISBN:	978-1-4577-0007-1

**Copyright © 2011 by the Institute of Electrical and Electronic Engineers, Inc
All Rights Reserved**

Copyright and Reprint Permissions: Abstracting is permitted with credit to the source. Libraries are permitted to photocopy beyond the limit of U.S. copyright law for private use of patrons those articles in this volume that carry a code at the bottom of the first page, provided the per-copy fee indicated in the code is paid through Copyright Clearance Center, 222 Rosewood Drive, Danvers, MA 01923.

For other copying, reprint or republication permission, write to IEEE Copyrights Manager, IEEE Service Center, 445 Hoes Lane, Piscataway, NJ 08854. All rights reserved.

***This publication is a representation of what appears in the IEEE Digital Libraries. Some format issues inherent in the e-media version may also appear in this print version.

IEEE Catalog Number: CFP1177N-PRT
ISBN 13:   978-1-4577-0007-1

Additional Copies of This Publication Are Available From:

Curran Associates, Inc
57 Morehouse Lane
Red Hook, NY 12571 USA
Phone: (845) 758-0400
Fax: (845) 758-2633
E-mail: curran@proceedings.com
Web: www.proceedings.com

2011 IEEE Applied Power Electronics Colloquium (IAPEC 2011)

Johor Bahru, Malaysia
18-19 April 2011

IEEE Catalog Number: CFP1177N-POD
ISBN: 978-1-45770-007-1

TABLE OF CONTENTS

A1. PHOTOVOLTAIC APPLICATIONS

Power Electronics for Photovoltaic Energy System of an Oceanographic Buoy 1
E. Behrouzian, A. Tabesh, F. Bahrainian, A. Zamani

Maximum Power Point Tracking for PV System under Partial Shading Condition via Particle Swarm Optimization 5
K. Ishaque, Z. Salam, H. Taheri, A. Shamsudin

Parameter Extraction of Photovoltaic Cell Using Differential Evolution Method 10
K. Ishaque, Z. Salam, H. Taheri, A. Shamsudin

Power Quality Improvement for Distributed Generation Employing Photovoltaic-Inverter 16
M. Hamide, A. Jusoh, M. Anwari

A Study of Maximum Power Point Tracking Algorithms for Stand-alone Photovoltaic Systems 22
M. Ngan, C. Tan

A2. POWER ELECTRONIC CONVERTERS

Robust Current-Mode DC Drive 28
A. Awan, M. Malik

Computer Simulation of Single-phase Control Rectifier using Single-phase Matrix Converter with Reduced Switch Count 34
R. Baharom, A. Idris, N. Hamzah, M. Hamzah

6-pulse Controlled Rectifier Synchronisation Method 40
B. Das, N. Watson, Y. Liu

A New Single-Phase to Three-Phase Converter using Quasi Z-Source Network 46
F. Khosravi, N. Azli, A. Kaykhosravi

B1. SWITCHING TECHNIQUES 1

Space Vector based Spread Spectrum Modulation Scheme for Three-Level Inverters 51
B. Jacob, M. Baiju

Optimization of Operating Parameters in a Unipolar PWM Inverter 57
T. Taufik, M. McCarty, M. Anwari, A. Prabuwono

A New Multi Carrier Based PWM for Multilevel Converter 63
B. Mohammed, K. Rao

B2. CONTROL 1

Parallel Configuration in Energy Management Control for the Fuel cell-Battery-Ultracapacitor Hybrid Vehicles 69
J. Wong, N. Idris, M. Anwari

Performance Comparison of SVPWM and Hysteresis Current Control for Dual Motor Drives 75
J. Lazi, Z.Ibrahim, M. Sulaiman, I. Jamaludin, M. Lada

A Bridgeless Cuk PFC Converter 81
M. Sahid, A. Yatim, N. Muhammad

C1. MODELING AND CONTROL

Development of a Doctor Following Mobile Robot with Mono-vision Based Marker Detection 86
M. Bakar, R. Nagarajan, A. Saad

Performance Improvement of Improved Practical Control Method for Two-Mass PTP Positioning Systems in the Presence of Actuator Saturation 92
M. Yakub, R. Akmeliawati

Automated Test Set-up for Reverse Recovery Characterization of Ultrafast Diodes 98
J. Stahl, D. Kuebrich, T. Duerbaum

Modeling of Lithium Ion Battery with Nonlinear Transfer Resistance 104
L. Yao, J. Aziz

Modeling and Simulation of Adaptive Neuro-Fuzzy Controller for Chopper-Fed DC Motor Drive 110
Y. Al-Mashhadany

C2. MACHINES AND DRIVES

Direct Torque Control of Induction Machines Utilizing 3-level Cascaded H-Bridge Multilevel Inverter and Fuzzy Logic 116
A. Mortezaei, N. Azli, N. Idris, S. Mahmoodi, N. Nordin

FPGA Based High Precision Torque and Flux Estimator of Direct Torque Control Drives 122
T. Sutikno, N. Idris, A. Jidin, M. Daud

Inter-Turn Stator Winding Fault Diagnosis and Determination of Fault Percent in PMSM 128
M. Nejad, M. Taghipour

High Reliability for Electric Machines Driving Critical Loads: A Review 132
J. Nandan, R. Gobbi

A Wavelet-Based Technique for Discrimination of Inrush Currents from Faults in Transformers Coupled with Finite Element Method 138
M. Jamali, M. Mirzaie, S. Gholamian, S. Cherati

D1. ACTIVE FILTERS AND HARMONICS

A New Schematic for Hybrid Active Power Filter Controller 143
E. Samadaei, S. Lesan, S. Cherati

Harmonic Study for MDF Industries: A Case Study 149
M. Yazdani-Asramil, S. Sadati, E. Samadaei

A Novel Hysteresis Bandwidth (NHB) Calculation To Fix the Switching Frequency Employed In Active Power Filter 155
H. Vahedi, A. Sheikholeslami, M. Bina

Simulation Single Phase Shunt Active Filter Based on p-q Technique using MATLAB/Simulink Development Tools Environment 159
M. Lada, O. Mohindo, A. Khamis, J. Lazi, I. Jamaludin

D2. CONTROL 2

Designing Dynamic Controller and Passive Filter for a Grid Connected Micro-turbine 165
R. Rahmani, M. Tayyebi, M. Majid, M. Hassan, H. Rahman

Impact of Double-Loop Controller on Grid Connected Inverter Input Admittance using Virtual Resistor 170
A. Rizqiawan, G. Fujita, T. Funabashi, M. Nomura

Design And Application Of A Novel Current Mode Controller On A Multilevel STATCOM 176
E. Najafi, A. Yatim

Design of a Current Mode PI Controller for a Single-phase PWM Inverter 180
S. Cherati, N. Azli, S. Ayob, A. Mortezaei

E1. SWITCHING TECHNIQUES 2

Comparison of Adaptive and Fixed-Band Hysteresis Current Control Considering High Frequency Harmonics 185
H. Vahedi, Y. Kukandeh, M. Kashani, A. Dankoob, A. Sheikholeslami

Pulse Density Modulated Soft-Switching Single-Phase Cycloconverter 189
T. Taufik, J. Adamson, A. Prabuwono

A Novel Soft Switching Bidirectional Coupled Inductor Buck-Boost Converter for Battery Discharging-Charging 195
A. Mirzaei, A. Jusoh, Z. Salam, E. Adib, H. Farzanehfard

Author Index

Power Electronics for Photovoltaic Energy System of an Oceanographic Buoy

Ehsan Behrouzian[1], Ahmadreza Tabesh[1], Farzad Bahrainian[2], Ahmadreza Zamani[2]

[1]Department of Electrical and Computer Engineering, [2]Subsea R&D Institute,
Isfahan University of Technology, Isfahan, Iran
Corresponding Email: e.behrouzian @ ec.iut.ac.ir

Abstract- **This paper reports on design of power electronics for photovoltaic (PV) energy system of an offshore remote sensing apparatus. Challenges in design of a PV energy system for a marine application are investigated and the design limitations compared to inland PV system are discussed. The designed system includes PV cells as the main source of energy, electric storage (battery), maximum power point tracking (MPPT) and protection circuitries. An MPPT algorithm based on measuring the slope of the PV power-voltage curves is presented which can be implemented with simple analog electronic circuits. The MPPT circuit uses Sepic converter as a core and it also includes a protection unit for maintaining the battery voltage in a safe range. The performance of the proposed MPPT algorithm in presence of measurement noises is verified using a circuit simulation software tool (PSCAD). Simulation results verify that the algorithm appropriately regulates the voltage of PV cells at MPP and it is robust against measurement noises for a signal-to-noise ratio above -2db.**

Keywords- **Stand-alone PV Systems, Tracking, Remote Sensing, Oceanographic Buoy, Sepic-Converter.**

I. INTRODUCTION

Oceanographic buoys are floating instruments that measure marine/metrological data and transmit them (via a wireless unit) to onshore stations for further analysis. Photovoltaic (PV) solar energy is the only reliable source of energy for wireless circuitry of a buoy [1, 2]. Design of PV energy systems for buoys compared to onshore systems is challenging due to a harsh marine environment. The limitations/requirements of the stand-alone PV energy system of a buoy compared with conventional inland PV systems are: 1) a buoy is often located far away of a shore and its cost of maintenance is huge, thus its PV energy system must be autonomous and robust, with additional electrical protections; 2) sea waves continuously fluctuate the structure of a buoy, therefore the topology of the PV power circuit of a buoy must be simple with minimum components to improve reliability; 3) the available surface area on a buoy to install PV panels is limited and panels are fixed which need efficient circuitry with a maximum power point tracking (MPPT) strategy. This paper deals with power electronics and system design aspects of a PV energy system for powering of *"Mowj-Negar"*, the oceanographic buoy of Isfahan University of Technology (Fig.1). Several analog and digital MPPT techniques have been proposed for inland stand-alone PV energy systems [3-5] which have been reviewed to select a suitable topology for a buoy. Then, the topology has been improved herein to address the specific requirements of the buoy. The designed PV energy system includes: 1) a power electronic circuit using Sepic

Fig. 1 Mowj-*Negar*, the oceanographic buoy of Isfahan University of Technology

dc/dc switching converter as a core; 2) a simple analog controller circuit for MPPT; and 3) the required protections for electrical units of the buoy. The proposed controller can be implemented with discrete analog electronic components which improves reliability of the circuit due to its simple structure as compared with DSP-based digital controllers. To verify the performance of the designed system, the electronic circuits and its controller were simulated using the electrical model of buoy's instruments and solar panels. The simulation investigates the MPPT capability of the designed circuit and robustness of the circuit with respect to measurement noises.

II. THE BUOY PV ENERGY SYSTEM

Fig. 2 shows the schematic diagram of the PV energy system of buoy. The input stage includes PV cells to capture energy of the sunlight irradiation. Next stage is a dc/dc converter which measures PV cells output voltages/currents and set the PV cells at their maximum power point. Instruments of a buoy periodically on/off during a day, however, we model the average power consumption of all instrumentations with a fixed load which is supplied with the buoy energy system.

978-1-4577-0007-1/11 $26.00 © 2011 IEEE

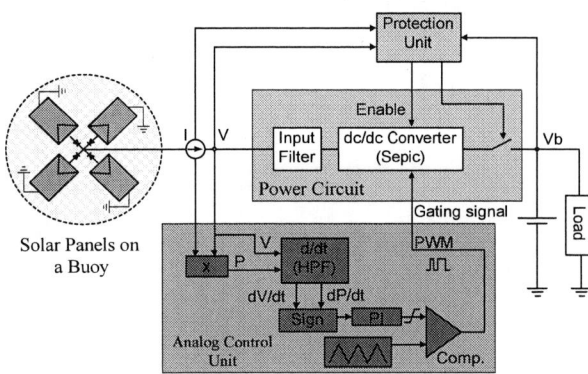

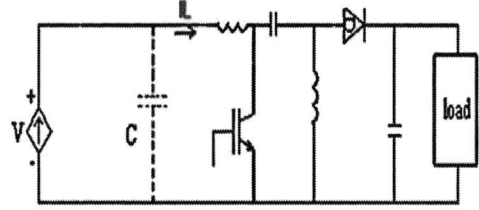

Fig. 2 The buoy PV energy system.

The dc/dc converter charges the battery during a sunny day and supplies power to the load. The battery must be protected against over charge. Thus, a protection unit is used to continuously monitor the battery voltage and disconnect it from converter at an over voltage condition.

The analog control unit in Fig. 2 consists of a high pass filter, a PI controller, a saw tooth signal generator, and a comparator. The detail function of this unit will be elaborated in the following sections.

III. ENERGY SYSTEM DESIGN FOR BUOY

The heart of MPPT circuit is a switch-mode dc/dc converter. Dc/dc convertors are widely used to convert an unregulated dc input into a controlled dc output at a desired voltage level. However, in a PV energy system, MPPT uses the converter to regulating the input voltage at the PV maximum power point. Input voltage regulation can be achieved via appropriate change in the duty cycle of converter.

A. Converter Selection and Design

A buck converter is a simple and efficient step-down dc/dc converter which is widely used in different applications. To implement the MPPT algorithm in the PV energy system of a buoy, we do need to measure terminal current and voltage of the PV cells. This necessitates using an additional low pass filters to mitigate the current pulsating and using high-side driver for the buck converter switch. Sepic converter is an alternative dc/dc converter for the PV energy system which has the buck-boost feature (Fig. 3). A Sepic converter has the merits of non-inverting polarity and easy-to drive switch. Furthermore, the input inductor of a Sepic converter will smoothen the measured current, hence, extra filter is not required. Integral characteristics of Sepic converter make it suitable for a low-power PV energy system and it may also be preferred for battery charging systems because of the blocking diode at the output. This diode prevents discharge of battery in PV cells during a night. Other advantage of Sepic over a simple buck converter is the capacitive isolation which protects the cells and battery against a switch failure.

Figure 3shows the schematic circuit of a Sepic converter connected to a PV array to provide a stepped-up-down voltage to the load. The input capacitor models the parasitic

Fig.3 PV array connected to a Sepic circuit.

capacitance of the array and any intentional capacitive filter at the input of the converter. The goal is to force I_L to track I_{Lmax}, the corresponding current to the MPP, independent of variations in the cells temperature and sunlight intensity.

B. The MPPT Control Strategy

The power-voltage characteristic of a PV cells yields:

$$\frac{dp}{dt} \times \frac{dv}{dt} > 0 \Rightarrow V < V^*$$

$$\frac{dp}{dt} \times \frac{dv}{dt} < 0 \Rightarrow V > V^* \qquad (1)$$

where V^* is the optimum voltage point and V is the PV terminal voltage. The duty cycle of converter (d) should be controlled such that eventually V approaches V^*. A method to control d is:

$$d = k \int \frac{dp}{dt} \frac{dv}{dt} dt \qquad (2)$$

where k is a constant gain. Based on (2) the voltage and duty cycle increase (decreases), respectively. The output voltage of the converter is fixed and equals the battery voltage and the input/output voltage relation is given by:

$$V_{in} = V_o \frac{1-d}{d} \qquad (3)$$

Since the output voltage V_o is constant at battery voltage level, any increase (decrease) in d leads to decrease (increase) in the input voltage. Thus, based on (1), we obtain the following control law for MPPT:

$$\frac{dp}{dt} \times \frac{dv}{dt} > 0 \Rightarrow V < V^* \Rightarrow d \text{ should decrease}$$

$$\frac{dp}{dt} \times \frac{dv}{dt} < 0 \Rightarrow V > V^* \Rightarrow d \text{ should increase} \qquad (4)$$

Eq. (4) shows that a negative k in (2) will lead to appropriate change in d such that V approaches V^*.

The control law can also be expressed in term of current by multiplying (4) with dp/di as:

$$d = k \int \frac{dp}{dt} \frac{di}{dt} dt \qquad (5)$$

Based on (5), MMP can be achieved via control of current, instead of voltage in (2), by changing d. However, in (5) k must be positive, since the current and voltage are inversely related in a PV array.

Implementations of control laws (2) or (5) have some practical limitations due to using derivatives of electrical quantities. To mitigate this problem, we suggest an alternative approach which uses sign of derivatives as discussed in the following section.

978-1-4577-0007-1/11 $26.00 © 2011 IEEE

C. Alternative Method

Based on (1) and discussion in [6], the signs of derivatives in (2) can be used instead of the derivatives. Thus, the control law can be modified as:

$$d = k \int sign\left(\frac{dp}{dt}\right) \times sign\left(\frac{dv}{dt}\right) dt \qquad (6)$$

Using the sign function limits the high amplitude of a derivative quantity due to a noise, which is a significant improvement to (2). The sign function can be readily realized using a simple logic gate, a saturated op-amp comparator, or it can be implemented with inexpensive synchronous demodulator integrated circuits (ICs).

The sign functions in the integrand of (6) are advantageous from a noise standpoint, though the integrand never asymptotically approaches zero. Another improvement to this method is to bind the integrand, such that law (2) is preserved if the derivatives are within a pre-specified small range.

D. Battery Protection

PV array size is designed to charge the battery and powered the output load during a day. When the battery is fully charged, it should be protected against overcharging to avoid damages to the battery. Several practical protection circuitries have been presented to control charging/discharging of a battery [7]. To design a protection circuit for buoy, we model the battery with a 1F super capacitor. Then, a simple circuit for protection of battery is shown in Fig. 4, using a Zener diode in series with an optocoupler. In this circuit, as soon as the voltage of battery reaches to a pre-specified upper level, the optocoupler will be activated and it can provide a control signal, for example to turn off the converter or disconnecting the battery from the cells. An improved charge controller circuit can be designed to use an output command based on a hysteresis loop which is adjusted corresponding to upper and lower limits of the battery voltage levels.

E. Differentiator Circuit

The basic differentiator circuit with single R and C (Fig. 5) is not widely used in practice because of sensitivity to a sensor measurement noise. Thus, we used the improved circuit of Fig. 6, which acts as a differentiator at low frequencies and limits the amplifying gain at high frequencies. Such a circuit provides a better noise rejection performance.

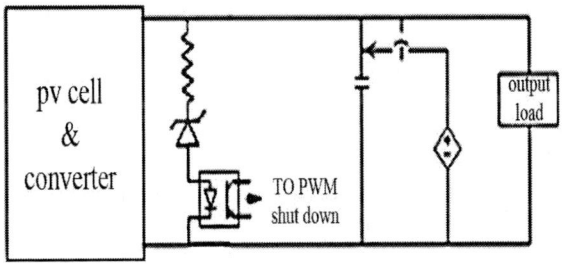

Fig. 4 Schematic of a battery protection circuitry

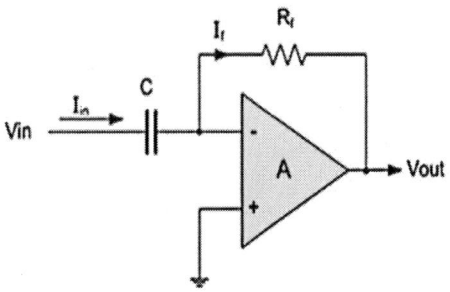

Fig.5 Basic single resistor and single capacitor differentiator circuit.

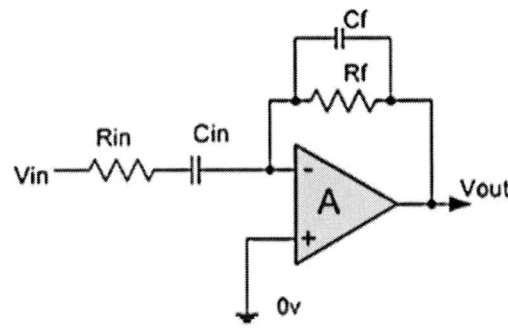

Fig.6 Improved differentiator amplifier circuit.

IV. DESIGN VERIFICATION

To verify the validity and performance of the suggest algorithm for MPPT in buoy PV energy system, the PV cell model, dc/dc converter, control loop based on (6), and the battery protection unit were simulated using PSCAD/EMTDC software tool as shown in Figs. 8 to 10. An analog multiplier was used for implementing the integrand of (6). To show robustness of the circuit against noise, a noise generator was used for adding noise with uniform distribution to the measured signals. The protection unit measured the output voltage and when the voltage exceeds its upper limit, the unit disconnects the PV cells path that charges the battery.

Fig. 7(a) shows the test signal (current) corresponding to a variation in sunlight radiation in the model of solar panels. Fig. 7(b) shows the capability of the control circuit to track the optimum operating point when the current increases as shown in Fig. 7(a). To evaluate the robustness of controller, a noise signal was added to the measured signals. The simulation results show that a sensor with signal-to-noise ratio (SNR) above -2 dB has no significant effect on the performance of the controller. However, SNR less than -2dB impacts on controller at low power operating points as shown in Fig.7(c).

Based on the results shown on Fig. 7, we conclude that the designed circuitry properly addresses the required MPPT specifications for a buoy with adequate robustness against noises.

V. CONCLUSION

Conceptual design of building blocks of a PV energy system for a buoy has been presented. It has been discussed that for offshore remote sensing applications, simplicity and robustness of the circuits are the key features of a successful design. Based on this fact, a simple algorithm for maximum power point tracking is presented and its performance is verified based on simulation. Simulation results show that the proposed algorithm can successfully track the maximum power point in presence of measurement noises.

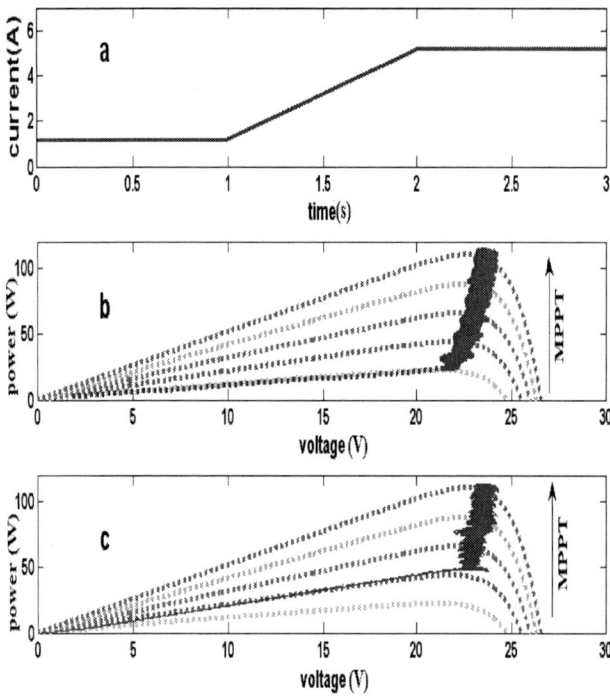

Fig. 7 Buoy's PV energy system simulation results: (a) variation of PV cell's current corresponding to sun light radiations; (b) MPPT performance with SNR>-2dB (c) Effects of noise with SNR<-2dB on MPPT capability.

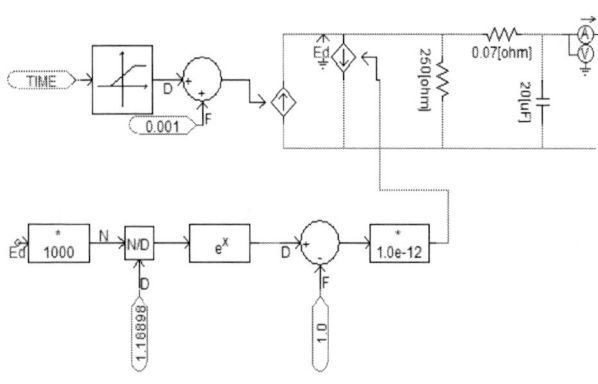

Fig. 8 Model of a PV cell with variable input radiation.

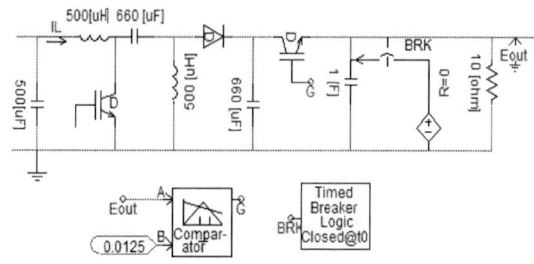

Fig.9 Schematic of electronics including Sepic converter, battery and load model, and the battery protection unit.

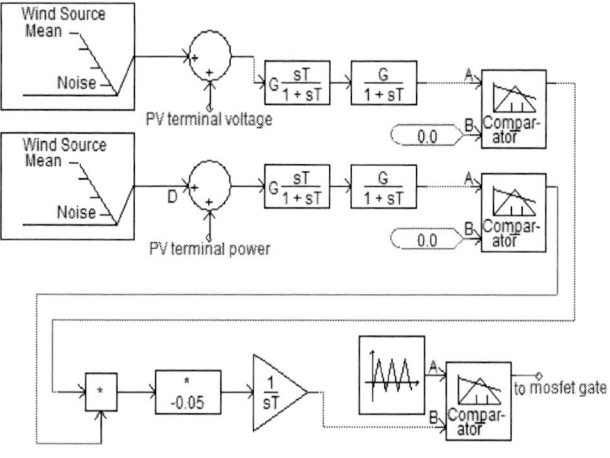

Fig.10 Analog control loop to produce appropriate duty cycle for Sepic converter.

REFERENCES

[1] C.Boccaletti, G.Fabbri, E.Santini," Innovative Solutions for standalone system powering" *29th International Telecommunications Energy Conference (INTELEC '07), 2007, pp. 294-301*

[2] B.Kroposki, R.De Blasio, "Technologies for the new millennium: photovoltaics as a distributed resource" *IEEE Power Engineering Society Summer Meeting, vol. 3, pp. 1798 – 1801, 16-20 July 2000.*

[3] J.A. Jiang, T.L. Huang, Y. T. Hsiao and C.H. Chen, "Maximum power tracking for photovoltaic power systems,"*Tamkang Journal of Science and engineering Vol. 8, No 2, pp. 147_153, 2005.*

[4] J. Seguel, S.Seleme. Jr, P.Garcia, L.Morais, P.Cortizo, M. Mendes, "Comparison of MPPT Approaches in Autonomous Photovoltaic Energy Supply System Using DSP" *Industrial Technology (ICIT), IEEE International Conference. pp.1149-1154, May ,2010.*

[5] T. Esram, and P.L. Chapman, "Comparison of photovoltaic array maximum power point tracking techniques," *IEEE Trans. on Energy Conversion, Vol. 22, No. 2, June 2007.*

[6] T. Esram, J.W. Kimball, P.T. Krein, P.L. Chapman, and Pallab Midya,"Dynamic maximum power point tracking of photovoltaic arrays using ripple correlation control,"*IEEE Trans. on Power Electronics, Vol. 21, No. 5, Sept. 2006.*

[7] P.Krein, R.Turnbull. R.Reppa, J.Kimball, "Dynamic Maximum Power Point Tracker for Photovoltaic Applications" *Power Electronics Specialists Conference, PESC '96 Record. 27th Annual IEEE., pp.1710-1716 , jun, 1996.*

Maximum Power Point Tracking for PV System under Partial Shading Condition via Particle Swarm Optimization

Kashif Ishaque, Zainal Salam, Hamed Taheri and Amir Shamsudin
Faculty of Electrical Engineering, Universiti Teknologi Malaysia, UTM 81310, Skudai,
Johor Bahru, Malaysia.
kashif@fkegraduate.utm.my, zainals@fke.utm.my

Abstract— **Performance of Photovoltaic (PV) system is greatly dependent on the solar irradiation and operating temperature. Due to partial shading condition, the characteristics of a PV system considerably change and often exhibit several local maxima with one global maxima. Conventional Maximum Power Point Tracking (MPPT) techniques can easily be trapped at local maxima under partial shading. This significantly reduced the energy yield of the PV systems. In order to solve this problem, this paper proposes a Maximum Power Point tracking algorithm based on particle swarm optimization (PSO) that is capable of tracking global MPP under partial shaded conditions. The performance of proposed algorithm is evaluated by means of simulation in MATLAB Simulink. The proposed algorithm is applied to a grid connected PV system, in which a Boost (step up) DC-DC converter satisfactorily tracks the global peak.**

Index Terms-- **differential evolution; genetic algorithm; parameter extraction; photovoltaic energy Introduction.**

I. INTRODUCTION

Large and small scale PV power systems have been commercialized in many countries due to its potential long term benefits, generous fed-in tariff schemes and other attractive initiatives provided by various governments to promote sustainable green energy. PV can be used in a wide range of applications from power supplies for satellite communications to large solar power stations feeding electricity into the grid [1]. The grid connected PV power system has a very large commercial potential. Despite its rapid growth, there remain several challenges that hinder the widespread use of PV power systems. A major problem is in PV system is the mismatching effects of the PV output due to broken parts or partial shading of the modules. The latter is of more critical, may be the result of sudden cloud changes in the sky, trees, poles, obstruction of buildings etc. An efficient PV system should be operated optimally in all conditions including during partial shading.

Generally, a grid connected PV system is operated in conjunction with a dc–dc power converter in order to track the instantaneous Maximum Power Point (MPP) of the PV source. Many MPP tracking techniques have been proposed in recent years. A well known Perturbation and Observation (P&O) works satisfactorily when the irradiance fluctuates very slowly [2]. However it often fails to track global MPP when irradiance changed suddenly. Another popular approach is the Incremental Conductance [3], offers better tracking performance but oscillation around the MPP may still occur.

Recently artificial intelligence methods which include Fuzzy logic (FL) [4] and Neural Network (NN) [5] have been applied to track the MPP. Although, better results can be sought but several fuzzy rules in FL and vast amount of training data in NN limit their utilization in the PV system.

Most of the MPP control algorithms mentioned above operate very satisfactorily under uniform irradiance conditions, in which only a single MPP is to be detected. If multiple MPPs exist due to the partial shading, the usefulness of the conventional MPPT algorithms diminishes rapidly. Since the MPP controller is not able to recognize the correct MPP (i.e. it detects the local MPP instead of the global MPP), efficiency of the PV system reduced significantly [6]. As a result, significant research has been carried out to reduce the effect of partial shading by improving the MPP capability of the controller [7].

In view of the excellent performance of multi-peak function optimization and global search of the evolutionary algorithm (EA), an EA technique known as particle swarm optimization (PSO) algorithm is applied to a grid connected PV system. The PSO is employed in order to track the MPP of the PV source. The remainder of the paper is organized as follows: the next section discusses about modeling of photovoltaic system under study including PV array model. This is followed by introducing the proposed particle swarm optimization (PSO) algorithm and then the simulation results are presented as well as discussed. Finally, the conclusions are made in the last section.

II. MODELLING OF PV MODULE

Among the modeling methods of PV module, the two diode model is known to be an accurate model as depicted in Fig. 1 [8]. The output current of the module can be described as:

$$I = I_{PV} - I_{d1} - I_{d2} - (\frac{V + IR_s}{R_p}) \qquad (1)$$

where

$$I_{d1} = I_{01}[\exp(\frac{V + IRs}{a_1 V_{T1}}) - 1] \qquad (2)$$

and

$$I_{d2} = I_{02}[\exp(\frac{V + IR_s}{a_2 V_{T2}}) - 1] \qquad (3)$$

Where I_{PV} is the current generated by the incidence of light; I_{o1} and I_{o2} are the reverse saturation currents of diode 1 and diode 2, respectively. The I_{o2} term is introduced to compensate for the recombination loss in the depletion region as described in [9]. Other variables are defined as follows: V_{T1} and V_{T2} (both equal to $N_s kT/q$) are the thermal voltages of the PV module having N_s cells connected in series, q is the electron charge ($1.60217646 \times 10^{-19}$ C), k is the Boltzmann constant ($1.3806503 \times 10^{-23}$ J/K) and T is the temperature of the p-n junction in Kelvin. Variables a_1 and a_2 represent the diode ideality constants; a_1 and a_2 represent the diffusion and recombination current component, respectively.

Although greater accuracy can be achieved using this model than the single diode model [9], it requires the computation of seven parameters, namely I_{PV}, I_{o1}, I_{o2}, R_p, R_s, a_1 and a_2.

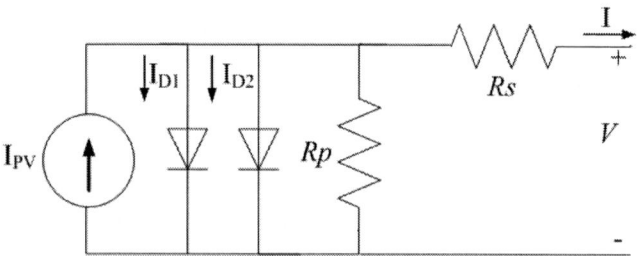

Fig. 1. Two diode model of PV cell

Recently, a fast and simple two diode model is proposed in [10]. The simplified current equation of [10] is given as:

$$I = I_{PV} - I_0\left(I_p + 2\right) - \left(\frac{V + IR_s}{R_p}\right) \qquad (4)$$

Where

$$I_p = \exp\left(\frac{V + IR_S}{V_T}\right) + \exp\left(\frac{V + IR_S}{(p-1)V_T}\right) \quad \text{and} \quad p = 1 + a_2$$

Fig. 2 shows the I-V curves for a commercial PV module Kyocera KC200GT [11], for different levels of irradiation (per unit $\lambda=1$ equivalent to 1000W/m²). Table I shows the parameter of the PV module. Fig.3 shows the I-V curves for the temperature variations for the same PV module.

TABLE III
PARAMETERS OF THE KC200GT PV MODULE AT 25 °C,
A.M1.5, 1000 W/m2

Peak Power(W), P_{mpp}	200
Peak power voltage(V), V_{mpp}	26.3
Peak power current(A), I_{mpp}	7.61
Open circuit voltage(V), V_{oc}	32.9
Short circuit current(A), I_{sc}	8.21
Temperature coefficient of current (mA/°C), K_i	0.003
Temperature coefficient of voltage (mV/°C), K_V	-0.123
Number of series cells, N_s	54

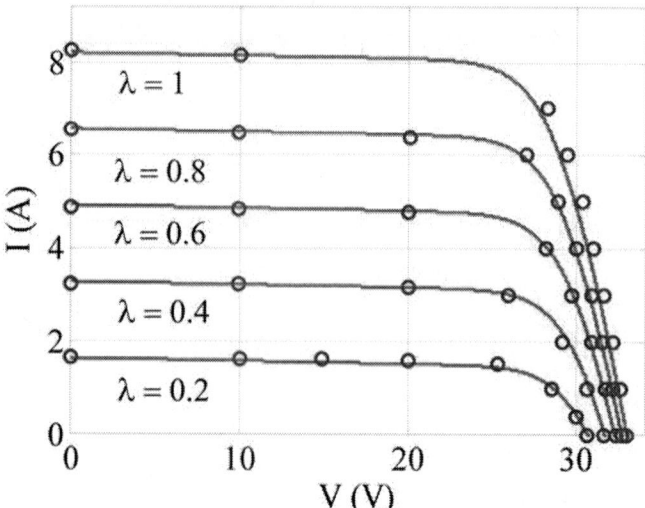

Fig. 2 I-V curves of the simplified two-diode model (solid) of the KC200GT PV module (circular points) for several Irradiation levels.

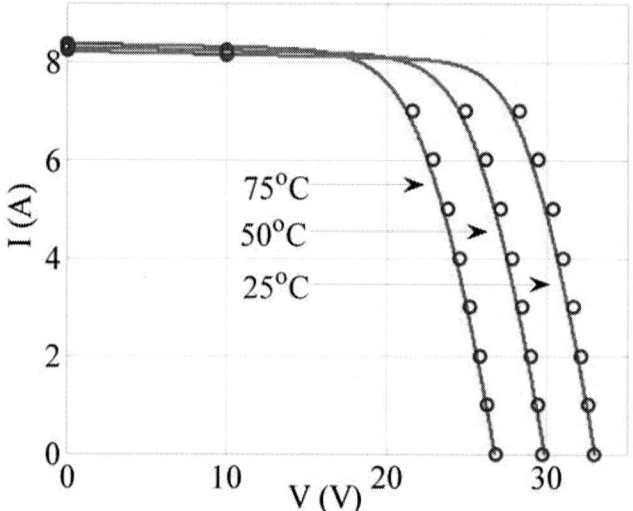

Fig. 3 I-V curves of the simplified two-diode model (solid) of the KC200GT PV module (circular points) for several temperature levels.

III. PV ARRAY SIMULATION

In a typical installation of a large PV power generation system, the modules are configured in a series parallel structure (i.e. $N_{ss} \times N_{pp}$ modules), as depicted in Fig. 4. To handle such cases, the output current equation given in (1) has to be modified as follows:

$$I = N_{pp}\left\{I_{PV} - I_o\left(I_p + 2\right)\right\} - \left(\frac{V + IR_s\Gamma}{R_p\Gamma}\right) \qquad (5)$$

where

$$I_p = \exp\left(\frac{V + IR_s\Gamma}{V_T N_{ss}}\right) + \exp\left[\frac{V + IR_s\Gamma}{(P-1)V_T N_{ss}}\right] \qquad (6)$$

$$\Gamma = \frac{N_{ss}}{N_{pp}} \qquad (7)$$

where I_{PV}, I_0, R_p, R_s, p are the parameters of the individual modules.

Fig. 5 shows the P–V curve for the 30×10 array configuration. The same KC200GT PV module is used here.

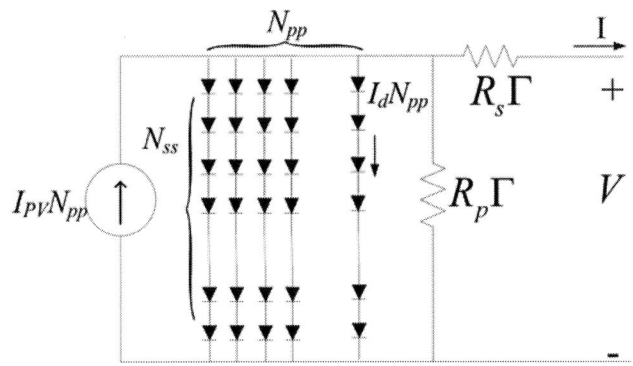

Fig. 4. 30×10 Series parallel combination of PV array.

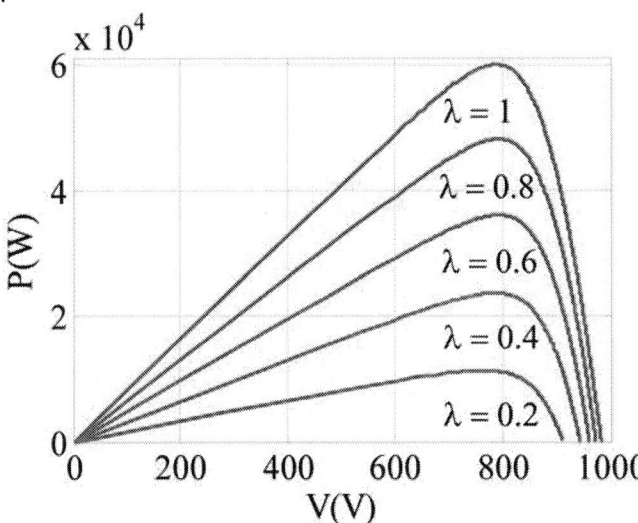

Fig. 5 P-V curves for KC200GT in 30×10 configuration for different irradiation levels.

As discussed earlier, partial shading condition significantly changes the characteristics of a PV array. Fig. 6 shows a typical partial shading condition. In this example, three shading patterns, i.e. λ=1, λ=0.75 and λ=0.5 are applied to the group of modules A, B, and C. Fig. 7 shows the resulting P-V curves for the above shading patterns.

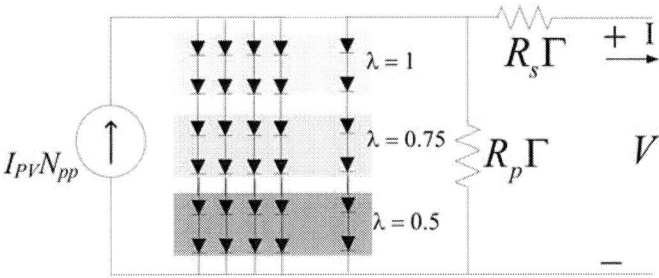

Fig. 6 A PV array illustration for the partial shading condition (λ=1000W/m^2).

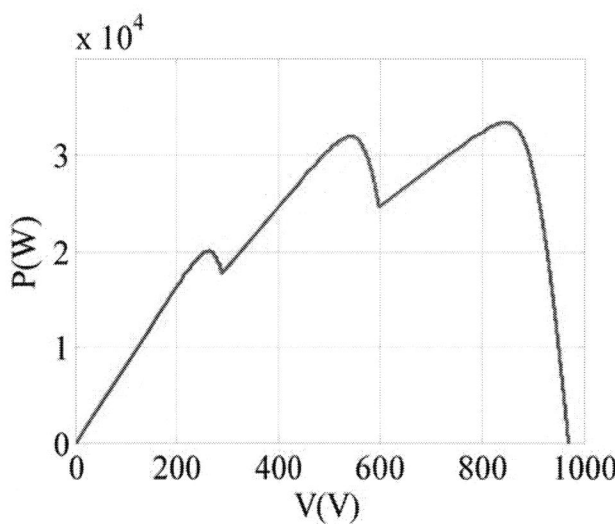

Fig. 7 P–V characteristics for the shading pattern of figure 6.

IV. PSO METHOD

Particle swarm optimization (PSO) is a stochastic, population-based search method, modeled after the behavior of bird flocks [11]. A PSO algorithm maintains a swarm of individuals (called particles), where each particle represents a candidate solution. Particles follow a simple behavior: emulate the success of neighboring particles, and its own successes achieved. The position of a particle is therefore influenced by the best particle in a neighborhood, as well as the best solution found by the particle. Particle position, x_i, are adjusted using

$$x_i(j+1) = x_i(j) + v_i(j+1) \qquad (8)$$

Where the velocity component, v_i, represents the step size. The velocity is calculated by

$$v_i(j+1) = wv_i(j) + c_1 r_1 \{y_i(t) - x_i(t)\}$$
$$+ c_2 r_2 \{\hat{y}_i(t) - x_i(t)\} \qquad (9)$$

where w is the inertia weight, c_1 and c_2 are the acceleration coefficients, $r_1, r_2 \in U(0, 1)$, y_i is the personal best position of particle i, and $\hat{y}_i$ is the neighborhood best position of particle i.

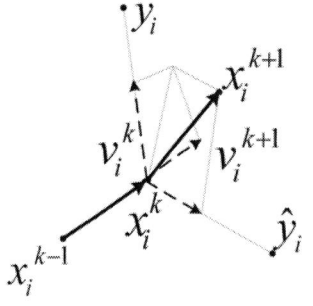

Fig. 8 movement of a PSO agent

V. PSO BASED MPPT

In this paper, a PSO algorithm is applied to track the maximum power of the PV array using the $P-I$ characteristic curve. In order to start the optimization (i.e. to find the global value of current in $P-I$ curve), a solution vector of global currents with Np particle can be defined as:

$$x_i^k = I_g = [I_{g1}, I_{g2}, I_{g3}, \ldots, I_{gj}]$$
$$j = 1, 2, 3, \ldots, Np \tag{10}$$

The objective function is defined as:

$$f(x_i^k) > f(y_i) \tag{11}$$

Where, f is the operating power of PV array.

Usually, power from PV array often changes due to partial shading. Hence, in such cases, the particles must be reinitialized to search the new global point (MPP). Therefore, following condition is used to reinitialize the particles.

$$\left| \frac{P(x_{i+1}) - P(x_i)}{P(x_i)} \right| > \Delta P \tag{12}$$

VI. RESULTS AND DISCUSSION

A. Simulation parameters

Table II shows the parameter of the PSO algorithm used in this work. Usually, in PSO algorithm, initialization of particles is achieved via random numbers. However, in this work predetermined initialization is used as can be seen in table III. Fig. 9 shows the PV system used in this work, in which the simulation of a grid connected PV system involving a boost-type DC–DC converter (with a MPPT controller) and an inverter is carried out. The PV modules are KC2000GT configured in a 6×2 array. Fig. 10 shows the $P-I$ characteristics curve for this configuration. The boost converter and inverter are designed using an averaging model [12]. For the boost converter, an input series winding resistance of 0.5Ω and an output current (source) of 30mA are used to model the conduction and switching losses, respectively. In the same manner, an output series winding resistance of 0.8Ω and an output current source of 40mA are used in the inverter to model the conduction and switching losses. A DC link capacitor (500μF) provides the energy storage necessary to balance the instantaneous power delivered to the grid. The capacitor value is calculated based on an 8% ripple in the V_{DC}. In the steady-state condition, I_{RMSref} is adjusted by the current controller to equalize $V_{DC} = V_{DCref}$. The error signal then goes to zero and the average power P_{ac} delivered to the AC grid matches the power generated by the PV array. A Simulink simulator [13] is used for the PV system simulation.

TABLE II
PARAMETERS VALUES OF PSO

Np	w	c_1	c_2	ΔP
3	0.4	1.2	1.6	0.15

TABLE III
INITIAL POSITION OF PARTICLES

Number of particles	1	2	3
Initial values	$0.8I_{sc}$	$0.4I_{sc}$	$0.6I_{sc}$

I_{sc}: Short Circuit Current

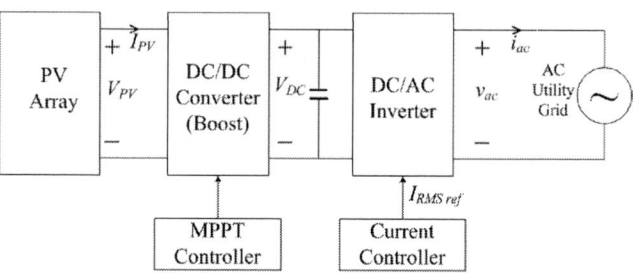

Fig. 9 Grid connected PV system

As can be seen in Fig. 11 (a)–(c), until $t = 0.05$ s, at which point shading occurs, the PSO based MPPT controller calculates the correct V_{mp} voltage (26.3×6 ≅ 157.8V) and I_{mp} current (7.6×2 ≅ 15.2A), respectively, corresponding to the maximum power point. Due to shading of the PV array (at $t = 0.05$ s), the output power of PV suddenly decreases from its optimal value owing to sudden change in operating current. This will tend to reinitialize the MPPT algorithm according to (12), and the new MPP (i.e. global MPP) will be search via PSO algorithm. It can be seen in Fig.11 (a)–(c), the MPPT controller accurately computes the new global I_{mp} current (11.6A), corresponding to the maximum power point (1213W). Moreover, the AC output power is almost equal to the input power as shown in Fig. 12(c). The expected 120Hz ripple (twice the mains frequency) at the output of the Boost converter is also evident.

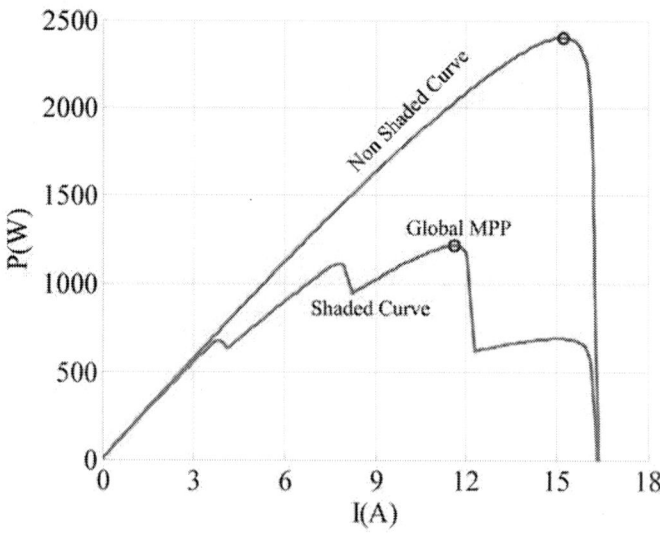

Fig. 10 $P-I$ curve for KC200GT in 6×2 configuration.

978-1-4577-0007-1/11 $26.00 © 2011 IEEE

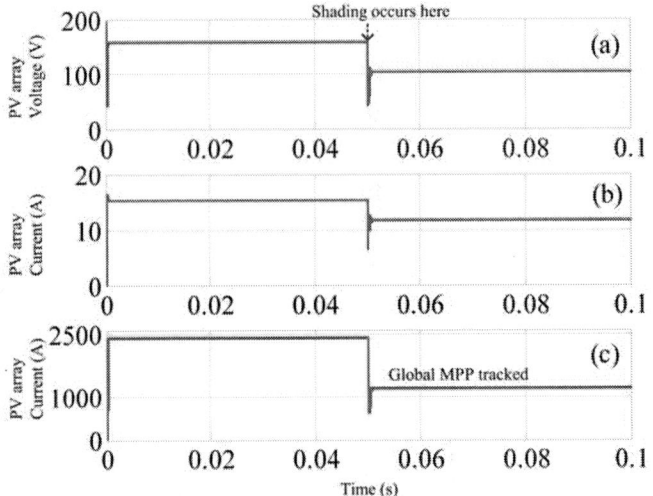

Fig. 11 (a)–(c) Output voltage, current, and output power from the PV array using PSO algorithm.

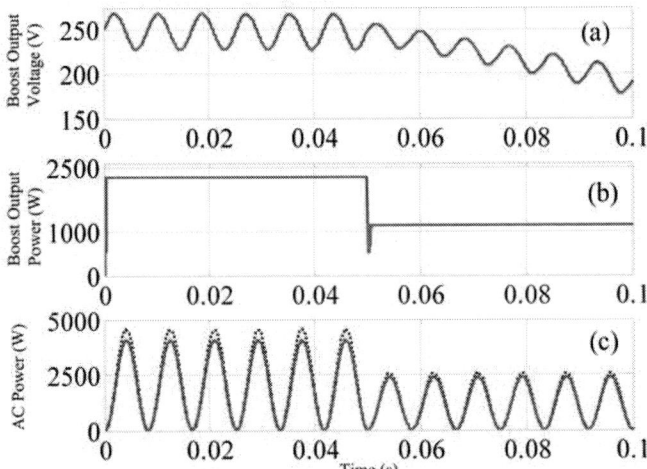

Fig. 12 (a)–(b) Output voltage and output power from the boost converter. (c) AC input (solid) and output power (dotted).

VII. CONCLUSION

In this paper, a particle swarm optimization (PSO) is used to track the maximum power point of PV system under partial shading condition. A grid connected PV system, designed in MATLAB Simulink was used for the validation, in which a DC–DC boost converter was used as a MPP tracker. The P–I characteristic curve was used for the MPP tracking. It was shown that the proposed MPPT algorithm satisfactorily tracks the global point for both non shading and shading condition.

ACKNOWLEDGMENT

The authors would like to thank Universiti Teknologi Malaysia for providing the facilities and research grant to conduct this research.

REFERENCES

[1] DJ. Flood, "Advanced space photovoltaic technology: Applications to telecommunication systems", Proc. of the 19th International Telecommunication Energy Conf., Melbourne, (1997) 647-651.

[2] C. Hua, J. Lin, and C. Shen, "Implementation of a DSP-controlled photovoltaic system with peak power tracking", *IEEE Trans. Ind. Electron.* 45(1) (1998) 99–107.

[3] K. H. Hussein, I. Muta, T. Hoshino, and M. Osakada, "Maximum photovoltaic power tracking: An algorithm for rapidly changing atmospheric conditions", in *Proc. IEE Generation, Transmiss. Distrib.* 142 (1995) 59–64.

[4] K. Ro and S. Rahman, "Two-loop controller for maximizing performance of a grid-connected photovoltaic-fuel cell hybrid power plant", *IEEE Trans. Energy Convers.* 13(3) (1998) 276–281.

[5] M. Veerachary, T. Senjyu, and K. Uezato, "Neuralnetwork-based maximum-power-point tracking of coupled-inductor interleaved-boostconverter- supplied PV system using fuzzy controller", *IEEE Trans. Ind. Electron.* 50(4) (2003) 749–758.

[6] Syafaruddin, E. Karatepe, T. Hiyama, "Polar coordinated fuzzy controller based real-time maximumpower point control of photovoltaic system", Renewable Energy, 34 (2009) 2597–2606.

[7] T. SHIMIZU, M. HIRAKATA, T. KAMEZAWA, WATANABE, "Generation control circuit forphotovoltaic modules", IEEE Trans. Power Electron.16 (3) (2001) 293–300.

[8] D. S. H. Chan, and J. C. H. Phang, "Analytical methods for the extraction of solar-cell single- and double-diode model parameters from I-V characteristics," *Electron Devices, IEEE Transactions on,* vol. 34, no. 2, pp. 286-293, 1987.

[9] K. Ishaque, Z. Salam, and H. Taheri, "Simple, fast and accurate two-diode model for photovoltaic modules," *Solar Energy Materials and Solar Cells,* vol. 95, no. 2, pp. 586-594, 2011.

[10] KC200GT High Efficiency Multicrystal Photovoltaic Module datasheet Kyocera.[Online].Available:http://www.kyocera.com.sg/products/solar/ /pdf/kc200gt.pdf.

[11] R. Eberhart, J. Kennedy. Micro Machine and Human Science, Proceedings of the Sixth International Symposium, pp. 39-43, 1995.

[12] Introduction to power electronics course notes, ECEN 2060 http://ecee.colorado.edu/~ecen2060/matlab.html, 2008.

[13] K. Ishaque, Z. Salam, and H. Taheri, "Accurate MATLAB Simulink PV System Simulator Based on a Two-diode Model," *Journal of Power Electronics,* In Press, Corrected Proof..

Parameter Extraction of Photovoltaic Cell Using Differential Evolution Method

Kashif Ishaque, Zainal Salam, Hamed Taheri and Amir Shamsudin
Faculty of Electrical Engineering, Universiti Teknologi Malaysia, UTM 81310, Skudai,
Johor Bahru, Malaysia.
kashif@fkegraduate.utm.my, zainals@fke.utm.my

Abstract—This paper proposes a new parameter extraction method of photovoltaic cell, based on the differential evolution (DE) technique. The proposed method requires very few control parameters and converges rapidly to a solution. Furthermore, it can fit the *I–V* curve very accurately irrespective of the values of the initial parameters guesses. The performance of DE is evaluated against the well known genetic algorithm (GA) using a synthetic and experimental I-V data set. It is found that the DE method fits the *I–V* curve better than GA, has a lower fitness function value and faster execution time.

Index Terms-- differential evolution; genetic algorithm; parameter extraction; photovoltaic energy Introduction.

I. INTRODUCTION

Photovoltaic (PV) power system is increasingly becoming a popular renewable energy with rapidly mature technologies. The main component of the system, i.e. the PV module utilizes standard semiconductor processes that can be fabricated with relatively minimum facilities [1]. Furthermore, the power conversion system (DC-DC converter and/or inverter) that interface the modules with the grid or batteries are well established. However, in a PV power generation, due to the high initial cost of the system, optimal use of the available photovoltaic energy has to be ensured. This necessitates an accurate simulation model of the PV system, particularly the PV cells.

Generally, two types of approach are used to obtain the cell parameters, namely the analytical [2]–[5] and numerical [6]–[13] extraction techniques. The simplicity of the analytical technique makes it very attractive, as it only requires information on several key points of the *I-V* characteristic curve (i.e., current and voltage at the maximum power point, short-circuit current, open-circuit voltage, and slopes). Computational-wise, it is very fast. However the accuracy of the extracted parameters is strongly influenced by the selected points on the *I-V* curve. Alternatively, the numerical extraction technique is based on algorithm to fit all the points on the *I–V* curve. The advantage of this approach is clear; more accurate cell parameters can be obtained because all points on the *I–V* curve are utilized. The shortcoming, however, is that it requires extensive computation effort. Furthermore the accuracy heavily depends on the type of fitting algorithm, the cost function and the initial values of the parameters to be extracted. It was reported that inappropriate selection of initial values may lead non-convergence of the algorithm [6,8].

Evolutionary algorithms (EA) techniques, particularly the genetic algorithm (GA), have attracted much attention because it can handle constrained and nonlinear functions without requiring derivatives information. Furthermore, it is able to deal with both continuous and discrete variables. The application of GA for PV cell parameter extraction is proposed by several researchers [8,14]. Despite the wide usage of GA in solving complex optimization problems, difficulties such as premature convergence, low speed, and degradation for highly interactive fitness function emerge [15]–[16]. Recently, another type of EA known as differential evolution (DE) was introduced. It is a stochastic optimization method that appears to be very efficient in optimizing real-valued multi-modal objective functions [17]. DE has three distinct advantages (1) able to locate the accurate global optimum irrespective of the initial parameter values (2) has rapid convergence (3) utilizes few control parameters, thus easy to use. This paper comprehensively demonstrates the ability of DE to extract the parameters through *I–V* data of a two-diode PV model [19]. The performance of DE is evaluated against the well known GA using a synthetic and experimental *I-V* data set.

II. TWO DIODE MODEL OF PV CELL

The two diode model is depicted in Fig. 1. The output current of the module can be described as

$$I = I_{PV} - I_{d1} - I_{d2} - (\frac{V + IR_s}{R_p}) \qquad (1)$$

where

$$I_{d1} = I_{01}[\exp(\frac{V + IRs}{a_1 V_{T1}}) - 1] \qquad (2)$$

and

$$I_{d2} = I_{02}[\exp(\frac{V + IR_s}{a_2 V_{T2}}) - 1] \qquad (3)$$

Where I_{PV} is the current generated by the incidence of light; I_{o1} and I_{o2} are the reverse saturation currents of diode 1 and diode 2, respectively. The I_{o2} term is introduced to compensate for the recombination loss in the depletion region as described in [20]. Other variables are defined as follows: V_{T1} and V_{T2} (both equal to $N_s kT/q$) are the thermal voltages of the PV module having N_s cells connected in series, q is the electron charge ($1.60217646 \times 10^{-19}$ C), k is the Boltzmann constant ($1.3806503 \times 10^{-23}$ J/K) and T is the temperature of the *p-n* junction in Kelvin. Variables a_1 and a_2 represent the diode ideality constants; a_1 and a_2 represent the diffusion and recombination current component, respectively.

978-1-4577-0007-1/11 $26.00 © 2011 IEEE

Although greater accuracy can be achieved using this model than the single diode model [20], it requires the computation of seven parameters, namely I_{PV}, I_{o1}, I_{o2}, R_p, R_s, a_1 and a_2.

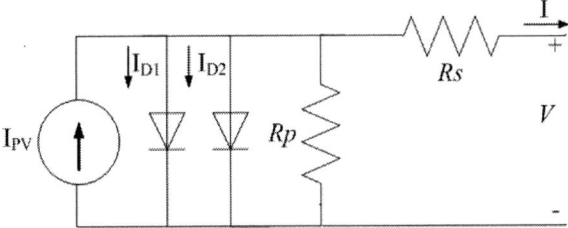

Fig. 1. Two diode model of PV cell

III. DE METHOD

The optimization process of DE involves similar operators to GA, such as crossover, mutation and selection. The population is randomly initialized within initial parameter boundaries. But unlike GA, which relies on crossover, DE primarily utilizes mutation operation (i.e. difference vector) as a search mechanism and selection operation to direct the search toward the prospective regions in the search space. DE process involves the following steps:

A. Initialization

In order to begin the optimization process, an initial population of *NP* *D* dimensional real-valued parameter vectors $X_{i,G} = \left[X_{1,i,G}, X_{2,i,G}, ..., X_{j,i,G}, ..., X_{D,i,G} \right]$ is created. Each vector forms a candidate solution to the multidimensional optimization problem. Initial parameter values are usually randomly selected uniformly in the interval $[X_L,\ X_H]$, where $X_L = \left[X_{1,L}, X_{2,L}, ..., X_{D,L} \right]$ and $X_H = \left[X_{1,H}, X_{2,H}, ..., X_{D,H} \right]$ are the lower and upper bound of the search space, respectively:

$$X_{j,i,0} = X_L + rand[0,1](X_H - X_L) \tag{4}$$

B. Mutation

For a given parameter vector $X_{i,G}$, three vectors ($X_{r1,G}$, $X_{r2,G}$, $X_{r3,G}$) are randomly selected in the range [1, *NP*], such that the indices i, r_1, r_2 and r_3 are distinct. A donor vector $V_{i,G}$ is created by adding the weighted difference between the two vectors to the third (base) vector as:

$$V_{i,G} = X_{r1,G} + F(X_{r2,G} - X_{r3,G}) \tag{5}$$

Where F is a mutation scaling factor, which is typically chosen from the range [0,1].

C. Crossover

The donor vector $V_{i,G+1}$ and the target vector $X_{i,G}$ are mixed to yield the trial vector

$$U_{i,G} = \left[U_{1i,G}, U_{2i,G}, ..., U_{ji,G}, ..., U_{Di,G} \right] \tag{6}$$

In this work, binomial crossover strategy is used which can be described as:

$$U_{j,i,G} = \begin{cases} V_{j,i,G}, & if\ (rand \leq CR\ or\ j = j_{rand}) \\ X_{j,i,G}, & otherwise \end{cases} \tag{7}$$

$j=1,2,.....,D$ and $i=1,2,.....,NP$

Where *CR* is known as the crossover rate and appears as another control parameter of DE just like *F*. $j_{rand} \in [1,2,...,D]$ is a randomly chosen index, which ensures that $U_{i,G}$ attains at least one element from $V_{i,G}$.

D. Evaluation and Selection

The selection operation at $G = G + 1$ is described as:

$$X_{i,G+1} = \begin{cases} U_{i,G} & if\ J\left(U_{i,G}\right) < J(X_{i,G}) \\ X_{i,G} & otherwise \end{cases} \tag{8}$$

$$i = 1,2,...,Np$$

Where, $J(X)$ is the objective function to be minimized. Thus, if the new trial vector acquires a lower value of the objective function, it swaps the corresponding target vector in the next generation; otherwise the target is preserved in the population. Hence, the population either gets better or remains the same in fitness status, but never declines.

The flow chart for the proposed DE method is shown in Fig. 2.

IV. PARAMETER EXTRACTION OF PV CELL USING DE

A. Methodology

The contribution of this paper is to apply the DE method to extract the parameters of the PV cell. To objectively evaluate the performance of the proposed method compared to others, a fitness value J is introduced. The cell parameters can be extracted with the following process: Given a set of *I–V* data of a particular cell, the DE method is applied to update the parameters until the *I–V* data is in close agreement to (1). Note that (1) is a transcendental equation which cannot be solved by a direct analysis. The proposed DE method does not require complex transformation of the original equation. The PV cell current equation can be re-written as:

$$y(I,V,\Phi) = I_{PV} - I_{01}[\exp(\frac{V+IR_S}{a_1 V_{T1}})-1] -$$
$$I_{02}[\exp(\frac{V+IR_s}{a_2 V_{T2}})-1] - (\frac{V+IR_s}{R_p}) - I \tag{9}$$

978-1-4577-0007-1/11 $26.00 © 2011 IEEE

Where $\Phi = \left[I_{PV}, I_{o1}, I_{o2}, R_s, R_p, a_1, a_2 \right]$ are the parameters to be extracted. In each iteration, the value of Φ is obtained by DE procedure as discussed in previous section. The procedure ends at intended maximum number of iteration has been reached.

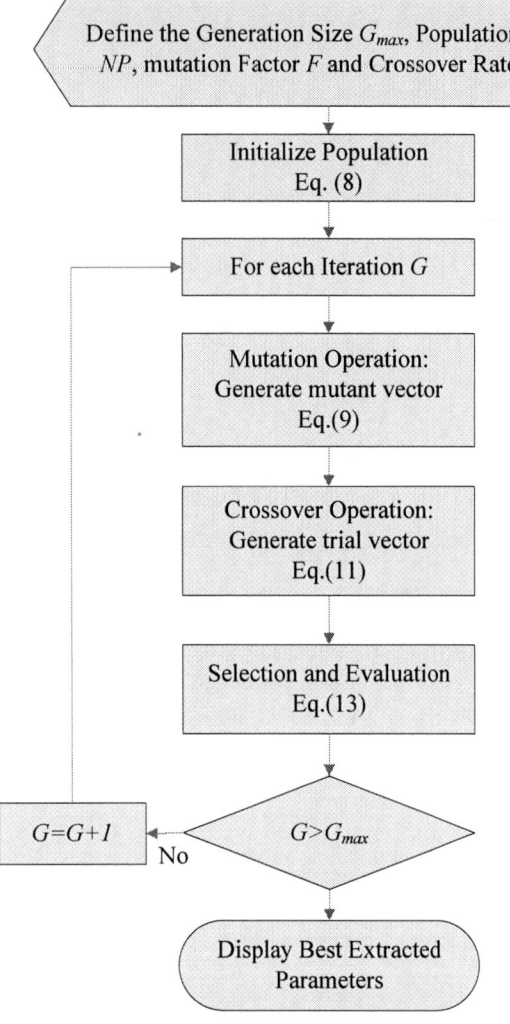

Fig. 2. Flowchart of DE

B. Fitness Function

In this work, the fitness function can be given by:

$$J = \sqrt{\frac{1}{N} \sum_{m=1}^{N} y(I_m, V_m, \Phi)^2} \qquad (10)$$

Where I_m and V_m are the data pair of I–V characteristics curve, respectively; N is the number of data. Hence, the extraction procedure using DE is based on the minimization of the objective function J with regard to $\Phi = \left[I_{PV}, I_{o1}, I_{o2}, R_s, R_p, a_1, a_2 \right]$. A smaller value of the objective function implies the least deviation between the I–V data and the DE I-V data. Ideally, a zero value of objective function J is desired. It can be seen from (11) that J is a nonlinear function. Therefore, no quadratic objective function is apparent which leads to several local minima with one global minimum. Conventional methods try to solve this problem iteratively which requires the gradient information. However, it is not easy to estimate when there are unobservable states and/or discontinuous types of nonlinearity exist in the model [21].

V. RESULTS AND DISCUSSION

A. With Synthetic Data

The parameter extraction is first evaluated using synthetic data obtained from [19]. The synthetic data comparison approach is also utilized by other researchers [21]. The data were calculated using the two diode model with the following seven parameters: I_{PV}=8.21A, I_{o1}=I_{o2}=4.218×10^{-10}, R_s=0.32Ω, R_p=160.5Ω, a_1=1, and a_2=1.2. The temperature and irradiance are at the Standard Test Conditions (STC) i.e. 1000W/m^2 and 298 Kelvin, respectively. Then the I–V curve fitting using DE is performed. For comparison, curve fitting using GA is also carried out. To ensure a fair evaluation for the two techniques, similar simulation conditions, i.e. population size, maximum generation number and search ranges are maintained. The extracted parameters obtained for these two methods are evaluated against the synthetic data.

The population size (NP), is chosen to be 70. This is a reasonable choice for NP; typical values are between $5D$ to $10D$. The maximum generation number (G_{max}) is set to 10000. The mutation scaling factor (F) is set at 0.8. There is no strict rule on the selection of F but in most cases, F>0.4 [22]. The crossover rate (CR) is chosen to be 1. Large value of CR intensifies the diversity of population, thus improving the convergence speed [22]. Moreover, a high value of CR is desirable as the parameters in the model are highly correlated [23]. The search ranges were set as follows: $I_{PV} \in [8.18, 8.25]$, $I_{o1} \in \left[4e^{-11}, 4e^{-6} \right]$, $I_{o2} \in \left[4e^{-11}, 4e^{-6} \right]$, $R_s \in [0.1, 0.6]$, $R_p \in [100, 200]$, $a_1 \in [0.95, 1.1]$, $a_2 \in [1.1, 2]$.

The DE/best/1/bin strategy is used in the proposed work. In this nomenclature, the word "best" defines the best vector from the current population, "1" specifies number of difference vector and "bin" describes the binomial crossover technique. For the GA implementation, a crossover rate, P_C = 0.8 and mutation rate, P_m = 0.2, was used. A GA function named as "ga" in MATLAB was used in this case.

Fig. 3(a) and (b) show the comparisons between the synthetic I–V data and the I–V characteristics derived by DE and GA, respectively. Clearly both methods fit the synthetic curve exactly −which indicate that both are appropriate to be used for cell parameter extraction. Fig. 4 illustrates the convergence performance of DE and GA for this particular example. As expected, DE outperforms GA in both convergence speed and fitness value J; a five order of magnitude lower J_{min} is obtained by the DE method. Moreover, GA method fails to converge to the global minimum.

Table I compares the parameters extracted from DE and GA in comparison to the synthetic data. Three evaluations are shown (1) difference between original and extracted

978-1-4577-0007-1/11 $26.00 © 2011 IEEE

parameters, (2) minimum fitness function values and (3) computation time. As can be seen, the extracted parameters using the DE are very close to the synthetic parameters. The relative errors between the extracted and original by using GA are much larger than those attained by DE for the same search ranges and population sizes. Moreover, the minimum fitness value J of DE method is much less than that of GA. The computation time demonstrates the speed of using the DE is six times faster than GA.

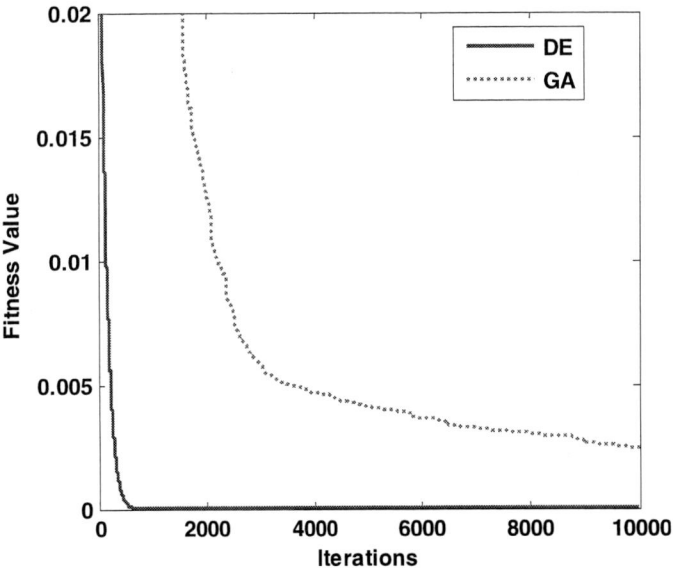

Fig. 4. Convergence performance of the fitness function with DE and GA

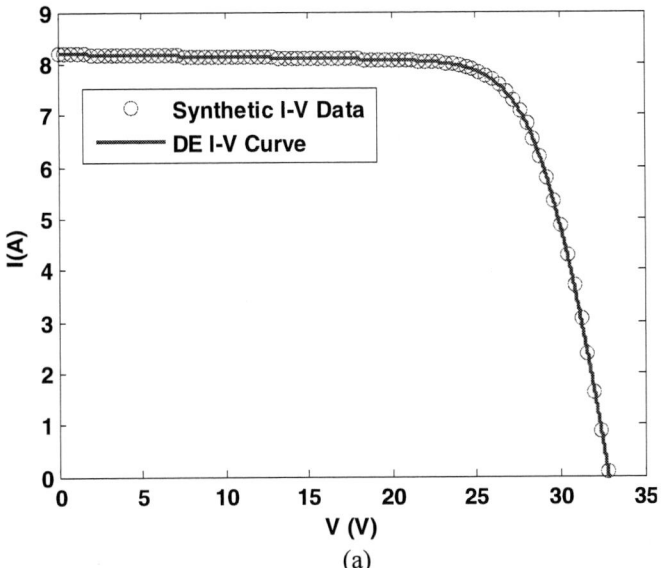

(a)

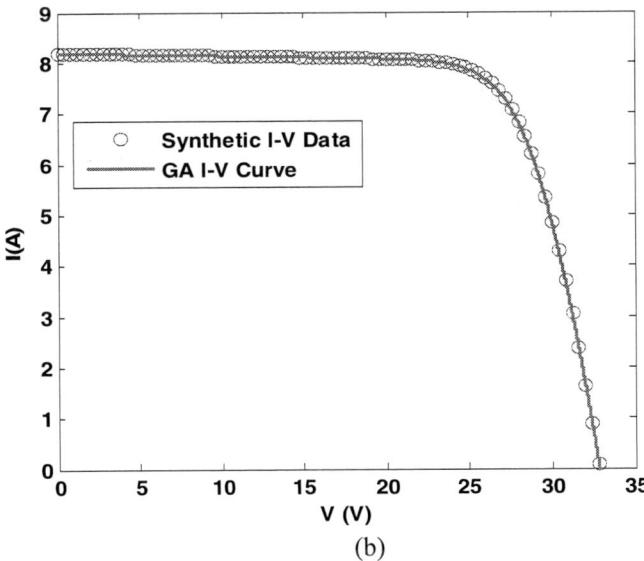

(b)

Fig. 3. *I-V* characteristics obtained from (a) DE and (b) GA methods

TABLE I. RESULTS OF PARAMETER EXTRACTION BY DE AND GA

PV Module Parameter	DE Method	GA Method	Relative Errors (%) DE	GA
I_{PV}	8.21A	8.20A	0	0.122
I_{o1}	4.1×10^{-10}A	4.1×10^{-9}A	0.2134	1.61
I_{o2}	4.1×10^{-10}A	1.5×10^{-8}A	0.2134	3.456×10^{3}
R_s	0.32Ω	0.316Ω	0	0.0625
R_p	160.4549	183.2 Ω	0.281	14.02
a_1	1	1	0	0
a_2	1.2	1.99	0	65.83
Time (s)	22.3	141.6	-----	-----
J_{min}	8.2×10^{-8}	2.5×10^{-3}	-----	-----

B. With Experimental Data

To further validate the performance of the cell extraction method using DE, curve fitting with experimental data is performed. The data is obtained from a commercial PV module (KC200GT) [24], measured at STC. Note that for experimental, the data is significantly fewer than synthetic data. Fig. 5(a) shows the comparisons between the experimental data and the *I-V* characteristics achieved by DE method. It can be seen that computation using DE fit the experimental data well, hence confirms the accurateness of the parameter extraction. Fig. 5(b) shows the *I-V* characteristics obtained from GA. It can be observed that the curve fitting of GA deteriorates slightly between the short circuit current (I_{sc}) and maximum current (I_{mp}). This does not happen to DE. The result implies that DE yields better accuracy with fewer *I-V* data.

Fig. 6 depicts the convergence of both methods. Clearly, the DE method offers a lower fitness value and faster convergence; DE converges in less than 500 iterations, whereas GA takes more than 1800 iterations. Table II shows the extracted parameters computed from DE and GA in relation to the experimental *I-V* data. Clearly, better results are achieved by the DE. As for computation time, DE is approximately five times faster than GA when the maximum generation is set to 10,000. Furthermore, its fitness value J_{min} is one order of magnitude lower.

TABLE II. RESULTS OF PARAMETER EXTRACTION BY DE AND GA

PV Module Parameter	DE Approach	GA Approach
I_{PV}	8.18A	8.22A
I_{o1}	1.72×10^{-9}A	3.622×10^{-9}A
I_{o2}	5.14×10^{-10}A	1.197×10^{-9}A
R_s	0.24Ω	0.30Ω
R_p	$147.2\ \Omega$	$161.13\ \Omega$
a_1	1.08	1.1
a_2	1.15	1.963
Time (s)	18.89	103.6
Fitness Value J	6.2×10^{-2}	0.12

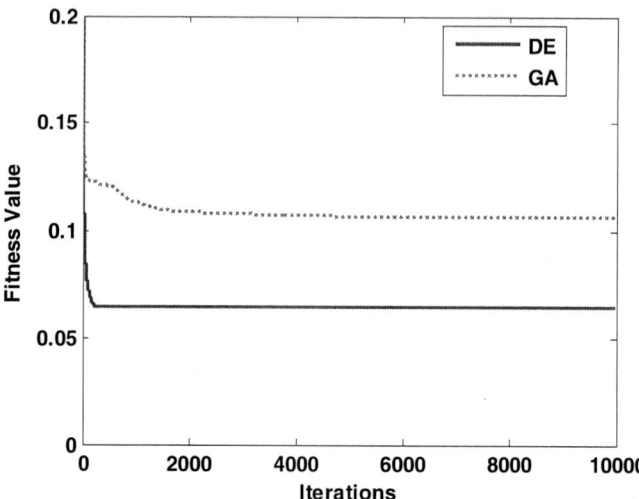

Fig. 6. Convergence performance of the fitness function with DE and GA

VI. CONCLUSION

This paper proposes a new parameter extraction technique for the PV cell using the Differential Evolution (DE) method. The results are compared with the Genetic Algorithm (GA). The feasibility of the proposed DE method has been validated by applying it to a synthetic and experimental *I-V* data. The DE method fits the *I–V* curve better than GA, has a lower fitness function value and faster execution time.

ACKNOWLEDGMENT

The authors would like to thank Universiti Teknologi Malaysia for providing the facilities and research grant to conduct this research.

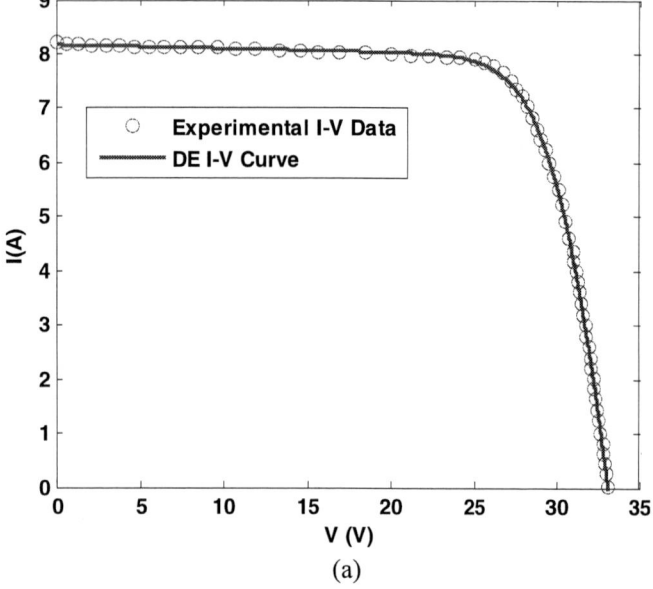

(a)

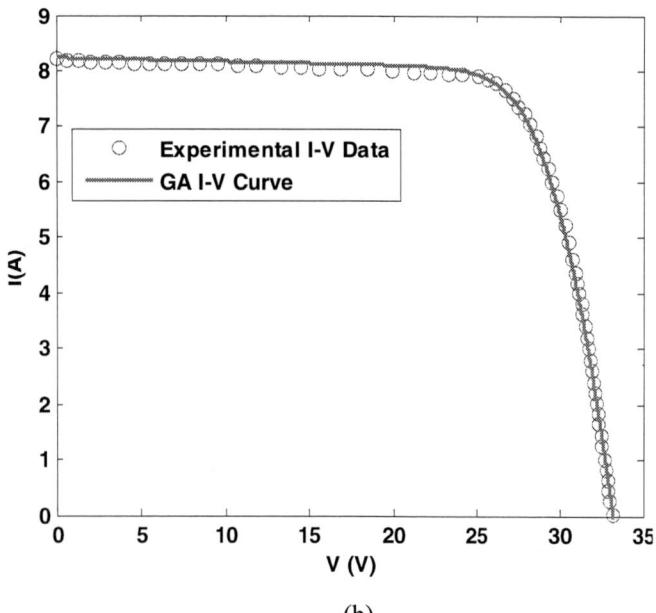

(b)

Fig. 5. *I-V* characteristics obtained from (a) DE and (b) GA methods

REFERENCES

[1] Roundy S, Wright P K and Rabaey J M 2004 Energy scavenging for wireless sensor networks: with special focus on vibrations, kluwer academic publishers, Norwell.

[2] Chan DSH and Phang JCH 1987 Analytical methods for the extraction of solar-cell single- and double-diode model parameters from I–V characteristics. *IEEE Trans Electron Dev* 34 286–93.

[3] Charles JP, Bordure G, Khoury A and Mialhe P 1985 Consistency of the double exponential model with the physical mechanisms of conduction for a solar cell under illumination. *J. Phys. D – Appl. Phys.*18 2261–2268.

[4] Ortiz-Conde A, Garcia Sanchez FJ and Muci J 2006 New method to extract the model parameters of solar cells from the explicit analytic solutions of their illuminated *I–V* characteristics. *Sol. Energy Mater. Solar Cells* 90 352–361.

[5] De Soto W, Klein S and Beckman W 2006 Improvement and validation of a model for photovoltaic array performance. *Solar Energy* 80 78–88.

[6] Gottschalg R, Rommel M, Infield DG and Kearney MJ 1999. The influence of the measurement environment on the accuracy of the extraction of the physical parameters of solar cells. *Meas Sci Technol* 10 796–804.

[7] Nakanishi F, Ikegami T, Ebihara K, Kuriyama S and Shiota Y 2000 Modeling and operation of a 10 kW photovoltaic power generator using equivalent electric circuit method. *In: Proceedings of 28th photovoltaic specialists conference* 1703–6.

[8] Jervase JA, Bourdoucen H and Al-Lawati A 2001 Solar cell parameter extraction using genetic algorithms. *Meas Sci Technol* 12 1922–1925.

[9] Chegaar M, Ouennoughi Z and Guechi F 2004 Extracting dc parameters of solar cells under illumination *Vacuum* 75 367–372.

[10] Haouari-Merbah M, Belhamel M, Tobias I and Ruiz J 2005 Extraction and analysis of solar cell parameters from the illuminated current–voltage curve. *Sol. Energy Mater. Solar Cells* 87 225–33.

[11] Chegaar M, Azzouzi G and Mialhe P 2006 Simple parameter extraction method for illuminated solar cells *Solid State Electron* 50 1234–1237.

[12] Duffie JA, Beckman WA 2006 Solar engineering of thermal processes. Third ed. Hoboken, NJ, USA: John Wiley & Sons.

[13] Liu CC, Chen CY, Weng CY, Wang CC, Jenq FL and Cheng PJ 2008 Physical parameters extraction from current–voltage characteristic for diodes using multiple nonlinear regression analysis *Solid State Electron* 52 839–843.

[14] Moldovan N, Picos R and Garcia-Moreno E 2009 Parameter extraction of a solar cell compact model using genetic algorithms *Proc. of Spanish conf. on Electron Devices* 379-382.

[15] Gaing ZL 2004 A particle swarm optimization approach for optimum design of PID controller in AVR system. *IEEE Trans Energy Convers* 19 384–391.

[16] Ji MJ, Jin ZH and Tang HW 2006 An improved simulated annealing for solving the linear constrained optimization problems *Appl. Math Comput.* 183 251–259.

[17] Storn R and Price K 1997 Differential evolution – a simple and efficient heuristic for global optimization over continuous spaces *J. Global Optim.* 11 341–359.

[18] da Costa W T , Fardin J F, Simonetti D S L and Neto L 2010 Identification of photovoltaic model parameters by Differential Evolution, *Industrial Technology (ICIT), 2010 IEEE International Conference* 931 - 936

[19] K. Ishaque, Z. Salam, H. Taheri, Simple, fast and accurate two-diode model for photovoltaic modules, Solar Energy Materials and Solar Cells, 95 (2011) 586-594.

[20] Sah C, Noyce R N and Shockley W 1957 Carrier generation and recombination in *p-n* junctions and *p-n* junction characteristics *in: Proceedings of IRE* 45 1228-1243.

[21] Meiying Y, Xiaodong W, and Yousheng X 2009 Parameter extraction of solar cells using particle swarm optimization *J. Appl. Phys.* 105 1-8.

[22] Gamperle R, Muller S D and Koumoutsakos P 2002 A parameter study for differential evolution *Adv. Intell. Syst. Fuzzy Syst. Evolut. Comput.* 293–298.

[23] H. Taheri H, Z. Salam, K. Ishaque, and Shafaruddin A Novel Maximum Power Point Tracking Control of Photovoltaic System Under Partial and Rapidly Fluctuating Shadow Conditions Using Differential, *2010 IEEE Symposium on Industrial Electronics & Applications* 82-87.

[24] KC200GT High Efficiency Multicrystal Photovoltaic Module Datasheet Kyocera. [Online]. Available: http://www.kyocera.com.sg/products/solar/pdf/kc200gt.pdf.

978-1-4577-0007-1/11 $26.00 © 2011 IEEE

Power Quality Improvement for Distributed Generation Employing Photovoltaic-Inverter

Muhammad Imran Hamid[a], Awang Jusoh[a], and Makbul Anwari[b]

[a]Faculty of Electrical Engineering, Universiti Teknologi Malaysia
81310 UTM Johor Bahru, Malaysia
[b]Electrical Engineering Department, Umm Al-Qura University
Makkah 21955, Saudi Arabia

Abstract- **Power quality problems such as variation of reactive power consumption and distorted current injection in distributed generation system can be compensated by a single-phase conditioner. The proposed conditioner works in feed forward mode and placed in parallel with the PV-inverters in the distributed generation system. A simple current extraction method is applied and an operating algorithm of the conditioner's hysteresis current controller is implemented. The proposed conditioner is simulated using Simulink/Matlab. The simulation has shown that the current detection method and the hysteresis current control based on the proposed algorithms are worked well and capable to improve the quality of injected currents to the main grid system.**

I. INTRODUCTION

Photovoltaic (PV) application for electricity shows a significant growth in recent years. This growth is indicated by more and more construction of photovoltaic power plant in various places in the world [1]. This condition is followed by the increased use and production of supporting components for these renewable generation systems. One of them is the PV-inverter, which is equipment that interfacing between photovoltaic conversion equipment into the electrical system.

Production of PV-inverters in wide-scale rating, from several hundred small-scale till several thousand watts has encouraged more widespread use of this equipment, particularly for distributed generation (DG) systems. However, the reality that PV-inverter is built based on a number of power electronics equipment as the characteristic of renewable energy generation system [2] and their interaction with others power system components and loads shows an impact to the power quality in the electrical system where they are connected [3] – [9].

Figure 1 shows injected current wave-shape of a group of PV-inverters connected to one phase of the grid. Besides showing current-shape imperfection during various level of generated power, the current-shape also shows a shifting phase angle between grid voltage and injected current which indicates the presents of reactive power consumption during conversion process.

To suppress the impact of power quality problem and to improve the power quality aspects in the operation of the PV-inverter many efforts have been reported. Several methods have been implemented to improve the quality of design on topology and control method with better power quality

parameters [10 – 18]. The current distortion caused by reactive power consumption is compensated by designing and adding a reactive power conditioner function integrated in the PV-inverter unit. In [10] and [11], this function is embedded in the current control and MPPT mechanism of the PV-inverters. In [12] and [13], PV-inverter is designed to enable it to role as a reactive power conditioner unit. In [14], a collaborative of two voltage source inverters are employed for controlling the quality of the current injected into the grid.

All the above methods focused to improve the internal functions of the PV-inverter. The internal functions are employed to respond the power quality deterioration signal entering them. These methods have generally succeeded in improving power quality performance of PV-inverters individually. However, the methods were not considered the distortion symptoms that arise as a result of interaction of several PV-inverters, in which variety of types, design, ratings, and different technology PV-inverters may be connected. So far the methods intended to improve the PQ of a number of inverters as a plant unit such as in PV-inverters in a PV-plant or BIPV DG are still lacking. These are described in the references [15] - [17].

This paper describes the implementation of a single phase active power conditioner works in feed forward scheme (Fig. 2) to improve the power quality output of single phase PV-inverters that work together to supply power as a unit of distributed generation into the main. The conditioner and its control are separated from the PV-inverter; its output is positioned on the point of common coupling (PCC) of inverters and the grid.

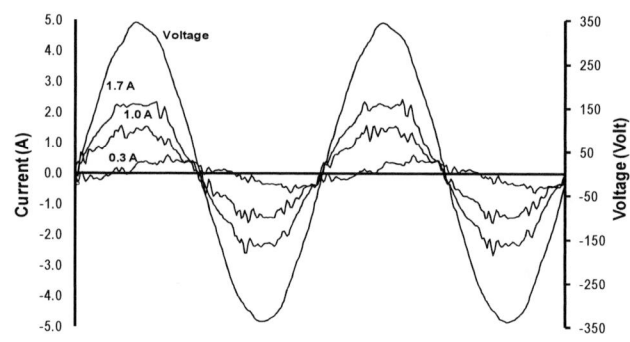

Fig. 1. Voltage and current wave-shape of a 700 Watt grid-tied PV-inverter in various levels of generated power

At this point, the compensated current from the conditioner is added to the distorted-current from PV-inverter(s), and as a result a unity power factor and sinusoidal current are produced and injected to the main. The conditioner works based on the conditions of the current output regardless of the external and internal conditions which may affect to the power quality of the inverters.

Conditioner current injection is carried out through a hysteresis current control where the reference signal for the compensation current generation is achieved by detecting and separating the fundamental and harmonics components of distorted output currents from PV-inverter(s) through multiplication with a sine and cosines signal. This method does not require any transformation techniques such as $p - q$ or $I_p - I_q$ transformation which been used in others detection methods, such as instantaneous power theory and its extension on single-phase system [18] and [19]. Besides, the orthogonal reference generator is also not required in order to get a 90° shifted signal of single phase systems as needed in detection method introduced in [22]. By separating the distorted current components, the determination of the components which need to be compensated such as reactive power and harmonics in line with the requirement to overcome the power quality problems faced in PV-inverter operation are easily implemented.

II. Reference Current Generation for Conditioner

Equation (1) shows the voltage at the terminal and PCC of PV-inverters and grid, whereas equation (2) shows the cumulative current generated by the inverters.

$$v(t) = \sqrt{2}\, V_{rms} \sin \omega_1 t \tag{1}$$

$$i(t) = \sum_{n=1}^{\infty} \sqrt{2}\, I_n \sin\left(n\omega_n t + \varphi_n\right) \tag{2}$$

Voltage is assumed ideally sinusoidal while the current equation is expressed as the summation of a number harmonics currents n where φ_n is the phase angle of each harmonic order with a reference of line to neutral voltage.

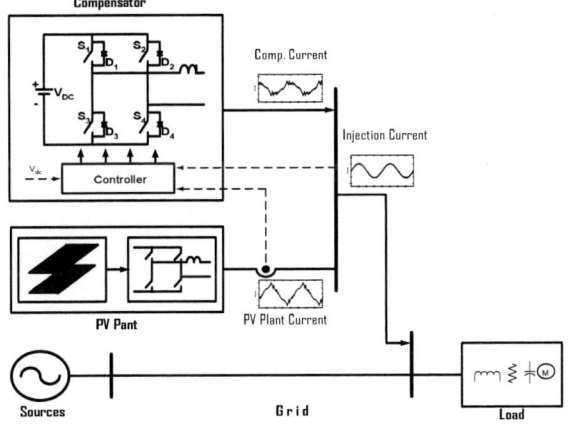

Fig.2. Configuration of proposed conditioner with PV-DG in an electrical grid

Then, by decomposing the distorted current, equation (2) into the fundamental and harmonic components, equation (3) is obtained:

$$i(t) = \sqrt{2}\, I_1 \sin\left(\omega_1 t + \varphi_1\right) + \sum_{n=2}^{\infty} \sqrt{2}\, I_n \sin(n\omega_n t + \varphi_n) \tag{3}$$

Equation (3) can be written as eq. (4) and (5):

$$i(t) = \sqrt{2}\, I_1 \cos\left(\varphi_1\right)\sin(\omega_1 t) + \sqrt{2}\, I_1 \sin\left(\varphi_1\right)\cos(\omega_1 t) + \sum_{n=2}^{\infty} \sqrt{2}\, I_n \sin(n\omega_n t + \varphi_n) \tag{4}$$

$$i(t) = I_{1p}(t) + I_{1q}(t) + I_h(t) \tag{5}$$

$I_{1p}(t)$ and $I_{1q}(t)$, are the fundamental active and reactive current components of $i_1(t)$, respectively. Their peak values are $I_p = \sqrt{2}\, I_1 \cos\left(\varphi_1\right)$ and $I_q = \sqrt{2}\, I_1 \sin\left(\varphi_1\right)$, which the DC components that do not change with time. Their magnitude depend on the angle of phase voltage and current (φ_1). By determining I_p and I_q, the instantaneous value of $I_{1p}(t)$ and $I_{1q}(t)$ can be obtained. These values then are used in equation (5) to find $I_h(t)$.

To determine the value of I_p, it is done by multiplying equation (4) with $2\sin(\omega_1 t)$ to obtain:

$$2\sin(\omega_1 t).i(t) = I_p - I_p \cos(2\,\omega_1 t) + I_q \sin(2\,\omega_1 t) + 2\sin(\omega_1 t)\sum_{n=2}^{\infty} \sqrt{2}\, I_n \sin(n\omega_n t + \varphi_n) \tag{6}$$

For I_q value, it is performed by multiplying equation (4) with $2\cos(\omega_1 t)$, which gives:

$$2\cos(\omega_1 t).i(t) = I_p \sin(2\,\omega_1 t) + I_q + I_q \cos(2\,\omega_1 t) + 2\cos(\omega_1 t).\sum_{n=2}^{\infty} \sqrt{2}\, I_n \sin(n\omega_n t + \varphi_n) \tag{7}$$

Equations (6) and (7) show that the DC fraction in active and reactive current component contained by $i(t)$ can be separated by multiplying the equation of current $i(t)$ with the factor $2\sin(\omega_1 t)$ and $2\cos(\omega_1 t)$. In a computer simulation or hardware implementation, these can be obtained by extracting the input voltage line. Input $v(t)$ as in equation (2) is divided by $\sqrt{2}$ to obtain the unity sinusoidal wave $\sin(\omega_1 t)$. The signal then be processed and multiplied by appropriate constant to obtain the factor $\cos(\omega_1 t)$. Other method that can be used is PLL. From the extraction process, according to equation (5), reference for conditioner current controller $i_{ref}(t)$ can be generated. Figure 3 shows Simulink block to determine $I_{1p}(t)$, $I_{1q}(t)$, and $I_h(t)$ from a distorted current $i(t)$ and wave shape of the signals.

III. Current Controller for Conditioner

Due to the behaviors of the current generated by the PV inverters, and also from calculation and simulation shown in Fig.3, it is noticed that the reference current for compensation

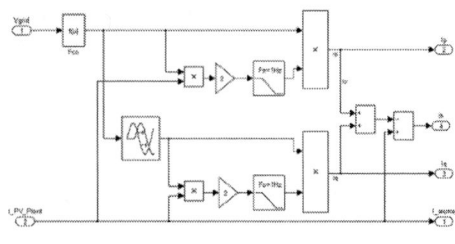

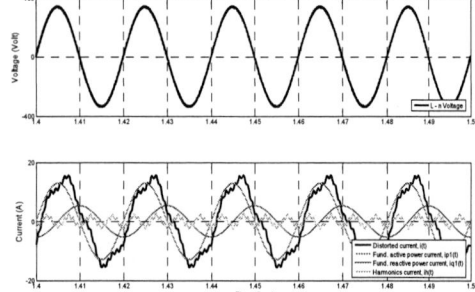

Fig.3. Simulink block for extracting $I_{1p}(t)$, $I_{1q}(t)$, and $I_h(t)$ from a distorted and lagging current $i(t)$ and wave shape of the signals

tend to change more frequently and random, and at several event there are sudden change (high value of $\frac{di_{ref}(t)}{dt}$) occurred. These conditions require a more responsive and robust current control method to face the high value of $\frac{di_{ref}(t)}{dt}$. Among the available of current control methods, hysteresis current controller has the capability to fulfill the requirements.

In hysteresis current controller, the output current of conditioner is forced to follow the reference current, deviation between them are limited by predetermined band h_B. Along with time, the actual current as the output of the conditioner swings from lower to upper limit of the band. If actual current reaches the upper limit, the switch component is switched off so that the current decrease till reach the lower limit. At this point, the switch component is switched on again and the actual current increases to upper limit. This process is repeated continually, following the path established by the reference band.

The upper and lower limits of reference band are determined based on the instantaneous value of the calculated compensated current reference. According to the value of $i_{ref}(t)$, the lower I_{LL} and upper limit I_{UL} of the bands are obtained as:

$$I_{LL} = i_{ref}(t) - h_B \quad \text{and}$$

$$I_{UL} = i_{ref}(t) + h_B \tag{8}$$

IV. RIPPLE CURRENT USING HYSTERESIS CURRENT CONTROL

The current condition under hysteresis current control of the single phase conditioner as shown in Fig.2 previously with

bipolar switching method can be analyzed to obtain important parameters of the controller. Figure 4 shows a small portion of output current during positive half cycle period. When current $i(t)$ increases on interval $t_1 - t_2$, switch S1 and S4 are ON, while switches S_2, S_3, and $D_1 - D_4$ are OFF. When current $i(t)$ decreases on interval $t_2 - t_3$, switch S_1, S_4, D_1, D_4 are OFF, while D_2 and D_3 are ON. Assumed conditioner resistance is small then $Ri_L(t)$ can be neglected, then during these periods, circuit equations are given in equation (9) and (10):

$$V_C - V_m \sin \omega t = L \frac{di_L^+(t)}{dt} \quad \text{or}$$
$$\frac{di_L^+(t)}{dt} = \frac{1}{L}(V_C - V_m \sin \omega t) \tag{9}$$

$$-V_C - V_m \sin \omega t = L \frac{di_L^-(t)}{dt} \quad \text{or}$$

$$\frac{di_L^-(t)}{dt} = -\frac{1}{L}(V_C + V_m \sin \omega t) \tag{10}$$

From the shape of the current as shown in Fig.4, the gradients of $i_L(t)$ during ON-OFF status of the switches are given as equation (11) and (12):

$$\frac{di_L^+(t)}{dt} t_{on} = \frac{di_{ref}(t)}{dt} t_{on} + 2h_B \quad \text{or}$$
$$\frac{di_L^+(t)}{dt} t_{on} - \frac{di_{ref}(t)}{dt} t_{on} = 2h_B \tag{11}$$

$$\frac{di_L^-(t)}{dt} t_{off} = \frac{di_{ref}(t)}{dt} t_{off} - 2h_B \quad \text{or}$$
$$\frac{di_L^-(t)}{dt} t_{off} - \frac{di_{ref}(t)}{dt} t_{off} = -2h_B \tag{12}$$

By adding equations (11) and (12) and consider that
$$t_{on} + t_{off} = T_s = \frac{1}{f_s} \quad \text{yield:}$$

$$\frac{di_L^+(t)}{dt} t_{on} + \frac{di_L^-(t)}{dt} t_{off} - \frac{di_{ref}(t)}{dt} \frac{1}{f_s} = 0 \tag{13}$$

By subtract equation (12) from (11), we get

$$\left(\frac{di_L^+(t)}{dt} t_{on} - \frac{di_L^-(t)}{dt} t_{off}\right) - \frac{di_{ref}(t)}{dt}\left(t_{on} - t_{off}\right) = 4h_B \tag{14}$$

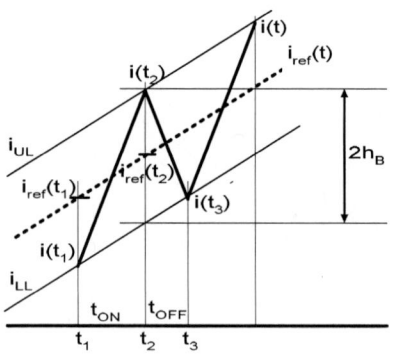

Fig.4. A portion of output current during positive half cycle of the grid voltage

978-1-4577-0007-1/11 $26.00 © 2011 IEEE

Substituting equations (9) and (10) into equation (13):

$$\left(t_{on} - t_{off}\right) = \frac{L}{f_s V_C}\left(\frac{1}{L}V_m \sin \omega t + \frac{di_{ref}(t)}{dt}\right) \tag{15}$$

Substituting equations (9), (10) and (15) into equation (14)

$$\frac{1}{L}\left(\left(t_{on} + t_{off}\right)V_C + \left(t_{off} - t_{on}\right)V_m \sin \omega t\right) - \\ \frac{di_{ref}(t)}{dt}\frac{L}{f_s V_C}\left(\frac{1}{L}V_m \sin \omega t + \frac{di_{ref}(t)}{dt}\right) = 4h_B \tag{16}$$

Expressing $\left(t_{off} - t_{on}\right)$ using other form of equation (15) and writing $\frac{di_{ref}(t)}{dt} = m$ and $\omega t = \theta$ on (16), yield equation (17), where h_B is hysteresis band.

$$h_B = \frac{V_C}{4 L f_s}\left[1 - \left(\frac{L}{V_C}\right)^2\left(\frac{V_m \sin \theta}{L} + m\right)^2\right] \tag{17}$$

Since the conditioner current is limited by the upper and lower of hysteresis band, the current value of the conditioner on the hysteresis band is the peak value of ripple current conditioner. The peak value of current in positive half cycle is:

$$\Delta i_{(+)} = i_{ref}(t) + h_B \tag{18}$$

And for the negative half cycle,

$$\Delta i_{(-)} = i_{ref}(t) - h_B \tag{19}$$

So, the peak-to-peak current ripple is obtained as:

$$\Delta i_{(p-p)} = \Delta i_{(+)} - \Delta i_{(-)}$$
$$\Delta i_{(p-p)} = \frac{V_C}{2 L f_s}\left[1 - \left(\frac{L}{V_C}\right)^2\left(\frac{V_m \sin \theta}{L} + m\right)^2\right] \tag{20}$$

The extreme value of $\Delta i_{(p-p)}$ for one cycle of grid voltage can be calculated by differentiating equation (20) with respect to θ, and equating it to zero: $\frac{\partial(\Delta i_{(p-p)})}{\partial \theta} = 0$. The extreme values of $\Delta i_{(p-p)}$ in one cycle of grid voltage are found at some points:

$$\theta = \begin{cases} \pm\frac{(2n-1)\pi}{2}, & n = 1,2,3 \dots \\ \sin^{-1}\left(-\frac{L m}{V_m}\right) \end{cases} \tag{21}$$

Equation (20) shows the relationship between ripple current $\Delta i_{(p-p)}$ to other conditioner variables, which are switching frequency, line voltage, DC link voltage, and gradient of the current control reference. From this equation, when the switching frequency f_s is fixed, the conditioner works in variable ripple current (variable band hysteresis current control mode). When the switching frequency is allowed to vary then the current control works in fixed band hysteresis current control mode.

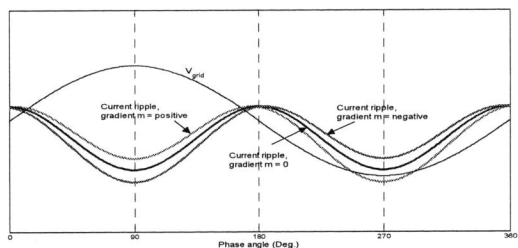

Fig.5. Variation of peak-to-peak current ripple $\Delta i_{(p-p)}$ at one level switching frequency with positive, zero, and negative direction of gradient m

Besides being a function of the phase angle θ, the equation (20) also shows that the current ripple is also depend on the value of the gradient of the current control reference. Figure 5 shows the influence of this variable to the current ripple for a level of given switching frequency. It is seen that for the given gradient, the resulting current ripple is different at different phase angle of the grid voltage. If the gradient is positive, a higher ripple appears on the positive cycle of the grid than negative one. The opposite condition occurs if the gradient is negative. Also it can be shown that the greater the value of the gradient, the deviation between the maximum and minimum ripple is also getting bigger.

V. SIMULATION AND DISCUSSION

In order to show the workability of the proposed compensator with respect to the performance of the current detection method, hysteresis current control operation and effect of the conditioner on the grid with a connected PV plant, a Matlab/Simulink simulation diagram is built as Fig.6. The PV Plant block which represents a number of PV-inverters and the conditioner model are connected on bus A, their connection to the grid are done trough breaker 1 and 2.

PV-inverters and conditioner are synchronized with phase-A of the grid voltage and are operated under current control scheme. This condition enable PV-inverters behave as a power source with a sinusoidal voltage terminal and allows the current to be non-sinusoidal as happened in the actual conditions in the grid-tied PV-inverter operation (see Fig.1). The current output of PV-inverters is sensed, and together with line to neutral grid voltage is used as control signals for conditioner circuit. The conditioned current then is injected to the grid trough bus C.

The second level of the conditioner block contains power and controller circuit. The sensed signals from line to neutral grid voltage and current from PV plant are scaled and filtered in the signal conditioner block. Then, using current extractor block, the fundamental current of active power $I_{1p}(t)$, reactive power $I_{1q}(t)$, and harmonics current component $I_h(t)$ from current output of PV plant are obtained, detail components of the block is shown in Fig.3. Compensator determination block is used to determine which components to be compensated: whether harmonics only or harmonics and reactive component. The output of this block is compensating reference signal, which to be used in hysteresis band calculator and hysteresis current controller.

978-1-4577-0007-1/11 $26.00 © 2011 IEEE

To produce low distortion grid injection current which is a summation of current originating from the PV-inverters and from the conditioner, then the current coming from the conditioner have to in low ripple and fixed for the entire cycle of grid voltage, thus ripple control is needed. To achieve this, the current controller of the conditioner was employed with fixed band mode. In order to generate current control that work in fixed band mode, equation (17) or (20) indicate that the switching frequency must be adjusted along the grid voltage cycle according to the value of phase angle θ and the reference current gradient $\frac{di_{ref}(t)}{dt} = m$.

Figure 7 shows the simulation of the current in phase-A on bus C. At the beginning, the existing linear load (I_L) is supplied entirely from sources in the grid (I_G). At t = 0.2, the PV plant is operated and work parallel with the grid supplying the linear load. Therefore, the current of the PV-plant (I_PV) is in the form of non-sinusoidal, causes the current from the grid supply to become distorted. Consequently the harmonic current is injected into the sources in the grid side. At t = 0.4, the conditioner is operated in the system. It supplying compensation current (I_C) and compensate for reactive or harmonic power components in the PV plant current.

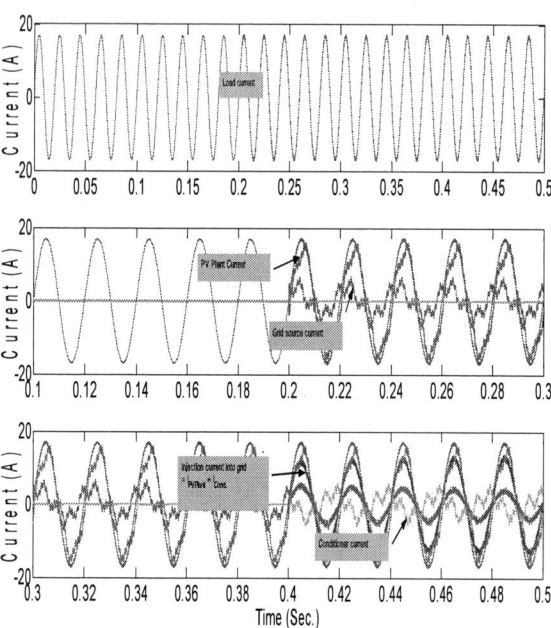

Fig.7. Phase-A current on bus C during simulation time

As a result, the supply current to the grid (I_(PV+C)) via bus B is improved. Furthermore sources in the grid are shielded from the harmonics current while they continue to work in parallel with the PV plant to supply the load.

Figure 8 shows the supply current from sources on the grid, load current, and the compensation current generated by conditioner in steady state condition for various PV plant output currents. Three different conditions are simulated and any changes in system currents are observed. The first condition is when the PV plant produces pure sinusoidal current in-phase with voltage grid (without compensation), Fig. 8.a. Second is when the PV plant produces distorted current in-phase with the grid voltage (harmonics compensation), Fig. 8.b, and the third condition is when the PV plant produces distorted current and lags from grid voltage phase (harmonics and reactive power compensation), Fig.8.c. Simulation results show that in all cases, the conditioner generates different forms of compensation current, by which each of them causes the sum of the conditioner and the PV plant current (I_(PV+C)) to be in-phase and sinusoidal. Thus, the current originating from sources on the grid will be always ideal. Furthermore, Fig. 9 shows harmonics spectrum of injected current to the grid (I_(PV+C)), with THD of 4.83 % . This indicates that the injection current has been successfully improved the plant output current to nearly ideal.

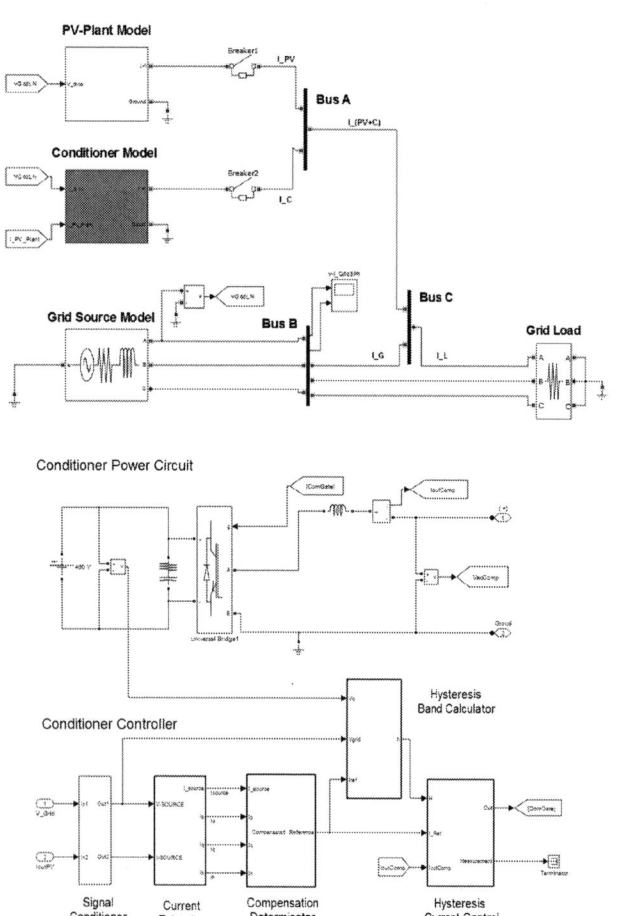

Fig.6. Simulation block diagram, main blocks on the first level (upper), and the second level of conditioner model block (lower)

VI. CONCLUSION

The implementation of the power conditioner to improve the power quality in a distributed generation system consisting of single-phase PV inverters has been described. It was found that the consumption of reactive power varies at different levels of power conversion. The distorted current generated by the PV inverters also contributes to the power quality problems of operating renewable energy based distributed generation. A simple method to compensate the reactive and distorted current using single-phase conditioner has been described and simulated in detail. It was noticed that the designed conditioner worked well and is able to compensate both the harmonics and the reactive power in various conditions of distributed generation plant power output.

The use of current extraction methods which extracting the fundamental components of active and reactive power, as well as the harmonics from a distorted current, provides potential advantages due to its simplicity in algorithm and resources. Furthermore, it can also accommodate the type of compensation used in distributed generation plant. The operation of hysteresis current control in fixed band mode gives the advantage in such a way that the current ripple can be controlled. The addition of current from the conditioner to the distributed generation plant does not cause an increased in ripple on the total injection current to the grid.

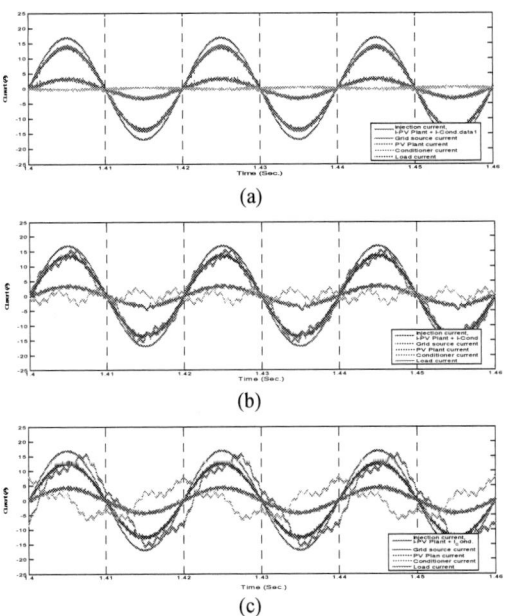

Fig.8. Compensated current in various PV plant output current condition

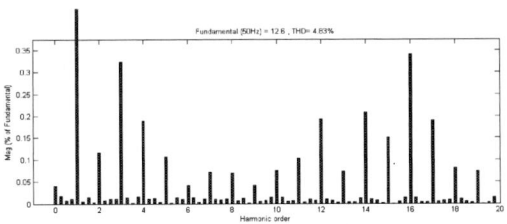

Fig.9. Harmonics spectrum of injection current into grid after compensated

VII. ACKNOWLEDGMENT

The authors thank the Malaysian Government, Ministry of Science, Technology and Innovation (MOSTI) for the Science Fund Grant, Project No. 01-01-06-SF0205 and Research Management Centre, UTM.

REFERENCES

[1] International Energy Agency Photovoltaic Power Systems. Trends in Photovoltaic Applications: Survey Report of Selected IEA Countries between 1992 and 2006. IEA-PVPS-T1-16:2007.

[2] F. Blaabjerg, Z. Chen, and S.B. Kjaer, "Power Electronics as Efficient Interface in Dispersed Power Generation Systems," *IEEE Trans. on Power Electronics*, vol.19 no. 5, pp. 1184-1194. Sept. 2004.

[3] T. Degner, C. Bendel, A. Engler, M. Viotto and C. Metzger, "Effect Renewable Energy Source on Power Quality – Recent Research Activities," in *Proc. The 17th International Conference on Electricity Distribution*, May 2003, pp. 1 – 4.

[4] A.D. Simmons and D.G. Infield, "Current Waveform Quality from Grid Connected Photovoltaic Inverters and its Dependence on Operating Conditions," *Progress in Photovoltaic Research and Application*, vol.8, pp. 411 - 420. John Wiley & Sons, 2000.

[5] A. R. Oliva and J.C. Balda, "A PV Dispersed Generator: A Power Quality Analysis within IEEE 519," *IEEE Trans, on Power Delivery*, vol. 18, no. 2, pp. 525-530, Apr. 2003.

[6] J.H.R. Enslin, P.J.M. Heskes, "Harmonic Interaction Between a Large Number of Distributed Power Inverters and the Distribution Network," *IEEE Trans. on Power Electronics*, vol. 19, no. 6, pp. 1586-1593, Nov. 2004.

[7] P.J.M. Heskes, "Power Quality Behaviour of Different PV-inverter Topologies," in *Proc. The 24th Int. Conference PCIM*, May 2003.

[8] A. Woyte, V.V. Thong, R. Belmans, and J. Nijs, "Voltage Fluctuations on Distribution Level Introduced by Photovoltaic Systems," *IEEE Trans. on Energy Conversion*, vol. 21 no.1. pp. 202-209, March 2006.

[9] D.G. Infield, P. Onions, A.D. Simmons, and G.A. Smith, "Power Quality from Multiple Grid-Connected Single Phase Inverter," *IEEE Trans. on Power Delivery*, vol. 19, no. 4, pp. 1983-1989, Oct. 2004

[10] D. Casadei, G. Grandi, and C. Rossi, "Single-Phase Single-Stage Photovoltaic Generation Based on Ripple Correlation Control Maximum Power Point Tracking," *IEEE Trans. on Energy Conversion*, vol. 21, no. 2, pp. 562-568, June 2006

[11] S.H. Ko, S.R. Lee, H. Dehbonei, and C.V. Nayar, "A Grid-Connected Photovoltaic System with Direct Coupled Power Quality Control," in *Proc. IEEE Ind. Electronics Conference - IECON*, 2006, pp. 5203-5208.

[12] T-F. Wu, H-S. Nien, C-L. Shen, and T.M. Chen, "A Single-phase Inverter System for PV Power Injection and Active Power Filtering with Nonlinear Inductor Consideration," *IEEE Trans. on Industry Application*. vol. 41, no. 4, pp 1075-1083, July/Aug. 2005.

[13] M.B. Bana Sharifian, "Single-Stage Grid Connected Photovoltaic System with Reactive Power Control and Adaptive Predictive Current Controller," *Journal of Applied Science*, vol. 9, pp 1503-1509, 2009.

[14] E.R. Cadaval, M.I.M. Montero, E.G. Romera, and F.B. González, "Power Injection System for Grid Connected Photovoltaic Generation Systems based on two Collaborative Voltage Source Inverters," *IEEE Trans. on Industrial Electronics*, vol. 56, no. 11, pp. 4389-4398, Nov. 2009.

[15] C.J. Hatziadoniu, E.N. Nikolov, and F. Pourboghrat, "Power Conditioner Control and Protection for DG and storage," *IEEE Trans. on Power System*, vol. 18, no. 1, pp.83-90, Feb 2003.

[16] C. Pica, M. Bollero, A. Bollero, A. Tenconi, and L. Limongi, "Single-phase Power Conditioner with Reduced Low-Frequency Current Ripple for Fuel Cells in Distributed Generation Applications," in *Proc. Optimization of Electrical and Electronic Equipment Conf. OPTIM 2008*, pp. 357–362.

[17] G. Grandi, D. Casadei, C. Rossi, "Dynamic Performance of a Power Conditioner Applied to Photovoltaic Sources," in *Proc. The 10th Int. Power Electronics and Motion Control Conf., 2002*, pp. P1-P10.

[18] T. Tanaka, H. Akagi, 1995, "A New Method of Harmonic Power Detection Based on Instantaneous power in Three-phase Circuits," IEEE *Trans. on Power Delivery*, vol. 10, no. 4, pp. 1737 – 1742, Oct. 1995.

[19] P.C. Tan, Z. Salam, A. Jusoh, "A Single-Phase Hybrid Active Power Filter using Extension p-q Theorem for Photovoltaic Application," in *Proc. IEEE PEDS Conf., 2005*, pp. 1250-1255.

[20] W. Hosny, B. Dobrucky, "Harmonic Distortion and Reactive Power Compensation in Single Phase Power System Using Orthogonal Transformation Strategy," *WSEAS Trans. on Power Systems*, vol. 3, no. 4, pp. 237-246, Apr. 2008.

A Study of Maximum Power Point Tracking Algorithms for Stand-alone Photovoltaic Systems

Mei Shan Ngan, Chee Wei Tan

Department of Energy Conversion, Faculty of Electrical Engineering
Universiti Teknologi Malaysia, 81300, Skudai, Johor, Malaysia.

Abstract- **The Photovoltaic (PV) energy is one of the renewable energies that attracts attention of researchers in the recent decades. Since the conversion efficiency of PV arrays is very low, it requires maximum power point tracking (MPPT) control techniques to extract the maximum available power from PV arrays. In this paper, two categories of MPPT algorithms, namely indirect and direct methods are discussed. In addition to that, the advantages and disadvantages of each MPPT algorithm are reviewed. Simulations of PV modules were also performed using Perturb and Observe algorithm and Fuzzy Logic controller. The simulation results produced by the two algorithms are compared with the expected results generated by Solarex MSX60 PV modules. Besides that, the P-V characteristics of PV arrays under partial shaded conditions are discussed in the last section.**

I. INTRODUCTION

In recent decades, the continuous growth of energy demand from all around the world has urged the society to seek for alternative energies due to the depletion of conventional energy resources. Among the available alternative energies, photovoltaic (PV) energy is one of the most promising renewable energy, PV energy is clean, inexhaustible and free to harvest [1]. However, there are two main drawbacks of PV system, namely the high installation cost and the low conversion efficiency of PV modules which is only in the range of 9-17% [1]. Besides that, PV characteristics are non linear and it is very much weather dependent. Fig. 1Fig. 2 show the I-V and P-V characteristics of a typical PV module for a series of temperatures and solar irradiance levels [2,3]. It can be noticed that PV output voltage greatly governed by temperature while PV output current has approximate linear relationship with solar irradiances. It can be seen from the P-V characteristic curve that there is only one peak operating point which is named as the maximum power point (MPP). Due to the high capital cost of PV array, maximum power point tracking (MPPT) control techniques are essential in order to extract the maximum available power from PV array in order to maximize the utilization efficiency of PV array. Therefore, a DC-DC converter is inserted between PV generator and load or battery storage. MPPT algorithms are used to control the switching of DC-DC converter by applying pulse-width modulation (PWM) technique [1,2].

This paper reviews and simulates the existing MPPT algorithms, namely MPPT under fully illuminated and MPPT for partial shaded conditions. The advantages and disadvantages of five different MPPT algorithms are discussed and simulations of PV system using Perturb and Observe (P&O) and Fuzzly Logic controller (FLC) algorithms are

described. Then, results and discussions of the simulations will be analyzed. The development of the PV system under partial shaded conditions will also be discussed. Finally, the last section concludes the review and the simulations of the studied MPPT algorithms.

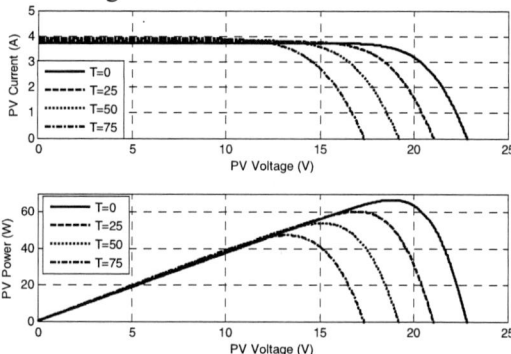

Fig. 1. I-V and P-V characteristics of a typical PV module for varied temperatures [2-4].

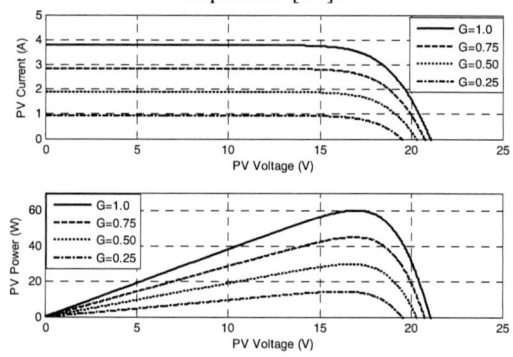

Fig. 2. I-V and P-V characteristics of a typical PV module for varied solar irradiances [2-4].

II. MPPT ALGORITHMS

There are many MPPT algorithms have been developed and implemented by researchers [1-3]. In general, MPPT techniques can be divided into two categories, namely direct and indirect methods [2]. Direct method of MPPT algorithms is independent from prior knowledge of PV modules' characteristics. The MPPT algorithms that include in this category are Perturb and Observe method (P&O), incremental conductance method (INCond.), feedback voltage or current, fuzzy logic method and neural network method. Indirect method requires prior evaluation of PV generator; it is based on mathematical relationship obtained from empirical data. Methods like look-up table, open-circuit PV voltage, short

978-1-4577-0007-1/11 $26.00 © 2011 IEEE

circuits PV current and other MPPT algorithms are included in indirect method.

A. Perturb and Observe (P&O) Method

The P&O method is most widely used in MPPT because of its simple structure and it requires only few parameters. Fig. 3 shows the flow chart of P&O method. It perturbs the PV array's terminal voltage periodically, and then it compares the PV output power with that of the previous cycle of perturbation [5-8].

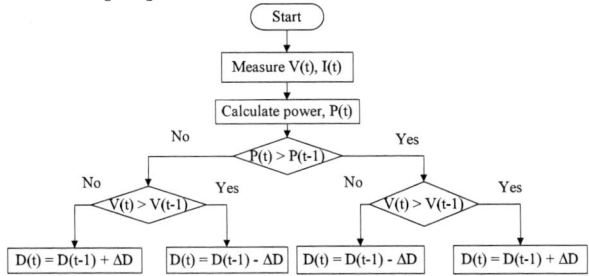

Fig. 3. Flow chart of perturb and observe method (P&O) [5].

Based on Fig. 3, when PV power and PV voltage increase at the same time and vice versa, a perturbation step size, ΔD will be added to the duty cycle, D to generate the next cycle of perturbation in order to force the operating point moving towards the MPP. When PV power increases and PV voltage decreases and vice versa, the perturbation step will be subtracted for the next cycle of perturbation [9]. This process will be carried on continuously until MPP is reached. However, the system will oscillate around the MPP throughout this process, and this will result in loss of energy. These oscillations can be minimized by reducing the perturbation step size but it slows down the MPP tracking system [7,8].

Neil S. D'Souza, et al. [10] employed a peak current control and instantaneous value to calculate the direction of next cycle of perturbation of the conventional P&O method. It increased the speed of response of the system and reduced the amplitude of power oscillations around the MPP. Youngseok.J, et al. [5] proposed an improved perturbation and observation method (iP&O). It satisfied the good dynamic response and steady-state performances. References [4,11] optimized the choice of ΔD for experiment and analysis. Literature [6] designed iP&O method with a current-mode controlled DC-DC step-up converter. It proved that MPP can be tracked accurately under both rapid and gradual changes of solar irradiance.

B. Incremental Conductance Method (INCond.)

The incremental conductance method is based on **(1)** with the derivative of PV output power with respect to the PV voltage. The derivation is zero at MPP, positive on the left side of MPP and negative on the right side of MPP (2) to (4) [8,12].

$$dP/dV = d(IV)/dV = I + V(dI/dV) = I + V(\Delta I/\Delta V) \quad (1)$$

By setting the result equals to zero,

$$\Delta I/\Delta V = -I/V \ (dP/dV = 0) \qquad \text{at MPP} \quad (2)$$
$$\Delta I/\Delta V > -I/V \ (dP/dV > 0) \qquad \text{left side of MPP} \quad (3)$$
$$\Delta I/\Delta V < -I/V \ (dP/dV < 0) \qquad \text{right side of MPP} \quad (4)$$

The MPP is tracked by comparing the instantaneous conductance (I/V) to the incremental conductance ($\Delta I/\Delta V$). Based on the flow chart in Fig. 4, V(k) and I(k) are the PV array output voltage and current at time k. The duty cycle, D of boost converter at which PV array is forced to operate at MPP. When the PV output voltage is constant but the output current increases, the duty cycle will increase. If the current decreases with the constant voltage, the duty cycle decreases [13,14]. The duty of the algorithm is to search a suitable duty cycle at which the incremental conductance equals to instantaneous conductance so that the PV system always operate at the MPP [12].

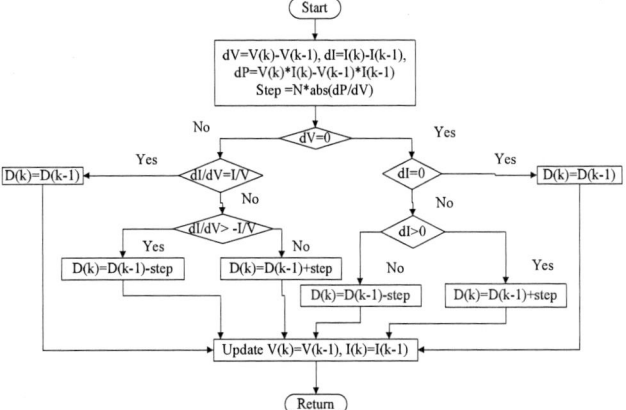

Fig. 4. Flow chart of Incremental Conductance method (INCond.) [8].

The incremental size, ΔD, determines how fast the MPP is being tracked. Fast tracking can be reached by greater incremental size, but it will make the system oscillates about its MPP. The main advantage of INCond method is that it can produce good results under rapidly changing environment. INCond. algorithm is able to achiev lower oscillation around MPP than P&O method. However, it requires two sensors to measure the instantaneous PV output voltage and current, which results in high cost and complex circuit of the system [13].

Jiyong.L, et al. [15] proposed a variable step size incremental conductance method and a space vector pulse width modulation (SVPWM) control scheme for three-phase voltage source PWM inverter. Fangrui. L, et al. [8] also designed a modified variable step size incremental conductance method with a simple constant voltage tracking (CVT) start program which may enable the smooth start process. Bangyin. L, et al. [13] has implemented the same method for experimentation and analysis. All these methods can improve the dynamic and steady-state performance of PV system simultaneously.

C. Open-Circuit Voltage Method

This is a method based on the linear relationship between output voltage of the PV array at the MPP, V_{MPP} and the PV array's open circuit voltage, V_{OC} in (5) under varying temperature and solar irradiance [2,16].

$$V_{MPP} \approx k_1 V_{OC} \quad (5)$$

Constant value of k_1 is dependent on the characteristics of PV array. Generally, it has to be computed empirically in order to determine the V_{MPP} and V_{OC} for varied temperatures and solar irradiances. The value of k_1 ranges from 0.73 to 0.80 for polycrystalline PV module over a temperature range of 0 to 60°C and solar irradiance range of 200 to 1000W/m² [2,16].

In Fig. 5, the PV array is temporarily isolated from MPPT, then the open circuit voltage, V_{OC} is measured periodically by shutting down the power converter momentarily. The MPPT calculates V_{MPP} from the pre-set value of k_1 and the calculated value of V_{OC}. Then, the array's voltage is varied until V_{MPP} is reached [3]. The shutdown of power converter periodically will incur temporary loss of power which results in power extracted will not be the maxima. Since (5) is an approximation, the PV array will never reach the MPP. Even though this technique is very easy and cheap for implementation, yet the constant value of k_1 is not valid in the presence of partial shading of the PV array [2,16].

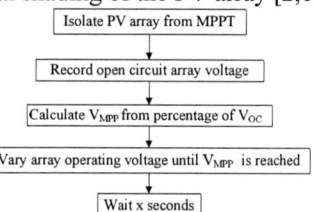

Fig. 5. Flow chart of constant voltage method [3].

D. Short-Circuit Current Method

This method is quite similar to the open circuit voltage method. It is based on the linear relationship between PV array output current at MPP, I_{MPP} and PV array short-circuit current, I_{SC} as shown (6) [2,16].

$$I_{MPP} \approx k_2 I_{SC} \tag{6}$$

where k_2 is the proportionality constant.

The constant value of k_2 also depends on characteristics of PV array. Generally, it ranges from 0.78 to 0.92 for polycrystalline PV modules under the same conditions like the one mentioned previously [2,16]. The way of determining k_2 is more complicated than a fixed value. After k_2 is obtained, the PV system remains with the approximation in **(6)**, until the next calculation of k_2.

The procedure of short-circuit current method is the same as that of open-circuit voltage method, the flow chart of this method can be referred to the same flow chart as shown in Fig. 5. An additional switch is added to the power converter and it is switched on momentarily to measure the short-circuit current, I_{SC} by using a current sensor, then the MPP current is calculated [3]. The output current of PV array is adjusted until the MPP current is reached. This process is repeated periodically. Like open-circuit voltage method, the MPP is never reached since (6) is an approximation, hence the output power produced will not be the maximum.

E. Fuzzy Logic Controller

Fuzzy Logic controller (FLC) works with imprecise inputs, it does not need an accurate mathematical model and it can handle nonlinearity well. It relies on the user's knowledge and experience rather than the technical understanding of the

system. FLC is divided into four categories, which include fuzzification, inference, rule based and defuzzification as shown in Fig. 6 [17,18].

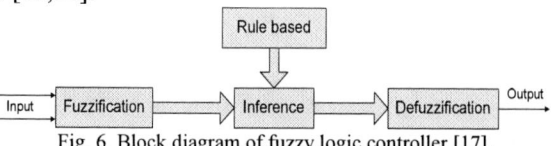

Fig. 6. Block diagram of fuzzy logic controller [17].

In this study, the inputs of FLC are error, E and change in error, CE at sample time k, which are defined by (7) and (8), while the output of FLC is the duty cycle, D [18,20].

$$E(k) = [P_{PV}(k) - P_{PV}(k-1)] / [V_{PV}(k) - V_{PV}(k-1)] \tag{7}$$

$$CE(k) = E(k) - E(k-1) \tag{8}$$

During fuzzification, the numerical input variables are converted into linguistic variables based on the membership functions as shown in Fig. 7 [18]. Five fuzzy levels are used for all the inputs and outputs variables: NB (negative big), NS (negative small), ZE (zero), PS (positive small), and PB (positive big). References [17,19,21] designed seven fuzzy levels to improve the control surface and to allow a smooth transition from transient to steady state. The user has the flexibility of choosing values of numerical variables of the inputs.

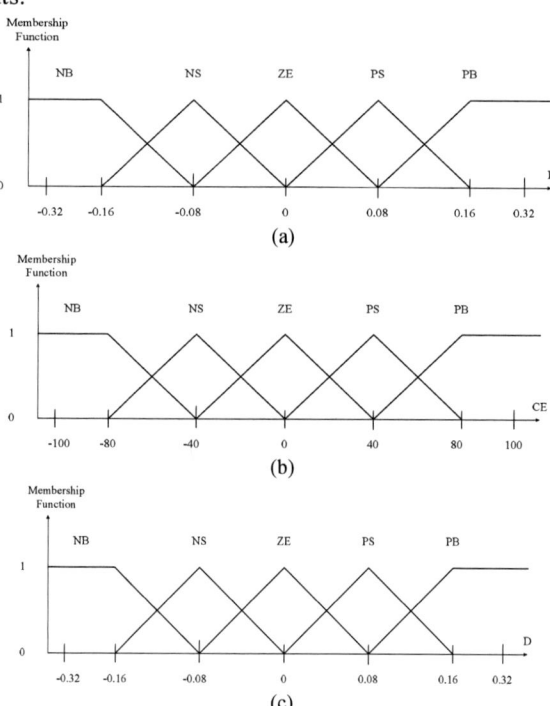

Fig. 7. Membership function for (a) input of error, E, (b) input of change of error, CE and (c) output of duty cycle, D.

Once E and CE are calculated, they are converted into linguistic variables and then the output, D is generated by looking up in a rule base table as shown in Table I, which consists of 25 rules. The FLC tracks the MPP based on master rule of "If X and Y, Then Z" [10,17]. Usually, weights are added to the rules to improve reasoning accuracy and to reduce undesirable consequent.

The inference method used to determine the output of fuzzy logic controller. Mamdani's inference method is the most common method used in engineering application [18,20]. It is employed in this study. Other inference methods include Compositional Rule of Inference (CRI), Generalized Modus Ponens (GMP) and Sugeno inference method [17].

TABLE I. Rule Base for Fuzzy Logic Controller.

E↓\CE→	NB	NS	ZE	PS	PB
NB	ZE	ZE	PB	PB	PB
NS	ZE	ZE	PS	PS	PS
ZE	PS	ZE	ZE	ZE	NS
PS	NS	NS	NS	ZE	ZE
PB	NB	NB	NB	ZE	ZE

During the process of defuzzification, FLC output is converted back to a numerical variable. This in turn, provides an analog signal to control the power converter. The centroid method is the most commonly used defuzzification methods because it has good averaging properties and it produces good results. Other defuzzification methods include bisector and Middle of Maxima (MOM) [17].

References [18-20] compared the results of fuzzy logic controller and P&O controller. It shows that FLC exhibits a better behavior than P&O method, which provides faster tracking of MPP and it presents smoother signal with less fluctuation in steady-state. Neil S. *et al.* [10] applied fuzzy logic for P&O method with peak current control and sampling of instantaneous values. It achieves very fast transient response and reduces the oscillations around MPP in steady-state. Kottas,T.L, *et al.* [22] proposed a method of fuzzy cognitive network by using fuzzy logic controller, which results in very good maximum power operation of PV array under varying change of conditions.

F. Other MPPT algorithms

B. Amrouche, *et al.* [23] proposed artificial neural network, (ANN) based modified P&O method to predict the power value during the next perturbation cycle so that the value of perturbation step can be adjusted for next perturbation cycle. Zhang. L, *et al.* [24] built a Genetic Algorithm trained Radial Basis Function Neural Network (GA-RBFNN) model to predict the reference DC bus voltage of the control system to maximize the output power. Veerachary. M, *et al.* [25] implemented a feed-forward MPPT scheme for coupled-inductor interleaved boost converter fed PV system by using fuzzy logic controller, while ANN is trained offline to estimate the voltage reference. Joe-Air Jiang, *et al.* [26] designed a three-point weight comparison method to avoid rapidly moving of the operating points of PV when it is under varying atmosphere conditions which could overcome the drawback of P&O method. References [27,28] proposed neural fuzzy network for MPPT control scheme. The neural network used to train sets of data off-line for inputs of fuzzy logic controller, while the fuzzy logic controller used to control the duty cycle effectively and hence the MPP can be tracked effectively.

III. SIMULATION

The Solarex MSX60, 60W PV module was chosen for modeling and simulation using MATLAB/SimuLink. The module has 36 polycrystalline cells which are connected in series. The electrical specifications are shown in Table II.

TABLE II. Specifications Of Solarex MSX60.

At temperature	25 °C
Open circuit voltage, V_{OC}	21.0 V
Short circuit current, I_{SC}	3.74 A
Voltage at maximum power, V_{MPP}	17.1 V
Current at maximum power, I_{MPP}	3.5 A
Maximum power, MPP	59.9 W

Fig. 8 shows the configuration of PV system built in Simulink. It consists of a PV module model, a boost converter, MPPT algorithm and PWM generator. The inputs of the PV module are the solar irradiance and the ambient temperature. The output produced by the PV module is the PV current, which acts as a controlled current source for the input of the boost converter. The input capacitance of the converter, C_{in} is 500 μF, the inductance, L is 25 μH and the output capacitance, C_{out} is set to be 5 μF. The load resistance, R is 20 Ω. The MPPT block consists of two MPPT algorithms, namely perturb and observe algorithm and fuzzy logic controller algorithm. The PWM generator provides a triangle waveform for pulse width modulation. The switching frequency, f_s was set to be 25 kHz in this simulation. The sampling time of the simulations is assumed to be 10 μs. Fig. 9(a) and Fig. 9(b) show the configurations of MPPT algorithms in Simulink according to the flow chart of P&O method explained in Fig. 3 and block diagram of FLC in Fig. 6.

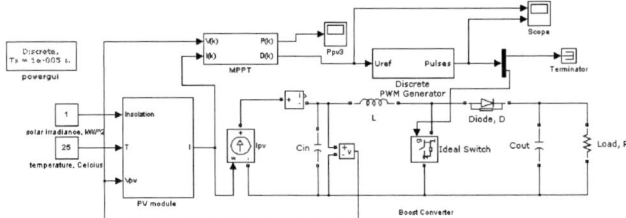

Fig. 8. Configuration of PV system in Simulink.

IV. RESULTS AND DISCUSSION

Fig. 10 shows the PV power waveforms simulated using the Perturb and Observe (P&O) algorithm and the fuzzy logic control, respectively. The incremental and the decremental perturbation step size of P&O algorithm are 0.00001 and 0.01, respectively. Both algorithms were tested at temperature of 25 °C and solar irradiance of 1 kW/m^2. The maximum PV power generated by both algorithms is 60 W, which is the expected maximum power, MPP of Solarex MSX60 as shown in TABLE II. Based on Fig. 10, it can be noticed that time for the PV power curve to reach at steady state for both algorithms is slightly different, it is about 0.005 s for P&O algorithm and 0.006 s for FLC algorithm. Meanwhile, the PV power curve generated by FLC algorithm has a lot of ripples during transient time as compared to that of P&O algorithm.

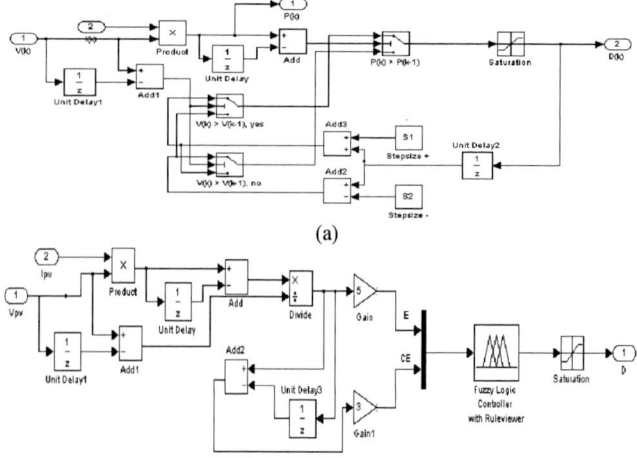

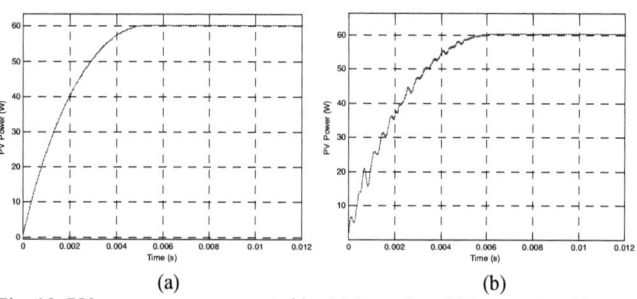

(b) Fig. 9. Configurations of MPPT algorithms in Simulink. (a) Perturb and Observe (P&O) MPPT algorithm and (b) Fuzzy Logic Controller (FLC) algorithm.

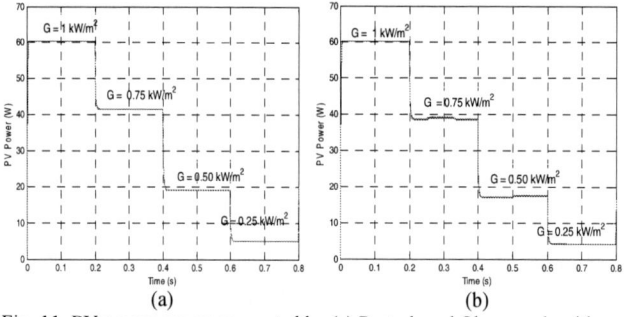

Fig. 10. PV power curves generated by (a) Perturb and Observe algorithm and (b) Fuzzy Logic controller at temperature 25 °C and solar irradiance 1 kW/m^2.

Besides that, the P&O and the FLC algorithms were tested under varying solar irradiance and at a constant temperature of 25°C, The solar irradiance changes from 1 to 0.25 kW/m^2 with every decremental of 0.25 kW/m^2. The PV power waveforms generated by the testing are shown in Fig. 11. It can be noticed that the PV power waveforms generated by P&O algorithm are smooth, while the PV power waveforms generated by FLC algorithm established a lot of ripples at each step.

Fig. 11. PV power curves generated by (a) Perturb and Observe algorithm and (b) Fuzzy Logic controller at temperature 25 °C and solar irradiance change from 1 kW/m^2 to 0.25 kW/m^2.

The P-V characteristic curves of both algorithms are shown in Fig. 12. The bolded curves show the maximum power that can be reached by using the algorithms under varied solar irradiance. It can be noticed the power curves of both algorithms failed to reach to the maximum when the solar irradiance start to drop from 0.50 to 0.25 kW/m^2. The PV

power values generated by these two algorithms are compared to the expected result as shown in Table III. The PV powers generated by P&O algorithm during low solar irradiance are very much less than the expected MPP, they are about 27% and 53% less than the expected MPP for solar irradiance of 0.5 kW/m^2 and 0.25 kW/m^2, respectively. Meanwhile, for the FLC algorithm even worst, the PV powers generated are 30% and 60% less than the expected MPP for solar irradiance of 0.5 kW/m^2 and 0.25 kW/m^2 respectively. This indicated that the P&O and the FLC algorithms are not able to perform well under low solar irradiance.

The simulations show that all these conventional algorithms are not able to perform well under varying solar irradiances. It is because of there is no modification or changes made on both of the P&O and FLC MPPT algorithms. Nevertheless, the comparison of the simulation results provides very useful information on the MPP.

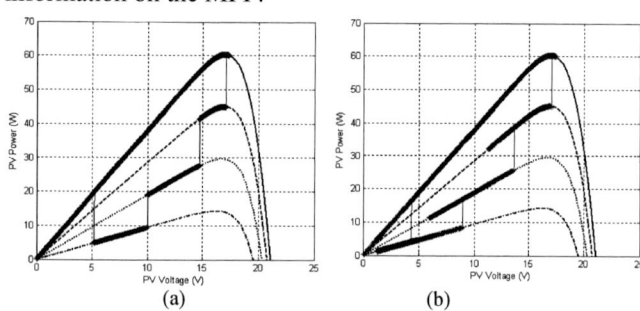

Fig. 12. The P-V characteristics curves generated by (a) Perturb and Observe algorithm and (b) Fuzzy Logic Controller when the solar irradiance changes from 1 to 0.25 kW/m^2.

TABLE III. Comparison results of PV power generated under Varied Solar Irradiance and Constant temperature.

Solar irradiance (kW/m^2)	PV Power (W)		
	Perturb and Observe algorithm	Fuzzy Logic Controller	Expected result
1.00	60.35	60.35	60.35
0.75	41.59	38.93	45.07
0.50	19.07	17.08	26.29
0.25	4.96	4.09	10.70

V. PV ARRAYS UNDER PARTIAL SHADED CONDITIIONS

In reality, PV arrays do not receive uniform insolations at all the time. Some part of the PV arrays might be shaded by heavy cloud, trees or nearby buildings. The shaded PV cells absorb a large amount of electric power generated by other PV cells that receive high insolation and convert it into heat. This situation is called the hot spot problem which may damage the PV cells with low insolation [29]. In order to relieve the stress on the shaded PV cells, bypass diodes are connected in parallel with each PV modules [30]. The inserted bypass diodes may cause multiple peaks are established in the P-V characteristic curves under non-uniform insolation as shown in Fig. 13. In general, most of the conventional MPPT algorithms that were mentioned previously cannot detect the global maximum peak effectively [31]. Therefore, further researches should be made on this situation so that the PV systems under partial shaded conditions can perform at the optimum condition.

978-1-4577-0007-1/11 $26.00 © 2011 IEEE

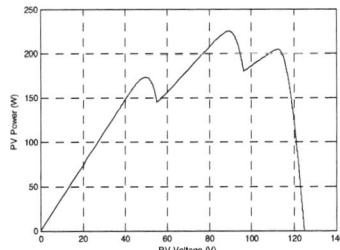

Fig. 13. P-V characteristic curve of three PV modules connected in series under non-uniform insolation.

VI. CONCLUSION

Two groups of MPPT control algorithms, namely direct and indirect methods were discussed and reviewed. The study presents a simulation comparison of the maximum power produced for P&O algorithm and FLC method. The P&O algorithm is the simplest method, which results in low cost of installation and it may be competitive with other MPPT algorithms. Hence, it should be further modified and improved to provide a good result. On the other hand, the FLC method provides a simpler way to arrive at a definite conclusion, but it highly depends on the user's knowledge of the process operation for the FLC parameter setting. It could also be further improved according to its flexibility of choosing numerical variables of input and output. The conventional MPPT algorithms are not capable of solving the problems of multiple peaks that established in the P-V characteristic curves of the PV systems due to partial shaded conditions. Therefore, further research should be done to extract maximum power effectively from the PV systems under non-uniform insolation.

REFERENCES

[1] Roberto Faranda, S.L., *Energy Comparison of MPPT Techniques for PV Systems.* WSEAS Trans. on POWER SYSTEMS, vol. 3, No.6.

[2] V. Salas, E.O., A. Barrado, A. Lazaro, *Review of the Maximum Power Point Tracking Algorithms for Stand-alone Photovoltaic Systems,* Solar Energy Materials and Solar Cells, 2006, p.p 1555–1578.

[3] Hohm, D.P. and M.E. Ropp, *Comparative Study of Maximum power point tracking algorithms,* Progress in Photovoltaics: Research and Applications, 2003, Vol.11, No.1, pp. 47-62.

[4] Nicola Femia, Giovanni Petrone, Giovanni Spagnuolo, and Massimo Vitelli, *Optimization of Perturb and Observe Maximum Power Point Tracking Method,* IEEE Trans. on Power Electronics, 2005, Vol. 20, NO. 4, p.p. 963-973.

[5] Youngseok, J., et al., *Improved Perturbation and Observation Method (IP&O) of MPPT Control for Photovoltaic Power Systems,* Photovoltaic Specialists Conference, IEEE 31st, 2005.

[6] Chee Wei, T., T.C. Green, and C.A. Hernandez-Aramburo, *A Current-mode Controlled Maximum Power Point Tracking Converter for Building Integrated Photovoltaics,* Power Electronics and Applications, 2007, European Conference.

[7] Al-Diab, A. and C. Sourkounis. *Variable Step Size P&O MPPT Algorithm for PV Systems,* Optimization of Electrical and Electronic Equipment (OPTIM) Conference, 2010.

[8] Fangrui, L., et al., *A Variable Step Size INC MPPT Method for PV Systems,* IEEE Trans. on Industrial Electronics, 2008, Vol.55, No. 7, p. p. 2622-2628.

[9] Marcelo Gradella Villalva, J.R.G., Ernesto Ruppert Filho, *Analysis and Simulation of The P&O Algorithm Using A Linearized PV Array Model,* Industrial Electronics Conference, IECON' 09, 2009, p.p. 231-236.

[10] Neil S. D'Souza, L.A.C.L.a.X.L., *An Intelligent Maximum Power Point Tracker Using Peak Current Control,* Power Electronics Specialist Conference, 2005, PESC'05, IEEE 36th.

[11] Femia, N., et al. *Optimizing Duty-cycle Perturbation of P&O MPPT Technique,* Power Electronics Specialists Conference, 2004. PESC 04.

[12] Dorofte, C., U. Borup, and F. Blaabjerg. *A Combined Two-method MPPT Control Scheme for Grid-connected Photovoltaic Systems,* Power Electronics and Applications, 2005.

[13] Bangyin, L., et al. *Analysis and Improvement of Maximum Power Point Tracking Algorithm Based on Incremental Conductance Method for Photovoltaic Array,* Power Electronics and Drive Systems, 2007. PEDS '07. 7th International Conference.

[14] Safari, A. and S. Mekhilef, *Simulation and Hardware Implementation of Incremental Conductance MPPT with Direct Control Method Using Cuk Converter,* IEEE Trans. Industrial Electronics, unpublished.

[15] Jiyong, L. and W. Honghua. *A Novel Stand-alone PV Generation System Based on Variable Step Size INC MPPT and SVPWM Control,* Power Electronics and Motion Control Conference, 2009.

[16] Trishan Esram, J.W.K., Philip T. Krein, Patrick L. Chapman, and Pallab Midya., *Dynamic Maximum Power Point Tracking of Photovoltaic Arrays Using Ripple Correlation Control.* IEEE Trans. on Power Electronics, 2006. Vol.21, No. 5.

[17] Kalantari, A., et al., *A Faster Maximum Power Point Tracker Using Peak Current Control,* IEEE Symposium on Industrial Electronics and Applications, 2009.

[18] M.S. Aït Cheikh*, C.L., G.F. Tchoketch Kebir and A. Zergueras, *Maximum power point tracking using a fuzzy logic control scheme,* Revue des Energies Renouvelables, 2007, Vol. 10, No. 3, p.p. 387 – 395.

[19] Subiyanto, A.M., M A Hannan, *Maximum Power Point Tracking in Grid Connected PV System Using A Novel Fuzzy Logic Controller.* Proceedings of 2009 IEEE Student Conference on Research and Development, 2009.

[20] F.Bouchafaa, D.B., M.S.Boucherit, *Modeling and Simulation of A Gird Connected PV Generation System With MPPT Fuzzy Logic Control,* 7th International Multi-Conference on Systems, Signals and Devices, 2010.

[21] Ahmed G. Abo-Khalil, D.-C.L., Jong-Woo Choi and Heung-Geun Kim, *Maximum Power Point Tracking Controller Connecting PV System to Grid,* Journal of Power Electronics, 2006, Vol.6, No. 3.

[22] Kottas, T.L., Y.S. Boutalis, and A.D. Karlis, *New Maximum Power Point Tracker for PV Arrays Using Fuzzy Controller in Close Cooperation with Fuzzy Cognitive Networks,* IEEE Trans. on Energy Conversion, 2006, Vol.21, No.3, p.p. 793-803.

[23] B. Amrouche, M.B.a.A.G., *Artificial Intelligence Based P&O MPPT Method for Photovoltaic Systems,* Revue des Energies Renouvelables ICRESD-07 Tlemcen, 2007, p.p11 – 16.

[24] Jiying, S., et al., *A Practical Maximum Power Point Tracker for The Photovoltaic System,* IEEE International Conference on Automation and Logistics, 2009.

[25] Veerachary, M., T. Senjyu, and K. Uezato, *Neural Network Based Maximum Power Point Tracking of Coupled Inductor Interleaved-Boost Converter Supplied PV System Using Fuzzy Controller,* IEEE Trans. on Industrial Electronics, 2003, Vol.50, No.4, p.p. 749-758.

[26] Joe-Air Jiang, T.-L.H., Ying-Tung Hsiao and Chia-Hong Chen, *Maximum Power Tracking for Photovoltaic Power Systems,* Tamkang Journal of Science and Engineering, 2005, Vol. 8, No 2, p.p. 147-153.

[27] He, S. and S.K. Starrett. *Modeling Power System Load Using Adaptive Neural Fuzzy Logic and Artificial Neural Networks,* North American Power Symposium (NAPS), 2009.

[28] Iskender, I., *A Fuzzy Logic Controlled Power Electronic System for Maximum Power Point Detection of A Solar Energy Panel,* The International Journal for Computation and Mathematics in Electrical and Electronic Engineering, 2005, Vol.24, No.4, p.p. 1164-1179.

[29] Young-Hyok Ji, et al., *A Real Maximum Power Point Tracking Method for Mismatching Compensation in PV Array under Partially Shaded Conditions,* unpublished.

[30] Ramaprabha Ramabadran, *MATLAB Based Modeling and Performance Study of Series Connected SPVA under Partial Shaded Conditions,* Journal of Sustainable Development, Nov. 2009, Vol.2, No.3, pp. 85-94.

[31] Y.-J. Wang and P.-C. Hsu, *Analytical Modeling of Partial Shading and Different Orientation of Photovoltaic Modules,* IET. Renew. Power Gener., 2010, Vol.4, Iss.3, pp 272-282.

Robust Current-Mode DC Drive

Aisha Akbar Awan,Mohammad Bilal Malik

Department of Electrical Engineering, College of Electrical and Mechanical Engineering Rawalpindi
National University of Sciences and Technology(Pakistan)
aishamaryum@hotmail.com, mbmalik@ceme.edu.pk

Abstract— **In this paper, we propose a robust controller that converts a conventional voltage-mode H-bridge into a current-mode drive. The design has also been physically implemented. This technique results into low-cost, high performance machine drive.**

Keywords-component; current mode drives, dc machine, output regulation, disturbance rejection, observer design.

I. INTRODUCTION

PWM based machine drives with current control loops have gained popularity. Although mainly work done on current mode control technique encompasses around inverters and ac drives. The current source controlled is particularly suited for drive systems working in high dynamic conditions such as servo drives for machine tools and robotics[1]. For current-mode controlled drives researchers have introduced various techniques which include hysteresis control, predictive control, adaptive control, ramp compensation, vector-based control techniques. A part from these techniques sliding mode control was being adopted by researchers to deduce current control scheme for dc motor drives[2-3].This technique required extensive calculations to calculate load parameters. Auto tuning technique was proposed in [3] which allows the algorithm to be applied without load information but this has made its implementation bit difficult.

Among these technique hysteresis mode control has gained lot of popularity because of good transient response and ease in implementation requiring minimum hardware [4-8].The main discrepancies of this method is wide variations in frequency, produces current ripples in steady state and is sensitive to phase commutation which subsequently results in generating PWM noise. For reducing current ripples and better steady state response vector-Control and predictive control method were used but their accurate and extensive calculations of parameters to assure good response made them quite complicated[9-11].Ramp compensation technique operates at fixed frequency and inductor current which directly controls the duty cycle of the switch. But these are prone to sub harmonic oscillations when dutycycle approach 50%. In our approach we have used regulator theory and observer design to control current drive. This method has eliminated lengthy calculations of load parameters[12-14].

Section –II will introduce the concept of current drives, role of conventional switches in h-bridge which drives the dc-machine. Section-III will describe interfacing of h-bridge and

dc machine. Section-IV will introduce the robust current mode output feedback regulation controller with its simulation results. Conclusion is presented in Section-V.

II. CURRENT MODE DC-DRIVE:

This section encompasses around the main idea of current mode controlled dc drive scheme. Current drive is preferred over conventional voltage-mode drives because it can directly control the torque of a dc-machine. The operation of current-controlled dc machine can be comprehended by figure-1.Here Q1, Q2, Q3 and Q4 are n-channel MOSFETs. 'M' is the dc machine whose output current is being monitored by the current sensor continuously. On the basis of output current and reference current error is being calculated. This error signal controls the PWM of the four MOSFETs.

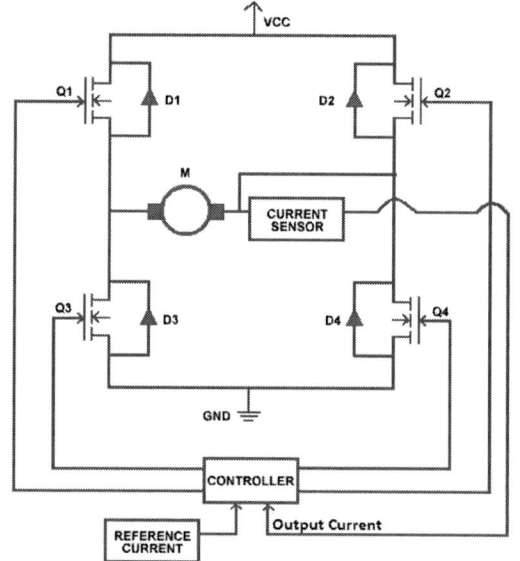

Figure.1 Schematic Diagram of Current Driven DC-Machines

A. Conventional H-bridge and its modes of operation

H-bridge is used for controlling direction, speed and operating modes of dc-machine. For simplification of model we have taken ideal switches having zero rise and fall time to avoid shorting at one side of bridge. In addition to switches diodes play an important role by connecting in anti-parallel direction to the switches. Whenever DC-Machine has been controlled with H-bridge it can operate in many different modes.

B. Mode-1

In this mode switch 1 and switch 4 are ON for ton. Left Side of the bridge gets connected to voltage source and other side gets connected to the ground. Energy will be flowing from source to the dc machine. During this time, current will increase from 0 to maximum; dc machine is absorbing electrical energy from supply and converting it into mechanical energy so it is working as motor in this mode. Average applied voltage is given by: $D(v_s - v_e)$

Here D is duty cycle vs is applied voltage and ve is the back EMF generated.

C. Mode-2

For current controlled machine devices, we have a choice of the recirculation path the current flows in "off-time".

In our model for "off-time" duration machine will send back the energy to power supply. Inductor has stored current in ton time and it will act as current source during off duration.

Inductor discharges the current depending on the time constant and current will flow through the anti-parallel diodes now. Here machine is sending back energy to power supply. The motor is forcing current right through its armature, through Q2'S diode then back to supply.

In this mode mechanical energy is being converted to electrical energy so it is acting as generator. H-bridge applies average applied voltage during "off time" will be given as:
$(1-D)(v_s + v_e)$

Average applied voltage to the dc machine is:
$$v_{ap} = D(v_s - v_e) - (1-D)(v_s + v_e)$$

Switch 1 and 4 are turned on for 30% of the time period and for rest of the time period all switches are turned off. Current will initially flow through MOSFET 1 and 4 forcing the machine to work as motor in "ON" duration. For rest of the time inductor will discharge its current through diode and back to the battery. Here in this duration it will work as generator. The net armature current has been shown in figure:2 for 30% on time.

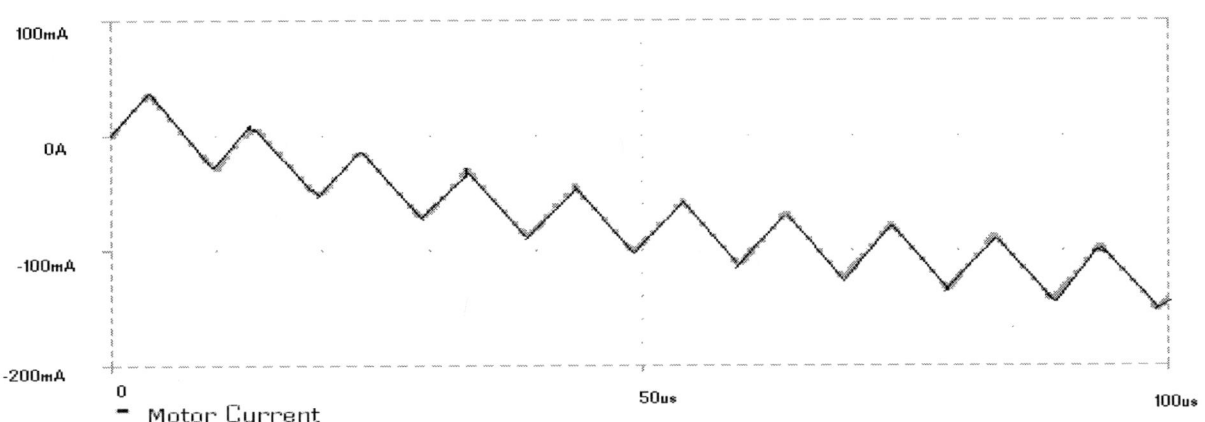

Figure2: Armature Current of DC-Machine

III. MACHINE MODEL:

In dc machines armature current varies depending on the load conditions and applied voltage. We represent variations of the armature current by the following first order differential equation.

$$v_{ap} = R_a i_a + L_a \frac{di_a}{dt} \qquad (1)$$

This equation represents dc-machine where Ra, ia, La represents armature resistance, armature current, armature inductance.

The average voltage seen by the machine will depend on duty cycle.

$$v_{ap} = D(v_s - v_e) - (1-D)(v_s + v_e) \qquad (2)$$

Whenever motor starts spinning because of change of flux it induces EMF which varies linearly with speed. This back-EMF has been taken as disturbance in the average applied voltage.

$$v_e = k_e \omega \qquad (3)$$

Let armature current be the state of the system.

$$x_1 = i \qquad (4)$$

$$\frac{dx_1}{dt} = \frac{v_{ap} - R_a x_1}{L} \qquad (5)$$

$$\frac{dx_1}{dt} = -ax_1 + bu - \frac{v_e}{L} \qquad (6)$$

Now here ;

$$a = -\frac{R}{L}; b = \frac{1}{L}$$

$$u = (2D-1)v_s$$

IV.   OUTPUT TRACKING CONTROL:

Traditional control theories don't model disturbances and reference signals. Utilizing the generalized output regulation technique mentioned in [15] improves the overall performance of current drive. The system is discretized through zero order hold equivalence. Our discrete system is represented by

$$x[k+1] = ax[k] + bu[k] + Pw[k] \qquad (7)$$

Here, x[k] represents the armature current, u[k] is average applied voltage seen by the dc machine. System is subjected to disturbance represented by Pw[k].

Where;

$$P= [0 \; b];$$

In our system we can model class of disturbance and reference signals by;

$$w[k+1] = Sw[k] \qquad (8)$$

Where;

$$S = \begin{pmatrix} 1 & 0 \\ 0 & 1 \end{pmatrix}; \; w[k] = \begin{pmatrix} w_1[k] \\ w_2[k] \end{pmatrix}$$

$w_1[k]$ is the reference signal to be tracked and $w_2[k]$ is the disturbance to be rejected.

$$v_e = w_2[k] = disturbance$$

Now the output of our system will be represented by this:

$$y[k] = Cx[k]; \qquad (9)$$

$$C = 1;$$

Tracking error has been given as;

$$e[k] = Cx[k] + Qw \qquad (10)$$

Where,

$$Q = \begin{pmatrix} -1 & 0 \end{pmatrix}$$

1.   When Eigen values of S are outside the unit circle.

2.   The pair(A,B) is stabilizable

If these two assumptions hold. Then the output regulation problem via full information feedback is solvable if and only if there exist matrices   and   which solve linear matrix equations.[15] We can then design a suitable tracking controller which makes the system stable.

$$\Pi_{1\times2} S_{2\times2} = A_{1\times1} \Pi_{1\times2} + B_{1\times1} \Gamma_{1\times2} + P_{1\times2}$$

$$0 = C_{1\times1} \Pi_{1\times2} + Q_{1\times2} \qquad (11)$$

Where;

$$\Pi_{1\times2} = \begin{pmatrix} 1 & 0 \end{pmatrix};$$

$$\Gamma_{1\times2} = \begin{pmatrix} \dfrac{1-a}{b} & -1 \end{pmatrix}$$

Π and Γ solved linear matrix equations so suitable feedback tracking control can be achieved which is given by;

$$u = K(x - \Pi w) + \Gamma w \qquad (12)$$

where K is an arbitrary feedback such that (A+BK) should be stable.

The input for suitable tracking controller which makes the system stable is given by:

$$u[k] = Kx[k] + Nr[k] - d[k] \qquad (13)$$

A. Observer Design

Using output feedback the objectives for closed loop state equation are that output signal should track any constant reference input. The basic idea behind this problem is to use observer so that it can generate asymptotic estimates of both the plant state and the disturbance.

The plant has been represented as :

$$\begin{pmatrix} x[k+1] \\ d[k+1] \end{pmatrix} = \begin{pmatrix} a & b \\ 0 & 1 \end{pmatrix}\begin{pmatrix} x[k] \\ d[k] \end{pmatrix} + \begin{pmatrix} b \\ 0 \end{pmatrix} u[k]$$

$$(14)$$

$$y[k] = \begin{pmatrix} 1 & 0 \end{pmatrix}\begin{pmatrix} x[k] \\ d[k] \end{pmatrix}$$

Observer design has been represented as

$$\begin{pmatrix} x[\hat{k}+1] \\ d[\hat{k}+1] \end{pmatrix} = A\begin{pmatrix} \hat{x[k]} \\ \hat{d[k]} \end{pmatrix} + Bu[k] + H(y - C\begin{pmatrix} x[\hat{k}] \\ \hat{d[k]} \end{pmatrix}) \quad (15)$$

Output feedback tracking control has been continuously rejecting the disturbance at the input. Our suitable output feedback tracking control has been given as:

$$u[k] = k\,\hat{x}[k] + Nr[k] - \hat{d}[k] \quad (16)$$

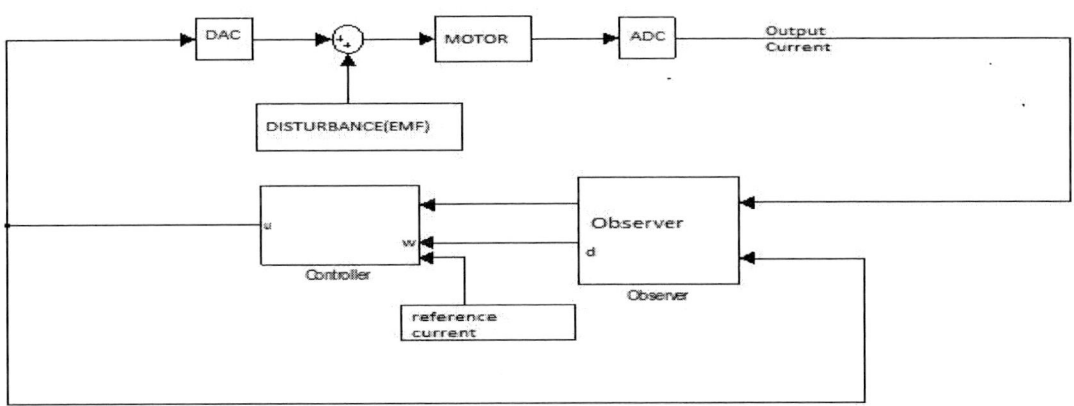

Figure 3: Output Feed Back Tracking Control

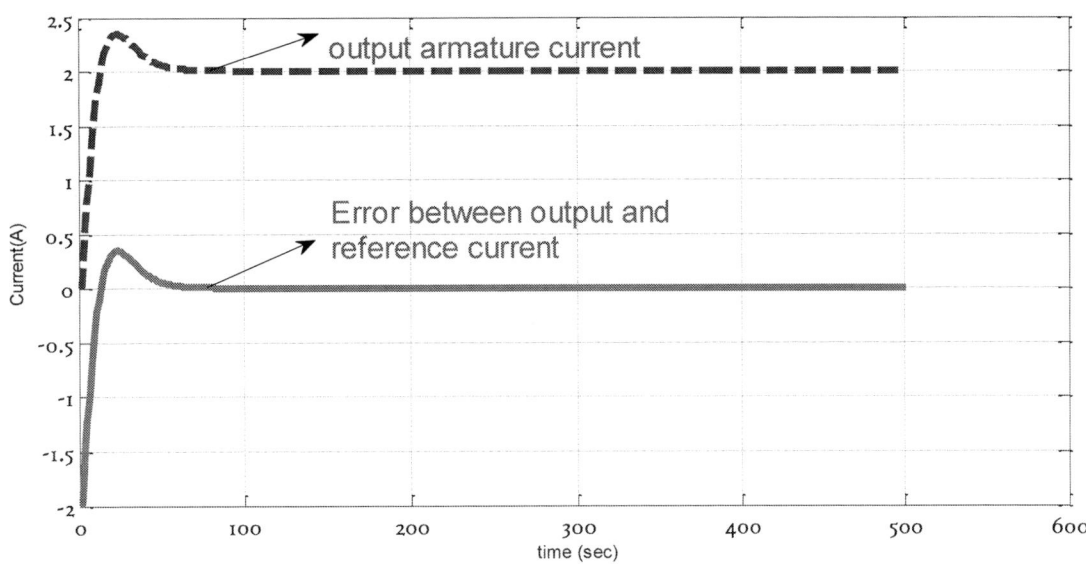

Figure 4:Simulated Results of Regulated Output Current and Error

We can see error of the system is going to be zero in a finite time. So system is asymptotically stable. System is achieving desired current in finite time by utilizing low cost sensor.

V. EXPERIMENTAL RESULTS:

The current-mode drive has been physically implemented. We have monitored output current of the dc- machine through current sensor.

For experimental verification of results we have used dc machine which has armature resistance and inductance given by these parameters. Ra = 1 Ω, La =2.1mH, Iref =1A . The duty cycle has been calculated which controlled the PWM of the H-bridge as per figure 5. On the basis of output current and reference current duty cycle has been established which ultimately reduces the error of the system to zero as per figure 7and achieves perfect tracking shown in figure.6

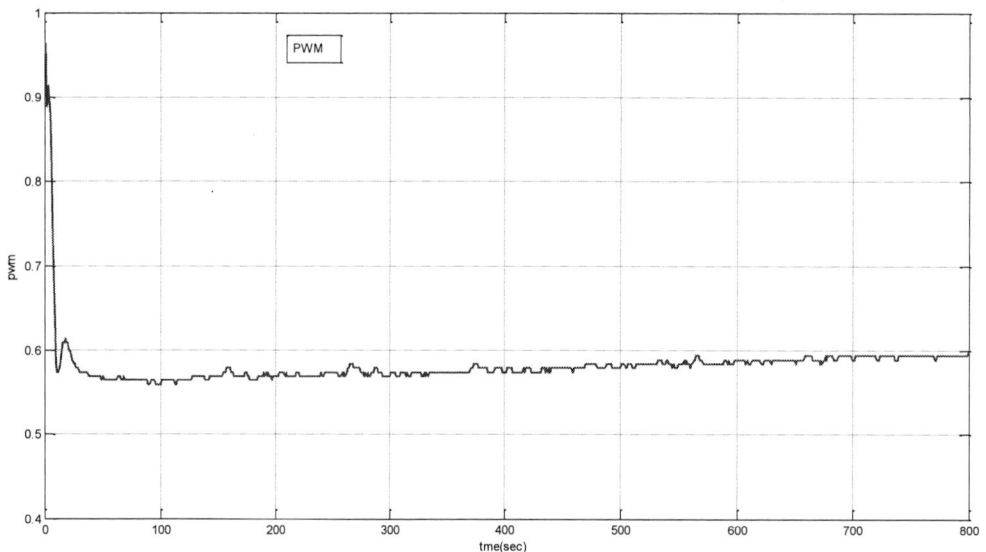

Figure 5:PWM of the Current Drive

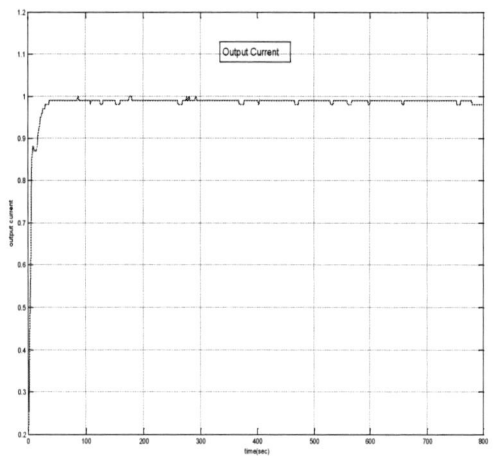

Figure 6: Output Current of Practical Systems

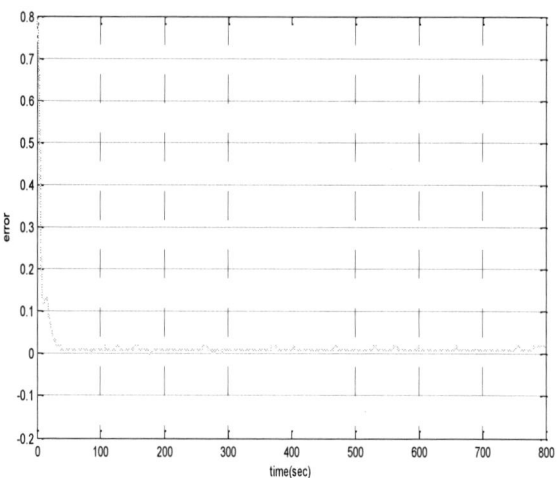

Figure 7: Error of the System

978-1-4577-0007-1/11 $26.00 © 2011 IEEE

VI.   CONCLUSION:

A robust current-mode control scheme for dc drives is presented in this paper. It has been shown that the proposed algorithm achieves asymptotic stability in a finite time. This scheme has utilized Generalized Output Regulation which modeled disturbances and references signals improved overall transient and steady state response of current drive. The disturbances are being continuously rejected which is improving the performance of the system. The experimental results depict the effectiveness of proposed scheme.

REFERENCES

[1] Andrzeg Sikroski and Tadeusz Citko," Current Controller Reduced Switching Frequency for VS-PWM Inverter Used with AC Motor Drive Applications" IEEE Transactions on Industrial Electronics, Vol.45, No.5 October 1998

[2] V.I.Utkin "Sliding mode control design principles and applications to electric drives" IEEE Trans.Ind Electron., vol. 40,pp. 23-34, Feb 1993

[3] Jessen Chen and Pei-Chong Tang "A Sliding Mode Current Control Scheme for PWM Brushless DC Motor Drives" IEEE Transactions on Power Electronics, Vol.14, No.3 May 1999

[4] Marian P. Kazmierkowski, and Luigi Malesani, "Current Control Techniques for Three-Phase Voltage-Source PWM Converters: A Survey" IEEE Transactions on Industrial Electronics, VOL. 45, NO. 5, October 1998

[5] Luigi Malesani, Paolo Mattavelli and Paolo Tomasin," Improved Constant-Frequency Hysteresis Current Control of VSI Inverters with

Simple Feedforward Bandwidth Prediction" IEEE Transactions on industry applications, VOL. 33, NO. 5, September/October 1997

[6] Luigi Malesani, PaoloTenti, Elena Gaio, and Roberto Piovan ,"Improved Current Control Technique of VSI PWM Inverters with Constant Modulation Frequency and Extended Voltage Range" IEEE Transactions on Industry Applications, VOL 21, NO 2, March/April 1991

[7] Luigi Malesani, PaoloTenti,"A Novel Hysteresis Control Method for Current –Controlled Voltage-Source PWM Inverters with Constant Modulation Frequency" IEEE Transactions on Industry Applications, VOL 26, NO 1, January/Feburary 1990

[8] Hoangle-Huy, "An adaptive Current Control Scheme for PWM Synchronous Motor Drives: Analysis and Simulation" IEEE Transactions on Power Electronics, VOL 4 October 1989

[9] AndrzejSikorski and TadeuszCitko," Current Controller Reduced Switching Frequency for VS-PWM Inverter Used with AC Motor Drive Applications" IEEE Transactions On Industrial Electronics, VOL. 45, NO. 5, OCTOBER 1998

[10]Jingquan Chen, AleksandarProndic ', Robert W. Erickson,andDraganMaksimovic ', "Predictive Digital Current Programmed Control" IEEE Transactions on Power Electronics, VOL. 18, NO. 1, January 2003

[11] StéphaneBibian, and Hua Jin, "High Performance Predictive Dead-Beat Digital Controller for DC Power Supplies" IEEE Transactions On Power Electronics, VOL. 17, NO. 3, MAY 2002

[12] Robert Sheehan," Understanding and Applying Current-Mode Control theory" Power Electronics Technology Exhibition and Conference October 30 – November 1, 2007 Hilton Anatole

[13] Dr. Ray Ridley,"A More Accurate Current-Mode Control Model"

[14] Kai Wan,and Mehdi Ferdowsi," Projected Cross Point – A New Average Current-Mode Control Approach"

[15]Ali Saberi, Anton A.Stoorvogel,ZongliLin,"Generalized Output Regulation for Linear Systems", Proceedings of the American Control Conference Albuquerque,New Mexico June 1997

Computer Simulation of Single-phase Control Rectifier using Single-phase Matrix Converter with Reduced Switch Count

R. Baharom; *Member, IEEE*, A. Idris; *Student Member, IEEE*; N. R. Hamzah, *Member, IEEE* and M. K. Hamzah;
Senior Member, IEEE
Faculty of Electrical Engineering
Universiti Teknologi MARA
Selangor, Malaysia.

Abstract- In this paper, single-phase control rectifier using single-phase matrix converter with reduced switch count is proposed. The proposed method is able to perform AC-DC converter operation using only six switches as perform by conventional single-phase matrix converter that uses eight main switches. The PWM technique was used to calculate the switch duty ratio to synthesize the output. Switch commutation arrangements were developed that allows dead time to avoid current spikes of non-ideal switches whilst providing a current path for the inductive load to avoid voltage spikes. The SPMC topology with reduced switch count can be easily control using suitable proposed switching algorithm. The feasibility of the proposed topology was verified by a computer simulation.

Keywords—Single-phase matrix converter (SPMC) with reduced switch count; AC-DC converter; Commutation strategy.

I. INTRODUCTION

In the recent past, matrix converters (MC) have marked their present as an advanced circuit topology that could perform a direct AC-AC power converter employing nine bidirectional switches. It is a modern power conversion topology, that was first proposed by Gyugyi [1] in 1976, followed by Alesina and Venturini in 1980 [2, 3], representing the circuit as a matrix of bi-directional power switches. It is a force commutated converter which uses an array of controlled bidirectional switches as the main power elements to create a variable output voltage system with unrestricted frequency.

There is presently considerable interest in matrix converter (MC) technology both in academic and industrial community [4]. One of the key benefits claimed for the matrix approach is the possibility of greater power density due to the absence of a DC link [5] in direct AC conversion. In addition, MC is able to operate in all the four quadrants of operation. These features make the matrix converter a suitable alternative to the traditional voltage source inverter [6]. Furthermore, it has the advantage of bidirectional power flow, high reliability and compact design. In comparison with conventional AC/DC/AC converter, it has the following merits [1]:

- No large energy storage components, such as large DC capacitors or inductors, are needed. As a result, a large capacity and compact converter system can be designed.

- Four-quadrant operation is straightforward, by controlling the switching devices appropriately, both output voltage and input current are sinusoidal with only harmonics around or above switching frequency.

- Clean input power characteristics with high input power factor.

- Increased power density with the possibility of operating at higher temperatures.

These ideal advantages can be fulfilled by MC, and this is the reason for the tremendous interest in the topology. However, the first 25 years in the development had been slow due to lack of understanding of the safe-commutation strategies.

The single-phase matrix converter (SPMC) has subsequently been investigated and has shown increased potential for use in converter designs. By manipulating its bi-directional switches additional new features could be designed and realized. The SPMC was first realized by Zuckerberger *et al.* [12] in 1997 with other works on AC-AC [7], DC-DC [8], DC-AC [9] and more recently the AC-DC operation [10-15].

The purpose of this paper is to present a switching algorithm for control AC-DC converter using single-phase matrix converter with reduced switch count. To verify the feasibility of the proposed topology, a simulation circuit using MATLAB/Simulink with suitable control switching technique was demonstrated.

II. SINGLE-PHASE MATRIX CONVERTER TOPOLOGY

The SPMC as shown in Figure 1 consists of a matrix of input and output lines with four bidirectional switches connecting the single-phase input to the single-phase output at the intersections. The matrix converter requires bidirectional switch capable of blocking voltage and conducting current in both directions. Figure 2 shows the bi-directional switch module using a common emitter anti-parallel IGBT with diode pair switch-cell to allow current flow in both directions, whilst at the same time blocking the voltage.

978-1-4577-0007-1/11 $26.00 © 2011 IEEE

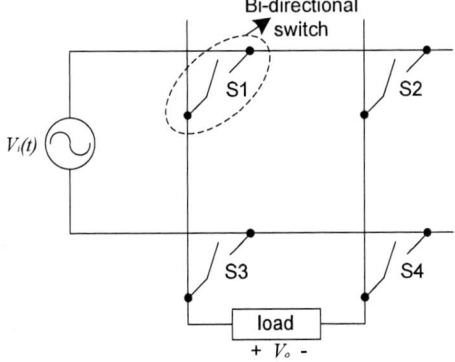

Figure 1: SPMC circuit configuration

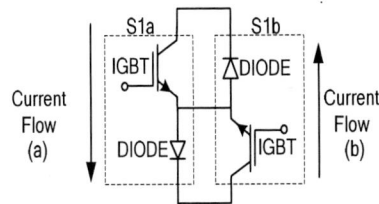

Figure 2: Bi-directional switches module (common emitter)

III. PROPOSED SPMC TOLOPOGY WITH REDUCED SWITCH COUNT

The proposed AC-DC converter using SPMC topology with reduced switch count is as shown in Figure 3. Two switches have been removed due to unused for all purpose of the operation for controlled AC-DC converter including safe-commutation strategy hence reduced the complexity of the conventional SPMC circuit.

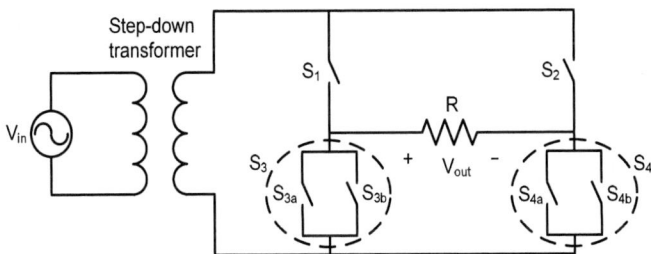

Figure 3: Proposed SPMC circuit topology with reduced switch count

The controllable amplitude of the DC output waveform is as shown in Figure 4. In this work the input frequency of power supply is set to the fundamental frequency of 50 Hz.

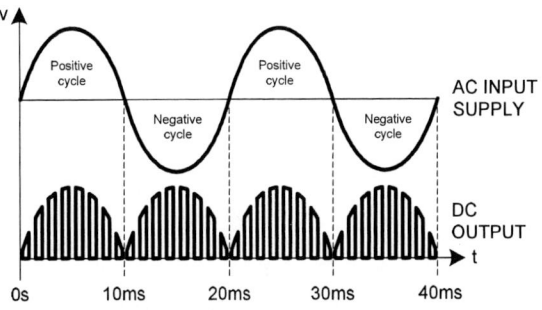

Figure 4: Waveform of AC-DC converter operation

i) Switching strategies

The operation of the proposed AC-DC converter using SPMC topology with reduced switch count were divided into positive and negative cycle operation in order to develop an output voltage as in Figure 5.

Positive cycle operation:

- At any time 't', only two switches $S1$ and $S5$ will be in 'ON' state and conduct the current flow during positive cycle of the input source.

Negative cycle operation:

- At any time 't' only two switches $S2$ and $S4$ will be in 'ON' state and conduct the current flow during negative cycle of the input source.

The two switching states defined, was illustrated in Figure 6.

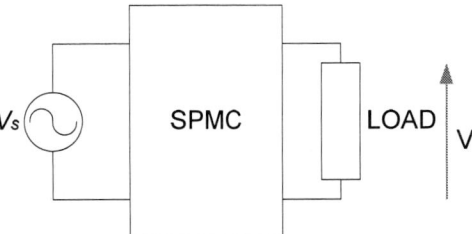

Figure 5: Output DC voltage for controlled rectifier

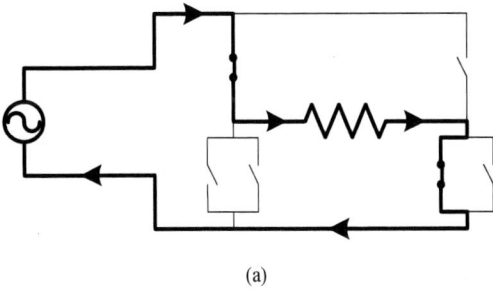

(a)

978-1-4577-0007-1/11 $26.00 © 2011 IEEE

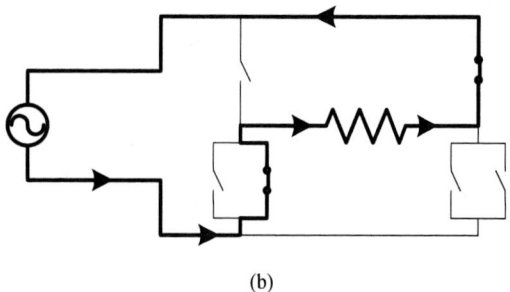

(b)

Figure 6: Basic control rectifier operation; (a) Positive cycle operation, (b) Negative cycle operation

ii) Control Implementation

Control implementation of the proposed AC-DC converter using SPMC with reduced switch count uses the pulse width modulation (PWM) technique due to its simplicity in operation that is function to calculate the switch duty ratio to synthesize the output waveform. The switching scheme follows those tabulated in Table 1.

Table 1: Sequence of switching controlled rectifier using SPMC with reduced switch count

Switches	Positive cycle	Negative cycle
S1	PWM	OFF
S2	OFF	PWM
S3a	OFF	OFF
S3b	OFF	PWM
S4a	PWM	OFF
S4b	OFF	OFF

For realization, a triangular carrier signal, A_c is compared with a reference signal, A_r of variable amplitude to control the modulation index, Ma, where,

$$M_a = \frac{A_r}{A_c} \qquad \ldots(1)$$

where: A_c = Amplitude of carrier signal

A_r = Amplitude of reference signal

The crossover points between these signals are used to determine the switching instants as shown in Figure 7. When the magnitude of the reference signal is higher than or equal to a triangular carrier signal a 'HIGH' output is produced from the comparator, otherwise a 'LOW' output is obtained. The output of PWM was separated into two parts; the first part called the positive cycle; operating from 0 to π degrees of sine wave and the other part called negative cycle; operating from π to 2π degrees as shown in Figure 8. Figure 9 illustrates the overall switching sequence of SPMC with reduced switch count for controlled rectifier operation based on those defined in Table 1.

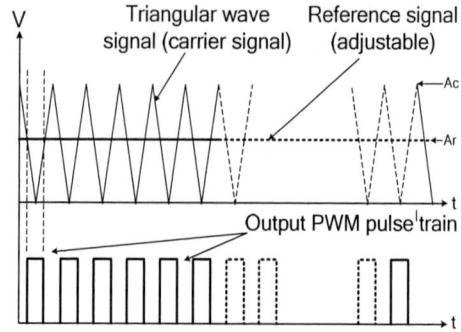

Figure 7: PWM waveform generation

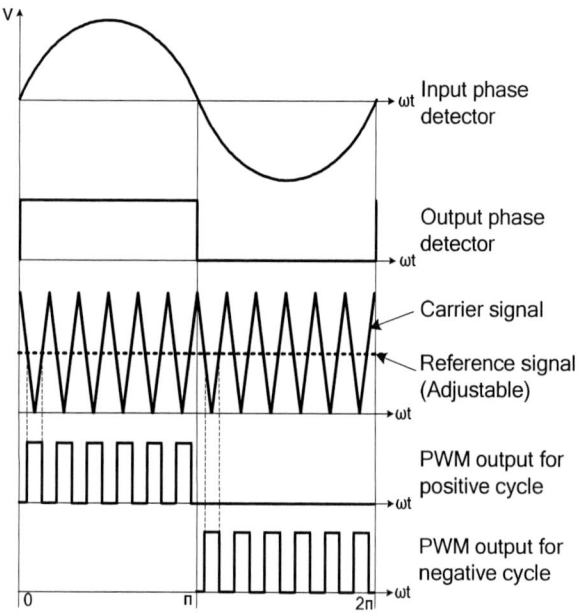

Figure 8: Separation of PWM waveform generation for positive and negative cycles

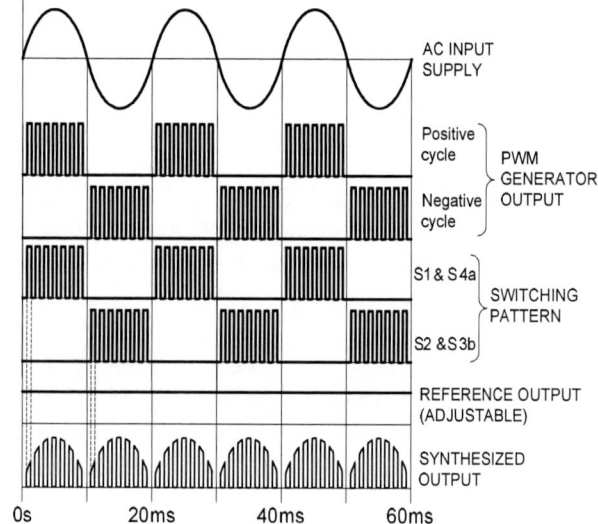

Figure 9: Waveform of switching pattern for controlled rectifier using SPMC with reduced switch count

978-1-4577-0007-1/11 $26.00 © 2011 IEEE

IV. SIMULATION VERIFICATION

The proposed operation was implemented using Matlab/Simulink (MLS) with SimPowerSystems to verify performance and operation. Figure 10 show the top level main model of the proposed AC-DC converter using SPMC with reduced switch count. Subsystem is used to improve clarity of modelling by breaking up large models into a hierarchical set of smaller models for structured implementation and are as shown in Figure 11 through 16. Figure 12 shows the switch model for S1 and S2 whilst Figure 13 shows the switch model for S3 and S4.

The controller unit implements for the operation of AC-DC converter using SPMC with reduced switch count is as shown in Figures 14 to 16. For the former, the sine wave is compared to zero using "Compare to Zero" block representing a phase detector function producing an 'ON' pulse for positive cycle operation.

A constant representing a straight line or reference signal used to vary the modulation index which is compared with the carrier (triangular waveform) to produce the required respective PWM output signal. This is implemented using the "Relational Operator" block. The signal is then locked to the positive and negative cycle by using the phase detector function.

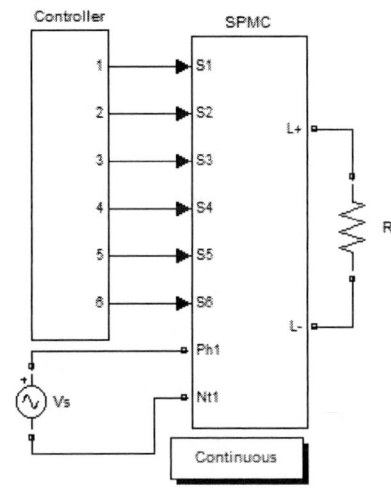

Figures 10: Top level main model of controlled rectifier using MLS

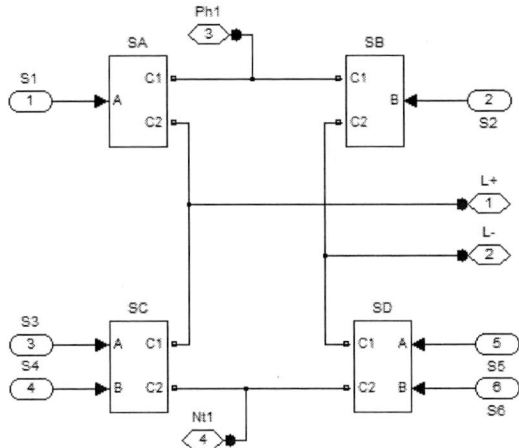

Figures 11: SPMC with Reduced Switch Count Model in MLS

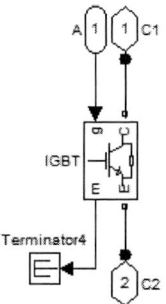

Figures 12: Switch Model for S1 and S2 in MLS

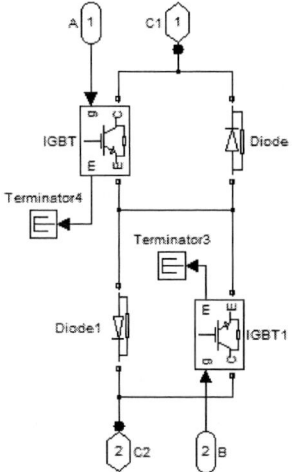

Figures 13: Bidirectional switch cell module (S3, S4, S5 and S6)

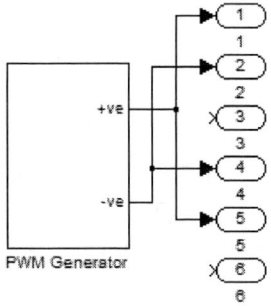

Figures 14: PWM generator model

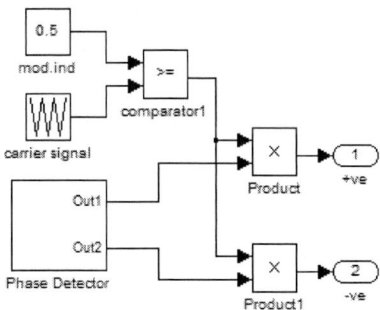

Figures 15: PWM generator model

978-1-4577-0007-1/11 $26.00 © 2011 IEEE

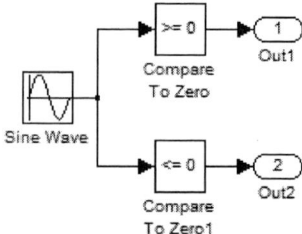

Figures 16: Phase detector model

V.   RESULT AND DISCUSSION

For the operation of controlled rectifier as described in the previous section, various loads had been implemented. Initially, pure resistive load was used to study the operational behaviour. An inductance is then inserted to examine associated problems such as spikes as described previously. A number of stages were implemented; a) using pure resistive load, and b) using resistive and inductive loads.

Figure 17 shows the simulation results of controlled AC-DC converter using SPMC with reduced switch count. A modulation index of 0.5 was used to illustrate sample results for a case of resistive load as presented in Figures 17 to 19 at a switching frequency of 5 kHz. A sinusoidal input voltage of 100V (pk-pk) at 50Hz and a resistive load of 50 ohm were initially used in this stage. The output patterns show good agreement for those obtained using the proposed topology and conventional SPMC topology. The introduction of 4mH inductor produces results as shown in Figure 18. It was observed that undesirable spikes seem to appear that requires elimination to avoid the damaging stress on switching device.

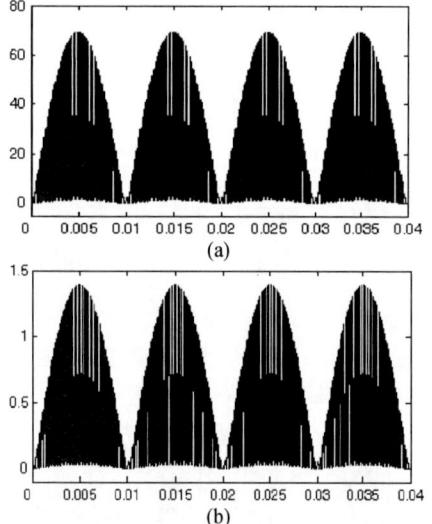

Figures 17: Simulation result (a) output voltage (b) output current

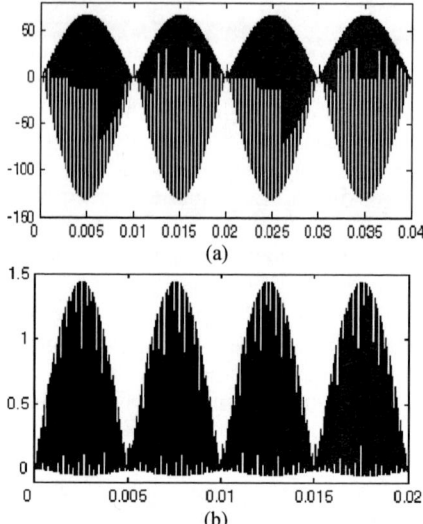

Figures 18: Simulation result (a) output voltage (b) output current

Figure 19 illustrate the result of investigation on effects of the proposed safe-commutation strategies obtained using 5 kHz switching frequency indicated that spikes as presented previously were successfully eliminated.

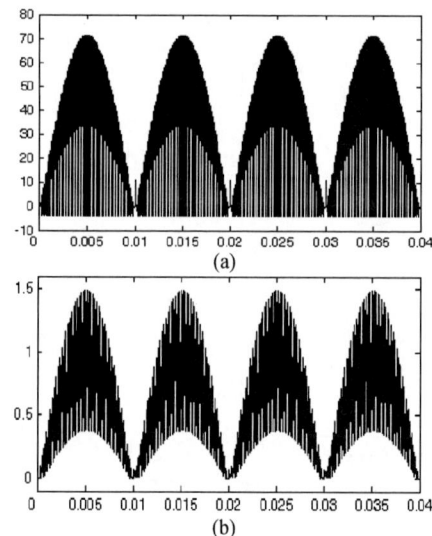

Figures 19: Simulation result (a) output voltage (b) output current

VI.   CONCLUSION

This paper briefly illustrates that the SPMC topology with reduced switch count could be operate as an AC-DC converter. Inherent commutation problems that lead to switching spikes have been presented with proposed safe commutation algorithm strategy. The simulation results were observed to confirm the predicted performance of the proposed topology.

978-1-4577-0007-1/11 $26.00 © 2011 IEEE

REFERENCES

[1] Gyugyi,L and Pelly,B.R, "Static Power Chargers, Theory, Performance and Application," John Wiley & Son Inc, 1976.

[2] Alesina, A., Venturini, M., "Solid-State Power Conversion: A Fourier Analysis Approach to Generalized Transformer Synthesis", IEEE Transactions on Circuits and Systems, April 1981, Vol. CAS-28, No. 4, Page(s): 319-330.

[3] Alesina, A., Venturini, M., "Analysis and Design of Optimum-Amplitude Nine-Switch Direct AC-AC Converters", IEEE Transactions on Power Electronics, January 1989, Vol. 4, No. 1, Page(s): 101-112.

[4] A. Daniels and D. Slattery, "New power converter technique employing power transistors", Proc. Inst. Elect. Eng., Feb. 1978, Vol. 125, No. 2, Page(s): 146-150.

[5] Bland, M.J., Wheeler, P.W., Clare, J.C., Empringham, L., "Comparison of Bi-directional Switch Components for Direct AC-AC Converters", 2004 IEEE 35th Annual Power Electronics Specialists Conference, 2004. PESC 04. Vol. 4, 2004, Page(s): 2905-2909.

[6] de Oliveira Filho, M.E, Ruppert Filho, E., Quindere, K.E.B., Gazoli, J.R., "A Simple Current Control for Matrix Converter", 41st IAS Annual Meeting. Conference Record of the 2006 IEEE Industry Applications Conference, 2006. Vol. 4, Oct. 2006, Page(s): 2090-2094.

[7] Firdaus, S., Hamzah, M.K.," Modelling and simulation of a single-phase AC-AC matrix converter using SPWM,", Student Conference on Research and Development 16-17 July 2002, SCOReD2002., pp286-289.

[8] Siti Zaliha Mohammad Noor, Mustafar Kamal Hamzah & Ahmad Farid Abidin, "Modelling and Simulation of a DC Chopper Using Single Phase Matrix Converter Topology" IEEE Sixth International Conference PEDS 2005, Kuala Lumpur, Malaysia.

[9] Noor, Siti Zaliha Mohammad; Hamzah, Mustafar Kamal; Baharom, Rahimi; Dahlan, Nofri Yenita; "A New Single-Phase Inverter with Bidirectional Capabilities Using Single-Phase Matrix Converter", IEEE Power Electronics Specialists Conference, 2007. PESC 2007. 17-21 June 2007, Page(s): 464 – 470.

[10] Baharom, R.; Hasim, A.S.A.; Hamzah, M.K.; Omar, M.F.; "A New Single-Phase Controlled Rectifier Using Single-Phase Matrix Converter", IEEE International Power and Energy Conference, 2006. PECon '06. Page(s): 453 – 458

[11] Baharom, R.; Hashim, N.; Hamzah, M.K.; "Implementation of controlled rectifier with power factor correction using single-phase matrix converter", International Conference on Power Electronics and Drive Systems, 2009. PEDS 2009. Page(s): 1020 - 1025

[12] Mohammad Noor, S.Z.; Baharom, R.; Hamzah, M.K.; Hamzah, N.R.; "Safe-commutation strategy for controlled rectifier operation using single-phase matrix converter", International Conference on Power Electronics and Drive Systems, 2009. PEDS 2009. Page(s): 1026 - 1029

[13] Baharom, R.; Hamzah, M.K.; Saparon, A.; Noor, S.Z.M.; Hamzah, N.R.; "A New Single-Phase Controlled Rectifier using Single-Phase Matrix Converter with Regenerative Capabilities", 7th International Conference on Power Electronics and Drive Systems, 2007. PEDS '07. Page(s): 1477 - 1482

[14] Baharom, R.; Hamzah, M.K.; Muhammad, K.S.; Hamzah, N.R.; "Boost rectifier using single-phase matrix converter", 3rd IEEE Conference on Industrial Electronics and Applications, 2008. ICIEA 2008. Page(s): 2205 – 2210

[15] Baharom, R.; Hamzah, M.K.; "A New Single-Phase Controlled Rectifier Using Single-Phase Matrix Converter Topology Incorporating Active Power Filter", IEEE International Electric Machines & Drives Conference, 2007. IEMDC '07. Volume: 1

6-pulse Controlled Rectifier Synchronisation Method

Bhaba Priyo Das, Neville Watson
Department of Electrical and Computer Engineering
University of Canterbury
Christchurch, New Zealand
E-mail: bhaba.das@pg.canterbury.ac.nz

Yonghe Liu
School of Information Engineering
Inner Mongolia University of Technology
Hohhot, China

Abstract— **The 6-pulse controlled rectifier circuit is very well known. For proper operation of the 6-pulse rectifier, it is necessary to synchronise the firing signals with the mains voltage, else the operation of the rectifier is affected. A firing angle controller, without using any phase lock loop (PLL), is described here. This scheme employs the dq-reference frame to estimate the magnitude and phase angle of the fundamental component of the mains voltage. This method offers structural simplicity and immunity to mains voltage unbalance, harmonics and voltage sag etc.**

Keywords- Phase Angle estimation, Rotating Reference Frame, Synchronisation .

I. INTRODUCTION

The 6-pulse controlled rectifier (Fig. 1) has a wide range of applications, from small rectifiers to large high voltage direct current (HVDC) transmission systems, FACTS devices - STATCOM, motor speed control and frequency converters etc. For proper operation of the 6-pulse rectifier, it is necessary to synchronise the gate signals with the mains voltage. Since low quality mains voltage is a common occurrence due to proliferation of power electronic devices, this poses a serious problem to correct mains synchronisation.

Synchronisation is usually carried out by detecting the phase angle of mains voltage and firing each thyristor with respect to a particular point of the phase angle. The signal used for synchronisation is not an ideal sine wave but is distorted by voltage unbalance, harmonics, sags, notches, phase angle and frequency step.

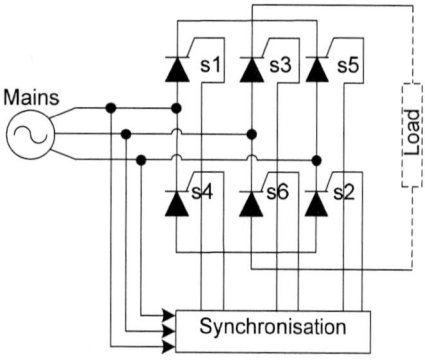

Figure 1.   General scheme of 6-pulse rectifier with synchronisation control.

Phase Lock Loops (PLL) has been widely used for synchronisation [1]-[2]. To deal with distortion of the mains voltage, various schemes have been proposed so far using modified PLLs [3]-[10]. Some schemes results in reducing the loop bandwidth, much lower than the frequency of the mains [11]. This leads to slower tracking. Although some other schemes achieve precision phase and frequency tracking, they result in highly complicated structures and practical realisation may become difficult [12]. Several other schemes have also been proposed apart from using PLLs. In [13], sinusoidal voltage at the mains is reconstructed in real time by measuring the distorted waveform at the load end. A neural network based approach is proposed in [14]. In [15] reference frame transformation is used and it drives a PLL. The requirement of 90° all-pass phase shifters prevents the use of this method if there is mains frequency variation. Adaptive filtering techniques use a specific "filtering" technique in order to obtain an un-distorted voltage signal [16].

In this paper, an alternative synchronisation scheme based on transforming the mains voltage into rotating reference (dq-reference) frame and generating the fundamental positive and negative sequence components is described. This scheme can operate under unbalanced, distorted and variable-frequency conditions of the mains voltage. This method offers structural simplicity which can be easily implemented both in hardware and software environments.

II. PROPOSED METHOD OF SYNCHRONISATION

Since higher order harmonics like 9th, 11th etc are insignificant for industrial power systems, this method is considered only upto the 7th harmonic. The DQ transformation is a transformation of coordinates from the three-phase (abc) reference frame to dq-reference frame. The transformation is given by:

$$
\begin{bmatrix} V_d \\ V_q \\ V_0 \end{bmatrix} = \frac{2}{3} \begin{bmatrix} \sin(\omega_0 t) & \sin(\omega_0 t - \frac{2\pi}{3}) & \sin(\omega_0 t + \frac{2\pi}{3}) \\ \cos(\omega_0 t) & \cos(\omega_0 t - \frac{2\pi}{3}) & \cos(\omega_0 t + \frac{2\pi}{3}) \\ \frac{1}{2} & \frac{1}{2} & \frac{1}{2} \end{bmatrix} \begin{bmatrix} V_a \\ V_b \\ V_c \end{bmatrix}
$$

$$(1)$$

where $V_{a,b,c}$: three phase mains voltage.

978-1-4577-0007-1/11 $26.00 © 2011 IEEE

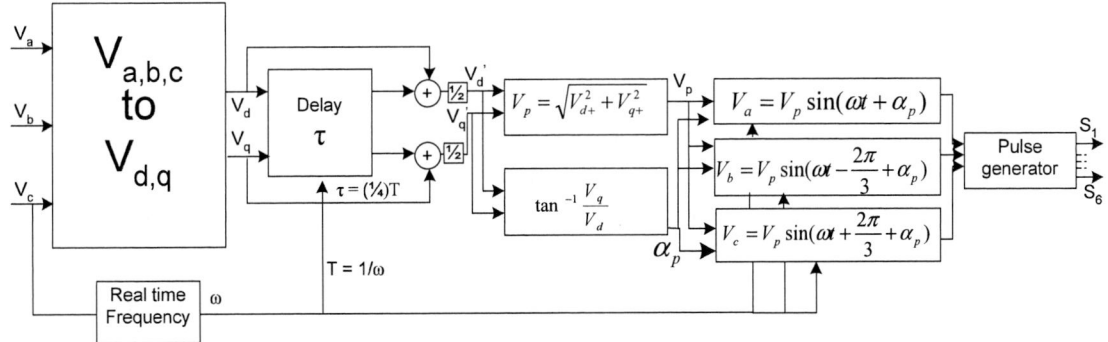

Figure 2. Proposed synchronisation scheme.

$\theta = \omega_0 t$, ω_0: fundamental angular frequency.

In the dq-reference the mains voltage is transformed into DC signals. Under the different corrupting sources abc to dq mapping is given as in Table I [1]:

TABLE I. COMPONENT MAPPING ABC AND DQ-REFERENCE FRAME

Components in ABC reference frame	Mapping into DQ reference frame
Any offset voltage, 3rd or multiple of 3rd harmonic present in all the three phases.	Harmonics having same frequency only in zero sequence.
Voltage unbalance in any or all the three phases	2nd harmonic in both d and q signal.
5th and 7th in all the three phases.	6th harmonic in both d and q signal.

The fundamental positive and negative sequence components are extracted from the dq-reference frame signal using Yao's method [17]. If $V_{dq}(t)$ represents the voltages in dq-reference frame, the dq signal is delayed by a time period (τ) equal to ¼ of the fundamental cycle. $V_{dq}(t-\tau)$ is obtained which has same amplitude as the original signal but is exactly $180°$ out of phase. Thus, by adding the delayed signal to the original signal cancellation of negative sequence, 5th harmonic, 7th harmonic in dq-reference frame is obtained. The amplitude is doubled which is divided by 2 to get the original amplitude. This is a very simple and fast method by which DC components in the dq-reference frame can be obtained. Once the magnitude of the positive sequence voltage and phase angle is obtained, a set of balanced three phase voltages which are in phase with the positive sequence of the mains voltage can be easily calculated by adding or subtracting $2\pi/3$ radians. Once the three sets of voltages are obtained, the first firing pulse is obtained at the zero crossing of V_{ca}, where V_{ca} is the phase to phase synchronised mains voltage. The remaining firing pulses are delayed by $60°$ from each other.

III. FUNDAMENTAL EXTRACTION

This section shows the results of fundamental extraction under various corrupting sources such as voltage harmonics, unbalance, sags, phase outage, frequency jump and phase jump.

A. Voltages in ABC reference frame are unbalanced.

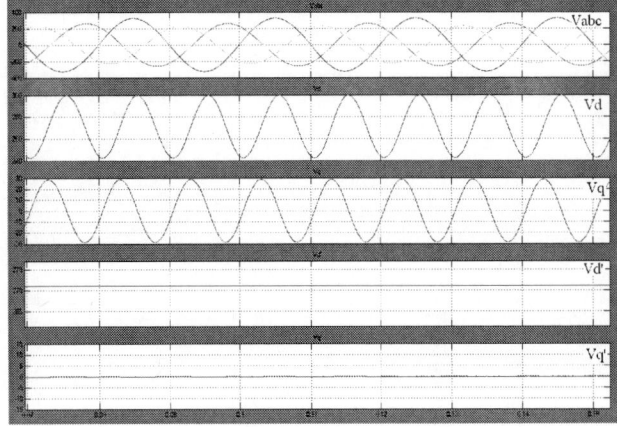

Figure 3. Fundamental extraction under voltage unbalance.

Fig. 3 shows the fundamental extraction under unbalanced mains voltage. The dq-reference consists of a DC + AC signal at twice the mains frequency. Using the method described in section II, the positive and negative sequence components are extracted after a delay of 5ms.

B. Voltages in ABC reference frame are unbalanced with 5th and 7th harmonic.

Fig. 4 shows addition of 5th and 7th harmonic to the unbalanced mains signal. Figs. 5 and 6 show that an additional 6th harmonic component is present in V_d and V_q along with the DC + AC (2nd harmonic) due to unbalance.

978-1-4577-0007-1/11 $26.00 © 2011 IEEE

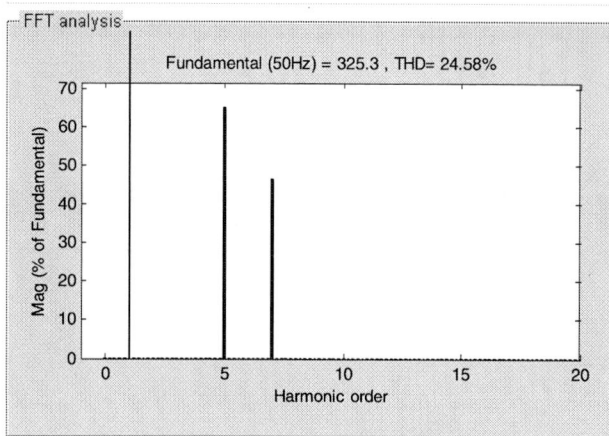

Figure 4.   Harmonic content of one phase.

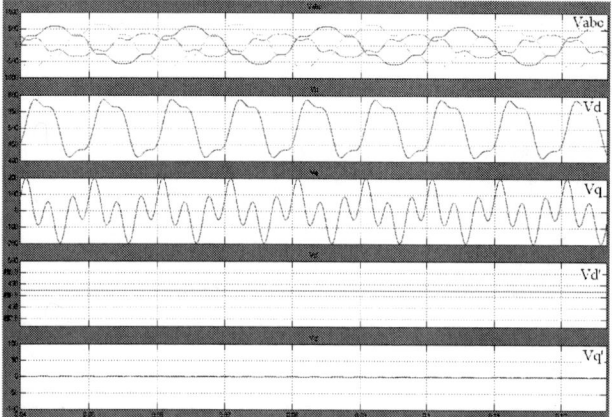

Figure 5.   Fundamental extraction under voltage harmonics and unbalance.

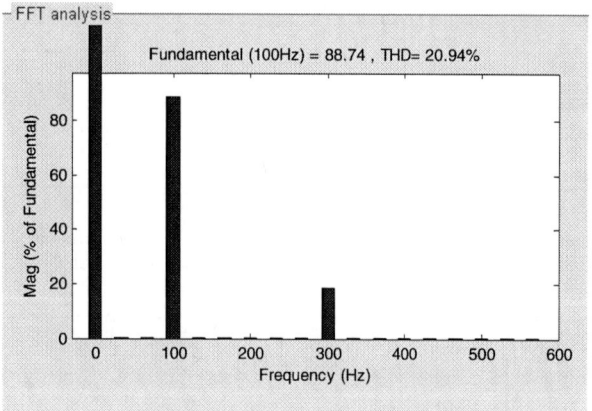

Figure 6.   Harmonic content of V_d.

Again, the positive and negative sequence components are extracted after a delay of 5ms.

C.   *Voltages in ABC reference frame are balanced with 5th and 7th harmonic.*

In this case, there is only DC + AC (6th harmonic) in V_d and V_q. Fig. 7 shows the extraction of the DC components as V_d' and V_q'.

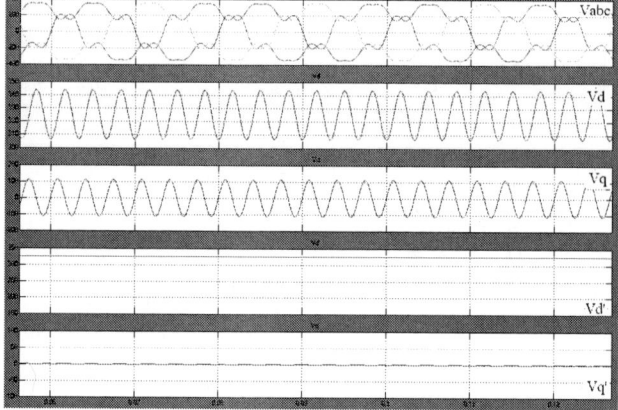

Figure 7.   Fundamental extraction under voltage harmonics.

D.   *Voltages in ABC reference with loss of phase A voltage at 0.1s.*

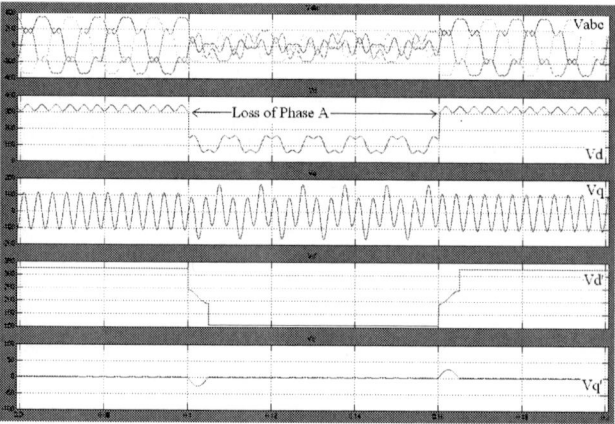

Figure 8.   Fundamental extraction under loss of phase A.

In this case, there is a sudden loss of phase A voltage from 0.10s until 0.16s. Due to loss of one phase, the fundamental positive sequence component V_d decreases, whereas the 2nd harmonic component appears in V_q. Fig. 8, again shows the extraction of the DC components as V_d' and V_q' within 5ms.

E.   *Voltages in ABC reference with 50% voltage sag at 0.1s.*

With voltage sag of 50 % at 0.1s, the fundamental positive sequence component V_d decreases. Fig. 9 shows that within 5ms, V_d' and V_q' are extracted.

F.   *Voltages in ABC reference with phase shift of 20° at 0.1s.*

Sudden phase change in mains voltage may occur if a large load is disconnected. This is assumed by applying a step change of 20° in the phase angle. As seen in Fig. 10, this step change is, again, tracked within 5ms with V_d' and V_q' settling to new DC values within the same timeframe.

G. Voltages in ABC reference with frequency shift of 1 Hz at 0.1s.

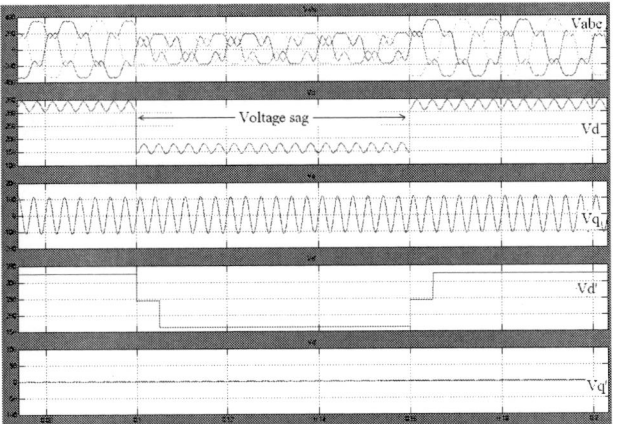

Figure 9.   Fundamental extraction under voltage sag.

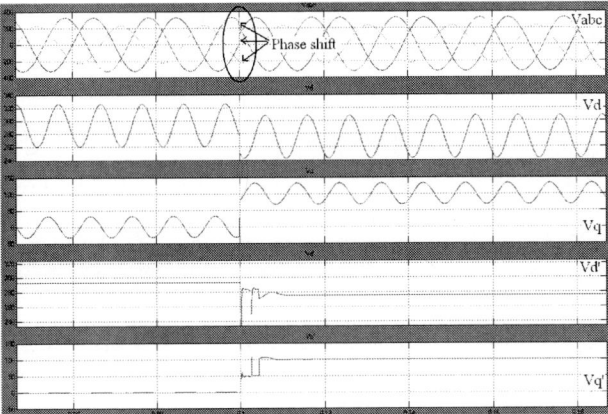

Figure 10.   Fundamental extraction under phase shift of 20°.

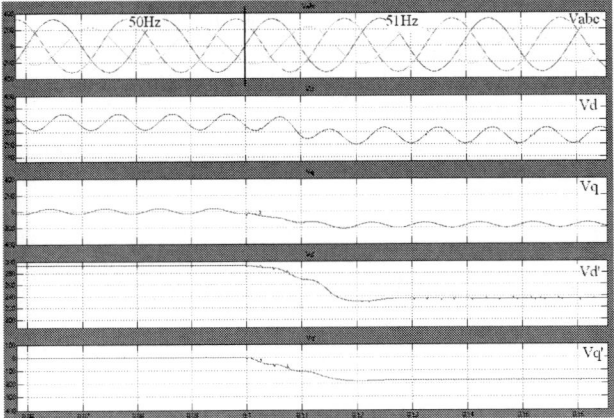

Figure 11.   Fundamental extraction under frequency shift of 1Hz.

Permitted frequency variation involves a change in frequency of ±1.5 % from the normally stable mains frequency of 50 Hz. To test the response to frequency variation, a step change of 1Hz is applied at 0.1s in the mains voltages. This is tracked by this synchronisation scheme within 17ms. This time is entirely dependent on the real time frequency estimation

principle used.   Frequency estimation plays a critical role in this scheme.

H. Voltages in ABC reference has an offset of 10V.

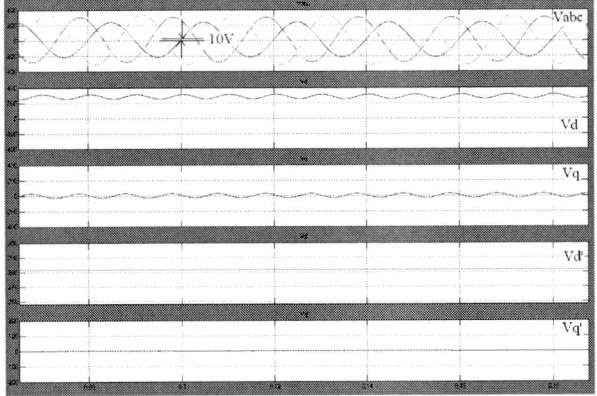

Figure 12. Fundamental extraction under voltage offset.

For digital implementation of such synchronisation schemes, all mains voltage measurements must be made after adding an offset in order to acquire bipolar signals for the analog-to-digital converter. Fig. 12 shows the results obtained after adding a 10V offset to the mains voltage. There is no effect of offset voltage in dq-reference frame as an offset voltage is mapped as zero sequence in dq-reference frame.

IV.   PULSE GENERATOR

As shown in section III, this scheme tracks the magnitude and phase angle of the positive sequence component (V_p and α_p) correctly in spite of corrupting sources of the mains voltage. Now a set of balanced three phase voltages, in phase with the positive sequence of the mains voltage, can be easily calculated by adding or subtracting $2\pi/3$ radians, as shown in Fig. 13.

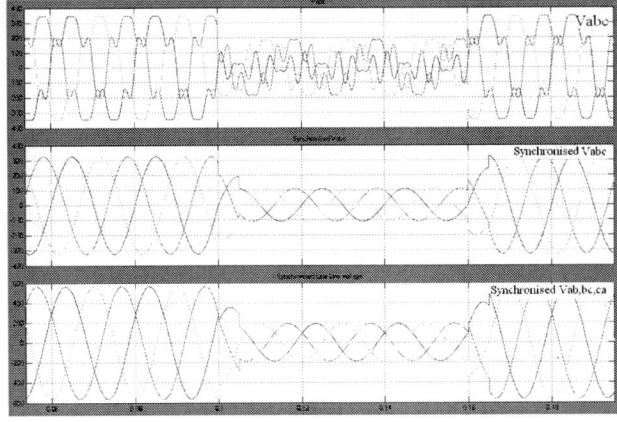

Figure 13. Line-to-neutral and line-line generation using proposed synchronisation method.

Loss of phase voltage or voltage sags usually have a phase-jump associated with it. This is quickly tracked as shown in Fig. 13.

978-1-4577-0007-1/11 $26.00 © 2011 IEEE

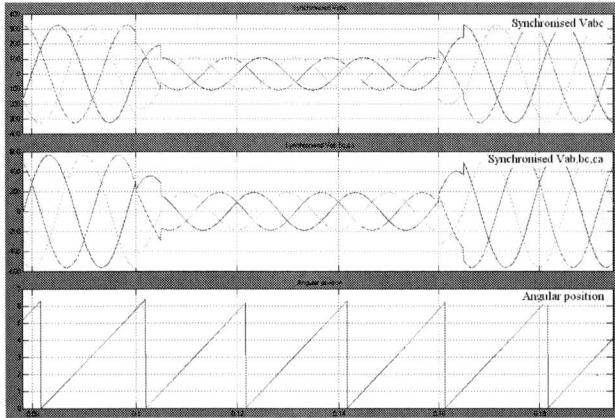

Figure 14. Phase angle determination using the proposed synchronisation method.

Once $V_{ab, bc, ca}$ is synthesized, a ramp starting at the falling zero crossing of V_{ca} is generated. Fig. 14 shows that correct phase angle is obtained even if there is a loss of phase A. This ramp is then compared with the desired firing angle delay to get the first firing pulse. Subsequently, the other 5 pulses are delayed by 60° from each other.

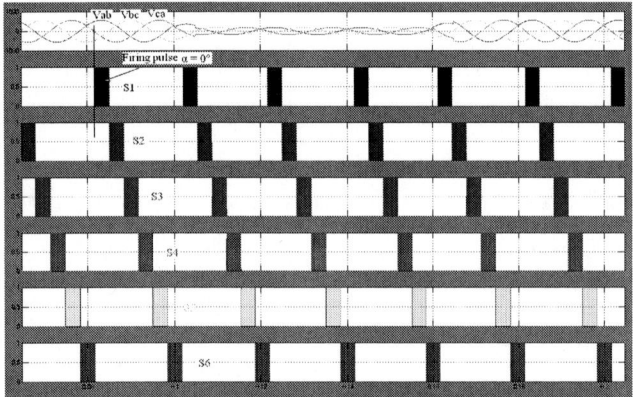

Figure 15. Firing pulses for S1 to S6 for α = 0°

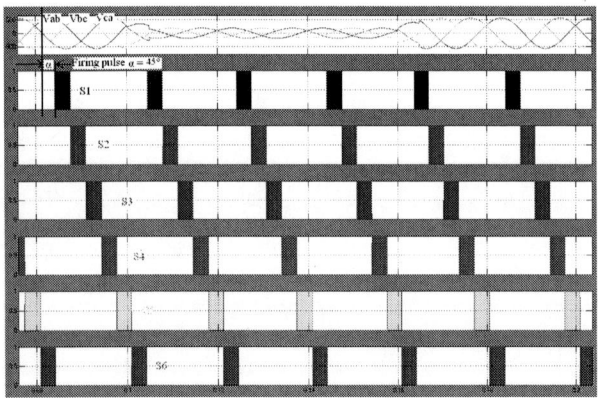

Figure 16. Firing pulses for S1 to S6 for α = 45°

V. COMPARISION WITH OTHER SCHEMES

This section provides a comparison of the proposed method with the existing synchronisation methods such as Low Pass Filter (LPF) based method, Space Vector (SV) based method, Extended Kalman Filter (EKF) based method, Least Squares (LS) based method, PLL based method and Enhanced PLL (EPLL) based method. The range of comparison is divided as:

Not applicable – 0   Poor – 1   Average – 2

Good – 3   Very good – 4   Excellent – 5

TABLE II.   COMPARISION BETWEEN DIFFERENT SYNCHRONISATION METHODS

	LPF	SV	EKF	LS	PLL	EPLL	Proposed
Harmonics	2	4	1	4	0	5	5
Unbalance	1	1	1	0	0	5	5
Phase step	5	5	5	5	5	5	5
Frequency step	0	1	2	5	5	5	5
Structure	5	4	3	3	4	3	5
Total	13	15	12	17	14	23	25

The LPF-based method is not capable of adjusting to frequency variations. Mains voltage unbalance affects its performance. The SV-based method performs better with respect to mains harmonics but frequency variation limits its operation. The EKF-based method cannot cope with mains harmonics and voltage unbalance. The main drawback of the three phase PLL method is that it cannot provide correct phase angle estimation under harmonics and unbalances. There is a trade-off between PLL bandwidth and response time. EPLL solves most of the problems but is complicated in structure.

VI. CONCLUSIONS

The simulation results obtained confirm that the proposed synchronisation method works very fast for fixed frequency applications. For variable frequency application, the response time depends on the real-time frequency detector. This method offers structural simplicity and immunity to voltage unbalance, mains harmonics, voltage sag, phase and frequency step and voltage offset. This scheme can easily be digitally implemented. This method generates correct firing pulses for the 6-pulse controlled rectifiers. The firing angle can be controlled precisely.

REFERENCES

[1] S. K. Chung, "A phase tracking system for three phase utility interface inverters," IEEE Trans. Power Electronics, vol.15, no.3, pp.431-438, May 2000.

[2] V. Kaura and V. Blasko, "Operation of a phase locked loop system under distorted utility conditions," Proc. Eleventh Annual Applied Power Electronics Conference and Exposition (APEC '96), vol.2, pp.703-708, Mar. 1996.

978-1-4577-0007-1/11 $26.00 © 2011 IEEE

[3] P. Rodriguez, L. Sainz, and J. Bergas, "Synchronous double reference frame PLL applied to a unified power quality conditioner," Proc. 10th International Conference on Harmonics and Quality of Power, vol.2, pp. 614- 619, Oct. 2002.

[4] A. V. Timbus, T. Teodorescu, F. Blaabjerg, M. Liserre, and P. Rodriguez, "PLL Algorithm for Power Generation Systems Robust to Grid Voltage Faults," Proc. 37th IEEE Power Electronics Specialists Conference (PESC '06), pp.1-7, June 2006.

[5] R. I. Bojoi, G. Griva, V. Bostan, M. Guerriero, F. Farina, and F. Profumo, "Current control strategy for power conditioners using sinusoidal signal integrators in synchronous reference frame," IEEE Trans. Power Electronics, vol.20, no.6, pp. 1402- 1412, Nov. 2005.

[6] P. Rodriguez, J. Pou, J. Bergas, J. I. Candela, R. P. Burgos, and D. Boroyevich, "Decoupled Double Synchronous Reference Frame PLL for Power Converters Control," IEEE Trans. Power Electronics, vol.22, no.2, pp.584-592, March 2007.

[7] D. Jovcic, "Phase locked loop system for FACTS," IEEE Trans. Power Systems, vol.18, no.3, pp. 1116- 1124, Aug. 2003.

[8] B. Han and B. Bae, "Novel phase-locked loop using adaptive linear combiner," IEEE Trans. Power Delivery, vol.21, no.1, pp. 513- 514, Jan. 2006.

[9] A. Ghoshal and V. John, "A Method to Improve PLL Performance Under Abnormal Grid Conditions," Proc. 3rd Bi-Annual National Power Electronics Conference (NPEC'07), Indian Institute of Science, Dec. 2007.

[10] Xu Ren-zhong, Liu Fei, Zhu Xiao-dong, Lv Hong-shui, Liu Yan-qin, "Study on the Three-Phase Software Phase-Locked Loop Based on d-q Transformation," Proc. 2010 Asia-Pacific Power and Energy Engineering Conference (APPEEC), pp.1-4, March 2010.

[11] B. P. Das, N. Watson and Y. Liu, "Evaluation of phase lock loops for power electronic applications," Unpublished. Submitted for 8th IEEE ICPE, Korea.

[12] M. Karimi-Ghartemani, H. Karimi, and M. R. Iravani, "A magnitude/phase-locked loop system based on estimation of frequency and in-phase/quadrature-phase amplitudes," IEEE Trans. Industrial Electronics, vol.51, no.2, pp. 511- 517, April 2004.

[13] R. Weidenbrüg, F. Dawson, and R. Bonert, "New synchronisation method for thyristor power converters to weak AC-systems," *IEEE Trans. Industrial Electronics*, vol. 40, no. 5, pp. 505–511, Oct. 1993.

[14] S. Väliviita, "Zero-crossing detection of distorted line voltages using 1-b measurements," *IEEE Trans. Industrial Electronics*, vol. 46, no. 5, pp. 917–922, Oct. 1999.

[15] Sang-Joon Lee, Jun-Koo Kang, Seung-Ki Sul, "A new phase detecting method for power conversion systems considering distorted conditions in power system," Proc. Thirty-Fourth IAS Annual Meeting Industry Applications Conference, vol.4, pp.2167-2172, 1999.

[16] O. Vainio, "Adaptive notch filtering in impulsive noise environment," Proc. IEEE Midnight-Sun Workshop on Soft Computing Methods in Industrial Applications (SMCia/99), pp.152-155, 1999.

[17] Ziwen Yao, "Fundamental Phasor Calculation With Short Delay," IEEE Trans. Power Delivery, vol.23, no.3, pp.1280-1287, July 2008.

A New Single-Phase to Three-Phase Converter using Quasi Z-Source Network

F. Khosravi[1], N. A. Azli[2], A. Kaykhosravi
Power Electronics and Drive Research Group (PEDG), Energy Research Alliance
Universiti Teknologi Malaysia, 81310 UTM Johor Bahru, Malaysia
farshadgoli@fkegraduate.utm.my[1]
naziha@ieee.org[2]

Abstract— This paper introduces a new cost effective structure of a 4-switch single-phase to three-phase converter using a Quasi Z-source (QZs) network for induction motor drive applications. In comparison to the traditional 6-switch structure, the proposed circuit reduces the cost of the system, switching losses and the complexity of the control method as well as the interface circuits used to make the trigger signals. In addition, The Quasi Z-Source network, similar to the Z-Source network uses a unique LC network with added advantages, such as; lower component ratings, reduced source stress, reduced component count and simplified control strategies for Adjustable Motor Drives (ASD) which require large range of gain. By controlling the zero shoot-through duty cycle, the converter can generate any desired output voltage, even greater than the supply voltage. Consequently, the converter is capable of extending the output voltage range, improving the power factor and reliability, reduces line harmonics and provides ride-through ability at the voltage sag interval. Simulation results on the proposed converter have indicated its efficiency and potential for further development.

I. INTRODUCTION

Over the years, three-phase motors, more than single-phase motors have been the main consideration in industries due to certain parameters such as; efficiency, torque ripples and power factor. In places like rural areas to the use of rolling mills, machine tools and in low-power industrial application for robotics, where by a three-phase utility may not be available, high performance converters must be used to run three-phase motor drives. Low losses and cost-effectiveness of these converters are very important.

Researchers have introduced different structures for the aforementioned converter with at least 6 switches for single-phase to three-phase converters [1-8]. Adjustable speed drives (ASD) for induction motors are widely used. By reducing the number of semiconductor power devices in the inverter structure, losses can be reduced to a desirable level. Component-minimized structures of the inverter have been introduced in literatures [9-11].

Investigations have been made on the extension of effective control methods for high efficiency ASD for induction motors [13-4]. The cost, simplicity, voltage sag mitigation and flexibility of the drive system are a few of the most significant parameters that have been taken into account. Majority of the introduced works on the 4-switch three-phase (4S3P) inverter for motor drives have not consider the closed loop control scheme, which is essential for high performance drives [12]. A traditional voltage source four-switch (B4) inverter has been presented in [12], which is shown in Fig. 1.

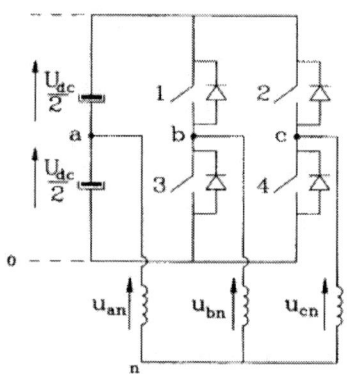

Fig.1. Traditional voltage source 4-switch inverter

In this structure, both switches of the same phase leg can never be gated ON at the same time. This is because at the zero shoot-through state, a short circuit on the DC bus will occur thus destroying the supply voltage. Furthermore, the maximum output voltage is limited by the input DC bus voltage.

II. QUASI Z-SOURCE CONCEPT

The Quasi Z-Source (QZs) structure can overcome the problems faced by traditional inverters. This circuit employs a unique LC network of different types which is connected between the loads and the power sources. The main circuit of a QZs inverter and its operating principle has been depicted in [15]. The voltage fed QZs inverter for continuous input current with six power switches is as shown in Fig. 2.

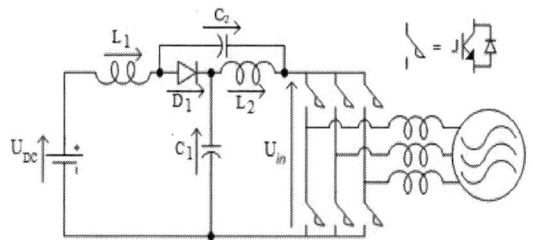

Fig.2. Voltage fed QZs inverter with continuous input current

The zero shoot-through state of the inverter is as shown in Fig. 3(a). In this interval, the output voltage and the voltage across the load are equal to zero. Fig. 3b shows the equivalent circuit of the inverter at the active state.

978-1-4577-0007-1/11 $26.00 © 2011 IEEE

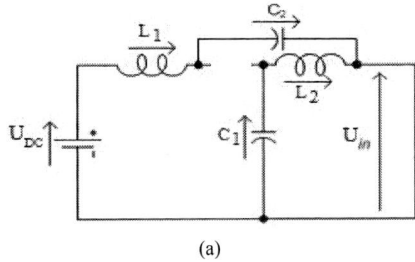

(a)

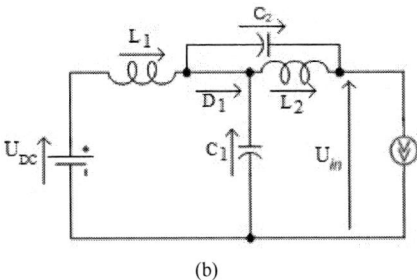

(b)

Fig.3. Equivalent Circuits of the qZSI in (a) Zero Shoot-Through State (b) Non Zero Shoot-Through State

The QZs inverter employs the zero shoot-through state to buck or boost the voltage of the DC bus to the desired output voltage by gating ON both the upper and lower switches of the same phase leg. Assuming the QZs network to be considered is a symmetrical circuit, it can be written for the network; $C_1 = C_2$ and $L_1 = L_2$.

From the above equivalent circuits and the main equations in [15], the required equations can be expressed as follows:

$$\frac{U_{in}}{U_{DC}} = S_D * \frac{1}{(1-2D)} \tag{1}$$

$$\frac{U_{C1}}{U_{DC}} = \frac{(1-D)}{(1-2D)} \tag{2}$$

where:

S_D, is the zero shoot-through switching function. S_D should be equal to 0 if the inverter is in the zero shoot-through state and 1 when the inverter is in the non shoot-through state.

D is the zero shoot-through duty cycle. If the shoot-through time interval is T_0 while the switching period is T_s, D can be expressed as:

$$D = \frac{T_0}{T_s} \tag{3}$$

III. THE PROPOSED STRUCTURE

As shown in Fig. 3(a) the voltage across the load will be zero, when the circuit is in the zero shoot-through state. Thus this state limits the maximum output voltage on the load [16]. For the proposed structure depicted in Fig. 4, one phase leg of the load is connected to the midpoint of two split capacitors while,

$$V_i = V_m \cdot \cos\theta \tag{4}$$

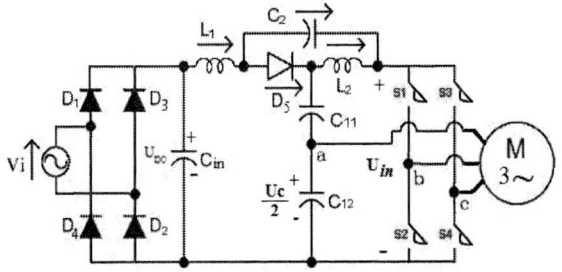

Fig.4. The proposed single-phase to three-phase QZs

Fig. 5 shows the equivalent circuit of the proposed structure in the zero shoot-through state for a wye-connected motor load. In this boost state, the network inductors will be charged by the network capacitors, over two separately LC loops.

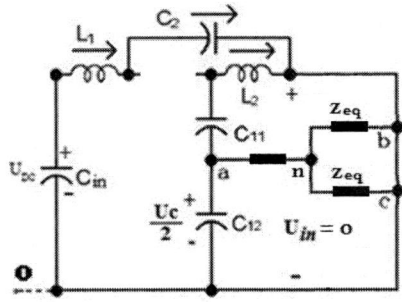

Fig.5. Equivalent circuit for the proposed structure at the zero shoot-through state

The average of the inductor voltage must be zero at steady state over one switching period while it is assumed that the average of the voltage across capacitors C_{11} and C_{12} has identical values. Thus by combining the relations (1) and (2), it can be written for the proposed structure that:

$$U_C = K * U_{in} \tag{5}$$

$$K = (1 - D) \tag{6}$$

$$K \leq 1$$

IV. SPACE VECTOR CONTROL METHOD FOR THE PROPOSED CIRCUIT

The voltage vectors are introduced as the line-to-neutral instantaneous motor voltages u_{an}, u_{bn}, u_{cn}, which can be obtained from the voltages u_{a0}, u_{b0}, u_{c0}, and u_{n0} as follows [17]:

$$u_{an} = u_{a0} - u_{n0} = \frac{2}{3} \cdot u_{a0} - \frac{1}{3} \cdot (u_{b0} + u_{c0})$$

$$u_{bn} = u_{b0} - u_{n0} = \frac{2}{3} \cdot u_{b0} - \frac{1}{3} \cdot (u_{a0} + u_{c0}) \tag{7}$$

$$u_{cn} = u_{c0} - u_{n0} = \frac{2}{3} \cdot u_{c0} - \frac{1}{3} \cdot (u_{b0} + u_{b0})$$

The proposed structure in Fig. 4 has four non shoot-through voltage vectors, $\vec{V}_1$, $\vec{V}_2$, $\vec{V}_3$, $\vec{V}_4$, and one zero shoot-through voltage vectors, $\vec{V}_5$. These voltage vectors are calculated according to switching states. In the analysis, switches S_1, S_2, S_3 and S_4 can be considered as IGBTs. When the switch is closed, it is represented by "1" while "0" indicates an open state for that switch.

978-1-4577-0007-1/11 $26.00 © 2011 IEEE   47

According to Clark's transformation, the components αβ of the aforementioned voltage vectors can be obtained as follows [18]:

$$\begin{bmatrix} u_\alpha \\ u_\beta \end{bmatrix} = \frac{2}{3} \begin{bmatrix} 1 & -\frac{1}{2} & -\frac{1}{2} \\ 0 & \frac{\sqrt{3}}{2} & -\frac{\sqrt{3}}{2} \end{bmatrix} \begin{bmatrix} u_{an} \\ u_{bn} \\ u_{cn} \end{bmatrix} \tag{8}$$

The five voltage vectors will be calculated as follows;

$$\vec{V}_i = \left(u_\alpha + j u_\beta \right) \quad , i = from\ 1\ to\ 5 \tag{9}$$

Table 1 shows all mentioned voltage vectors in details. Switching state also depicted in the table.

Table 1- Voltage space vectors for all switching states

S_1	S_2	S_3	S_4	u_{a0}	u_{b0}	u_{c0}	u_{an}	u_{bn}	u_{cn}	Vectors
1	0	0	1	$\frac{k}{2} \cdot U_{in}$	U_{in}	0	$\frac{(k-1)}{3} \cdot U_{in}$	$\frac{(4-k)}{6} \cdot U_{in}$	$-\frac{(k+2)}{6} \cdot U_{in}$	$\vec{V}_1$
0	1	1	0	$\frac{k}{2} \cdot U_{in}$	0	U_{in}	$\frac{(k-1)}{3} \cdot U_{in}$	$-\frac{(k+2)}{6} \cdot U_{in}$	$\frac{(4-k)}{6} \cdot U_{in}$	$\vec{V}_2$
1	0	1	0	$\frac{k}{2} \cdot U_{in}$	U_{in}	U_{in}	$\frac{(k-2)}{3} \cdot U_{in}$	$\frac{(2-k)}{6} \cdot U_{in}$	$\frac{(2-k)}{6} \cdot U_{in}$	$\vec{V}_3$
0	1	0	1	$\frac{k}{2} \cdot U_{in}$	0	0	$\frac{k}{3} \cdot U_{in}$	$-\frac{k}{6} \cdot U_{in}$	$-\frac{k}{6} \cdot U_{in}$	$\vec{V}_4$
1	1	1	1	$\frac{k}{2} \cdot U_{in}$	0	0	$\frac{k}{3} \cdot U_{in}$	$-\frac{k}{6} \cdot U_{in}$	$-\frac{k}{6} \cdot U_{in}$	$\vec{V}_5$

where:

$$\vec{V}_1 = \left(\left(\frac{k-1}{3} \right) + j \left(\frac{1}{\sqrt{3}} \right) \right) \cdot U_{in}$$

$$\vec{V}_2 = \left(\left(\frac{k-1}{3} \right) - j \left(\frac{1}{\sqrt{3}} \right) \right) \cdot U_{in}$$

$$\vec{V}_3 = \left(\frac{k-2}{3} \right) \cdot U_{in} \tag{10}$$

$$\vec{V}_4 = \frac{k}{3} \cdot U_{in}$$

$$\vec{V}_5 = \frac{k}{3} \cdot U_{in}$$

The voltage vectors are dependent on the value of "k". Equations (3) and (6) have illustrated that the voltage vectors are also dependent on "T_0" which is the zero shoot-through time. For $T_0 = 0.3 T_s$ and k = 0.7, the location of the voltage vectors in the αβ coordinates are as shown in Fig. 6.

Considering a symmetrical switching pattern, the average of the space vectors shown in this figure and calculations of the switching states are obtained for $1/2 \cdot T_s$ as follows:

$$\bar{V}_s \cdot (T_s/2) = \vec{V}_1 \cdot t_1 + \vec{V}_2 \cdot t_2 + \vec{V}_3 \cdot t_3 + \vec{V}_4 \cdot t_4 + \vec{V}_5 \cdot T_0 \tag{11}$$

where:

$$t_1 + t_2 + t_3 + t_4 + T_0 = T_s/2 \tag{12}$$

and t_i is half of the ON gating time for the ith switch over one switching cycle.

To obtain a set of identical three-phase voltages on the motor load, and for calculating the switching times, the following equations must be met.

$$u_\alpha = M \cdot U_{in} . \cos \theta$$
$$u_\beta = M \cdot U_{in} \cdot \sin \theta \tag{13}$$

Where "M" is the so called coefficient for amplitude tuning and it can be adjusted by the maximum magnitude of the space vector.

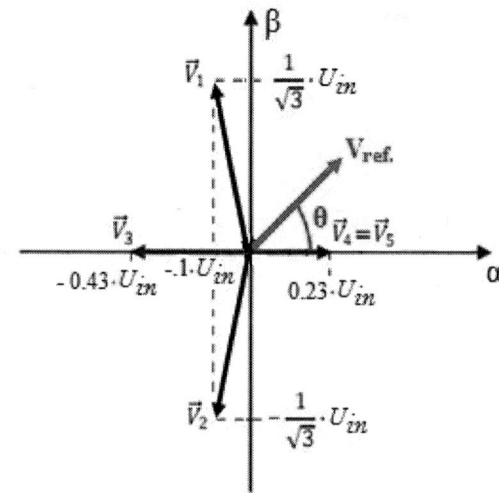

Fig.6. Arrangement of voltage vectors in αβ coordinates for k = 0.7

According to the mentioned five vectors and in order to have a one change status for all switches in a switching cycle, calculation of the switching times will be done for the two zones as follows:

A. Zone One: $\quad 0 \le \theta < 180$

In this zone the space vector is calculated with four vectors, all vectors except $\vec{V}_2$.

$$t_2 = 0$$

Thus:

$$t_1 = \sqrt{3} M \sin \theta \cdot (T_s/2)$$
$$t_3 = \left(k - M \left(3 \cos \theta + \sqrt{3} \sin \theta \right) \right) \cdot (T_s/4) \tag{14}$$
$$t_4 = \left((T_s/2) - T_0 - t_3 - t_1 \right)$$

B. Zone Two: $\quad 180 \le \theta \le 360$

In this zone the space vector is calculated with four vectors, all vectors except $\vec{V}_1$.

$$t_1 = 0$$

Thus:

$$t_2 = -\sqrt{3}M\sin\theta \cdot (T_s/2)$$

$$t_3 = \left(k + \sqrt{3}M\left(\sin\theta - \sqrt{3}\cos\theta\right)\right) \cdot (T_s/4) \qquad (15)$$

$$t_4 = \left((T_s/2) - T_0 - t_3 - t_2\right)$$

The pulse patterns of switches for both zones are shown in Fig. 7. The switching frequency has been fixed to 10 kHz.

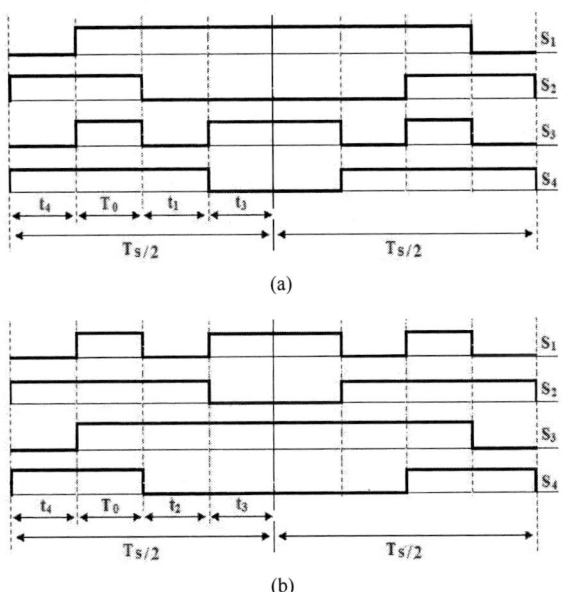

(a)

(b)

Fig.7. Pulse patterns for switches in the proposed structure in (a) Zone one (b) Zone two

V. SIMULATION RESULTS

In order to capture the advantages of the proposed structure and its control method, it is connected to a three-phase motor load with parameters as given in the Appendix of [17]. Simulations have been conducted in Matlab/Simulink environment and results have been obtained to show the effectiveness of the structure.

The QZs network has been simulated by considering the following parameters:

$L_1 = L_2 = 700\ \mu H$, $C_2 = 220\ \mu F$, $C_{11} = C_{12} = 440\ \mu F$

$C_{in} = 220\ \mu F$

Fig. 8 shows the waveforms of the three-phase current that flows through the three-phase motor. The motor currents are not perfectly symmetrical because of voltage variation at the midpoint connection by the phase leg "a", as illustrated in Fig. 14.

Fig. 9 depicts the total harmonic distortion (THD) of the motor phase current. In comparison to the traditional structure [12], the proposed converter has a smaller current percent THD of 4.01%. Fig. 10 shows the line to line three-phase voltage waveform across the motor load.

A V/f closed-loop control method is employed in controlling the motor speed. Fig. 11 shows the speed response of the control system for the QZs converter at a load torque of 0.4 N. m. The speed response is found to be better than that of which is based on the traditional structure.

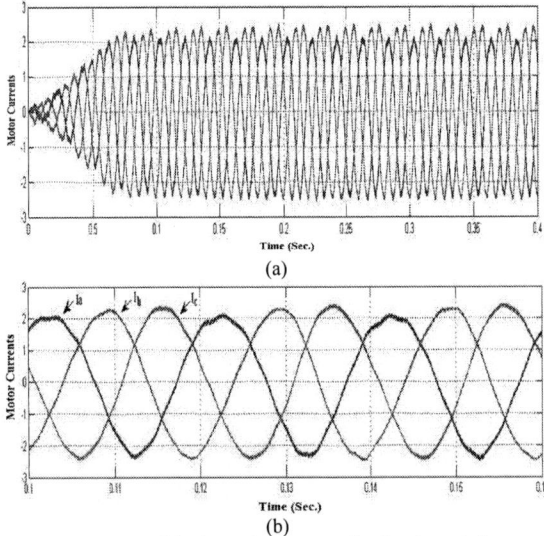

(a)

(b)

Fig.8. Waveforms of the three-phase current flowing through the motor (a) normal view (b) zoomed

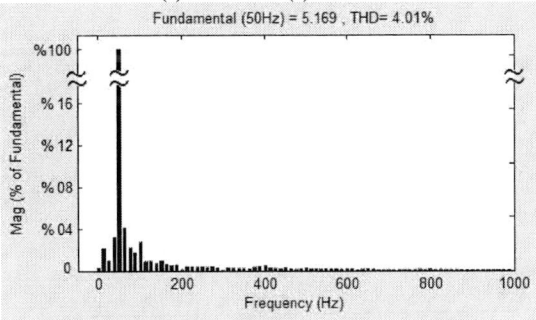

Fig.9. Harmonic spectra of the motor phase current

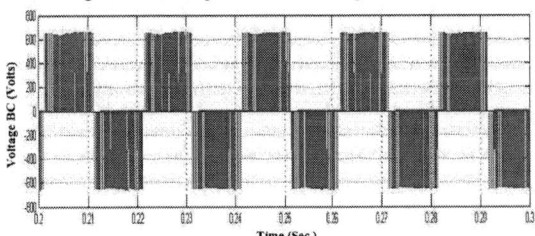

Fig.10. Line to line voltage waveform across the motor load

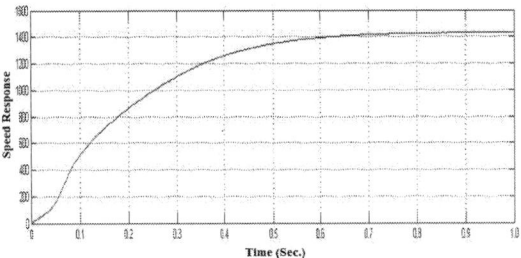

Fig.11. Speed response of the proposed structure

Furthermore another advantage of the proposed structure is the voltage sag mitigation. By controlling the zero shoot-through duty cycle, the DC bus voltage will remain constant while the magnitude of the input voltage is reduced. Fig. 12 and Fig. 13 show the voltages of the DC bus for two states, D = 0 and D = 0.2.

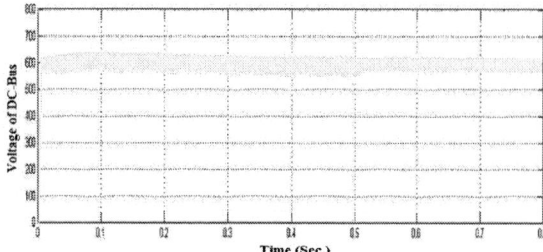

Fig.12. Voltage of DC bus for D = 0

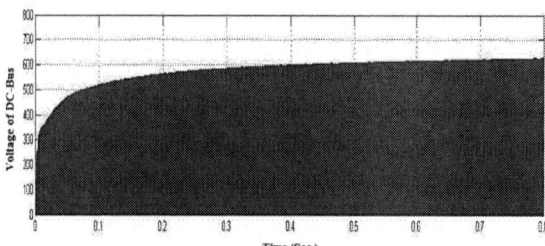

Fig.13. Voltage of DC bus for D = 0.2

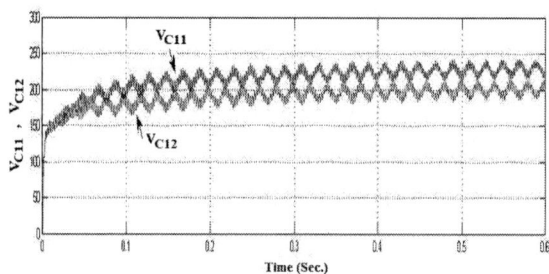

Fig.14. Voltages of the QZs network capacitors

VI. CONCLUSION

The results in this paper have proven that the proposed Quasi Z-source single-phase to three-phase converter with only four switches is a more efficient, more reliable and cost efficient converter. It can be used as an adjustable speed drive for a three-phase induction motor. Current THD in this structure is less when compared to the traditional circuit. Any desired output voltage can be generated, even if it is greater than the input voltage.

The proposed converter is also capable of mitigating voltage sag as it can provide ride-through during its occurrence without any additional structures. In addition it is expected that with this structure, the inrush current can be reduced while the desired speed response is achieved.

REFERENCES

[1] N. Mohan, T.M. Underland and R.J. Ferraro, "Sinusoidal Line Current Rectification with a 100 kHz B-SIT Step-up Converter," *IEEE PESC Conf. Rec,* pp. 92-98, 1984.

[2] P. Enjeti and A. Rahman, "A New Single-Phase Three-Phase Converter with Active Input Current Shaping for Low Cost AC Motor Drives," *IEEE. Tran. On IA*, Vol. 29, No. 4, 1990.

[3] S.B. Bekiarov and A. Emadi, "A New On-Line Single-Phase to Three-Phase UPS Topology with Reduced Number of Switches," *IEEE Conference, Proc. PESC3*, pp. 451-456, 2003.

[4] M.D. Bellar, M. Aredes, J.L. Siklva, L.G.B. Rolim, F.C. Aquino and V.C. Peterson, "Comparative Analysis of Single-Phase to Three-Phase Converters for Rural Electrification," *IEEE Conf. on IEs*, Vol. 2, pp. 1255-1260, 2004.

[5] H. Haga, K. Ohishi and K. Sumida, "Unity Power Factor Control Method of Single-Phase to Three-Phase Power Converter without Reactor and Electrolytic Capacitor," *IEEE, EPE Conference*, pp. 1-7, 2005.

[6] X. Yang, R. Hao, X. You and T.Q. Zheng, "A New Topology for Operating Three-Phase Induction Motors Connected to Single-Phase Supply," *IEEE Conference, ICEMS*, pp. 1391-1394, 2008.

[7] X. Yang, R. Hao, X. You and T.Q. Zheng, "A New Single-Phase to Three-Phase Cycloconverter for Low Cost AC Motor Drives," *Proc. ICIEA 3rd IEEE Conf.*, pp. 1752-1756, 2008.

[8] E.C. Santos, C.B. Jacobina, J.A.A. Dias, "Active Power Line Conditioner Applied to Single-Phase to Three-Phase Systems," *IEEE Conference on Industrial Electronic*, pp. 148-153, 2009.

[9] H.W. Broeck and H. Ch. Skudelny, "Analytical Analysis of the Harmonic Effects of a PWM AC Drive," *IEEE Trans. Power Electronic, Vol. 3, No. 2, pp. 216-223, 1988.*

[10] P.N. Enjeti, A. Rahman and R. Jakkli, "Economic Single-Phase to Three-Phase Converter Topologies for Fixed and Variable Frequency Output," *IEEE Trans. Power Electronic,* Vol. 8, No.3, pp. 329-335, 1993.

[11] F. Blaabjerg, S. Freysson, H.H. Hansen and S. Hansen, "A New Optimized Space-Vector Modulation Strategy for a Component-Minimized Voltage Source Inverter," *IEEE Transaction on Power Electronics*, Vol. 12, No. 4, 1997.

[12] M.N. Uddin, T.S. Radwan and M.A. Rahman, "Performance Analysis of a Four Switch 3-Phase Inverter Fed IM Drives," *IEEE Conference,* pp. 36-40, 2004.

[13] B.K. Bose, *Power Electronics and AC Drives*, Englewood Cliffs, NJ: Prentice Hall, 1986.

[14] M. N. Uddin, T.S. Radwan and M.A. Rahman, "Performances of Fuzzy Logic Based Indirect Vector Control for Induction Motor Drive," *IEEE Trans. On Industry Applications*, Vol. 38, No. 5, pp. 1219-1225, 2002.

[15] J. Anderson and F.Z. Peng, "Four Quasi Z-Source Inverters," *IEEE Power Electronic Conf.*, pp. 2743-2749, 2008.

[16] Fang Z. Peng, M. Shen and Z. Qian, "Maximum Boost Control of the Z-Source Inverter," *IEEE Proc. PESC*, pp. 255-260, 2004.

[17] F. Blaabjerg, H.H. Hansen and S. Hansen, "A New Optimized Space-Vector Modulation Strategy for a Component-Minimized Voltage Source Inverter," *IEEE Trans. On Power Electronics*, Vol. 12, No. 4, 1997.

[18] P.Q. Dzung, L.M. Phuong, P.Q. Vinh, N.M. Hoang and T.C. Binh, "New Space Vector Control Approach for Four Switch Three Phase Inverter (FSTPI)," *IEEE PEDS Conf.*, pp. 1002-1008, 2007.

Space Vector based Spread Spectrum Modulation Scheme for Three-Level Inverters

Biji Jacob, M.R.Baiju
College of Engineering, Trivandrum, India
biji @ece.cet.ac.in, mrbaiju@ece.cet.ac.in

Abstract- **A Spread Spectrum Modulation scheme based on Space Vector for three level inverter is proposed in this paper. The scheme disperses the power spectrum of the inverter output voltage as a wide-spectrum noise. The vector space of inverter reference space vector region is divided into voronoi regions to find out switching vectors. A new method is presented to code these voronoi regions with Vector Quantization using instantaneous reference phase amplitudes without using lookup table. The space-vector diagram of three-level inverter is simplified to two-level space-vector diagram by the principle of mapping. The switching vectors are determined for two-level inverter using the proposed scheme and these switching vectors are then translated to the actual switching vectors of the three-level inverter by reverse mapping. The proposed scheme naturally selects the outer vectors in the over-modulation condition and hence results in a smooth transition from linear to over-modulation region. The scheme is implemented for 2-HP three phase induction motor driven by three-level cascaded inverter topology.**

I. INTRODUCTION

Three-level inverters are widely used in industrial medium-voltage adjustable speed drives. In variable voltage, variable frequency drives, the dc link can be kept constant and voltage control is achieved by varying the duty ratio of inverter switches [1]–[3]. Three-inverters have improved total harmonic distortion and reduced stress on switching devices compared to two level inverters [2]–[6]. Commonly used modulation and control strategies in multilevel inverters are classified into carrier based sinusoidal Pulse Width Modulation (SPWM); selective harmonic elimination PWM (SHEPWM); and space vector PWM (SVPWM) [3]–[8]. The power spectrum of inverters using constant switching frequency PWM schemes tends to be concentrated around the switching frequency and its harmonics [9]-[12]. This results in the electromagnetic interference radiation from the inverter and acoustic noise generated by electric machines driven by these inverters [9]-[12].

To spread the harmonic energy over a large frequency range instead of being concentrated at few discrete frequencies variable frequency switching schemes can be used. Different variable frequency switching schemes used in two level inverters are frequency modulation of the system clock, random or quasi-random modulation of the system clock frequency, delta or sigma-delta modulation, chaotic control and hysteresis control [10]–[22]. Randomized switching time results in high frequency switching and narrow switching pulses. Spread spectrum scheme for multilevel inverters are not yet investigated extensively.

In this paper, a variable switching frequency scheme for multilevel inverters using sigma delta modulation is proposed for spreading the power spectra. The motivation for adopting the principle of sigma delta modulation in the case of multilevel inverter is that the switching converters can be viewed as analog-to-digital converters [20]–[22]. Sigma delta modulators are used to reduce quantization noise in over sampling analog-to-digital converters [23]–[24]. The switching frequency in sigma delta modulator varies randomly under the constant sampling frequency, resulting in the spreading of the output spectra.

Sigma delta modulation with scalar quantizer is used for power control in 2-level voltage source inverters [14]–[19]. For efficient quantization in digital communication and data compression, the concept of Vector Quantization is used instead of scalar quantization [25]–[26]. Space Vector based Sigma Delta Modulator with Vector Quantization for two level inverter is proposed in [27]. The space-vector diagram of three-level inverter can be simplified to two-level space-vector diagram by the principle of mapping [28]–[29]. The present work proposes to extend the sigma delta modulation to three-level inverters and adopt the principle of Vector Quantization in the quantizer of sigma delta modulator.

This paper proposes a new approach to spread spectrum modulation for three-level voltage source inverter. The proposed scheme uses a space vector based sigma delta modulation. For quantizing reference space vector in the sigma delta modulator, the principle of Vector Quantization is used. In the present paper, sixty-degree coordinate system is used for the representation of space vector to eliminate fractional arithmetic and the computational overhead instead of conventional orthogonal coordinate system [37]–[39]. The proposed scheme is experimentally verified for three-level, four-level, five-level and six-level inverter topologies driving 2-HP three phase induction motor and experimental result are presented.

II. PRINCIPLE OF THE PROPOSED SCHEME

The Space-Vector diagram of a Three-Level Inverter is shown in Fig.1(a). This can be visualized as a hexagon formed by seven small subhexagons as shown in Fig. 1(b) [28]–[29]. The subhexagon at the centre (inner subhexagon) corresponds to the space-vector diagram of a two-level Inverter having the vector 000 as the center. There are six outer subhexagon

978-1-4577-0007-1/11 $26.00 © 2011 IEEE

around the centre subhexagon. By placing the centers of outer six subhexagon on the vertex of inner subhexagon, space-vector diagram of a three-level inverter is obtained (Fig. 1).

Depending on where the tip of the reference voltage space-vector lies, there are three regions of operations namely Two-level operation, Three-level operation, and over modulation region of operation. If the tip of the reference vector is confined to the inner subhexagon { | Vref | < √3/4 V_{DC}}, it is Two-level operation mode. If the tip of the reference vector lies in the outer subhexagons {√3/4 V_{DC} < | Vref | < √3/2 V_{DC}}, it is Three-level operation mode. If the tip of the reference vector lies outside the hexagons { | Vref | > √3/2 V_{DC}}, it will be in the over modulation region.

If the outer subhexagons are shifted towards the inner subhexagon center, the space vector diagram of a three level inverter is simplified to that of a two-level inverter. The shifting of outer subhexagons in the space vector diagram of three-level inverter towards the zero vector 000 involves the mapping of sectors of outer subhexagons to sectors of inner subhexagon. This is done by subtracting the vector at the center of the outer subhexagon from its other vectors.

The position of the reference voltage space-vector located anywhere in the three-level space vector diagram can be mapped to inner subhexagon. Mapping of reference voltage space-vector to two-level space vector diagram is achieved by subtracting the subhexagon centre vector in which the reference voltage space-vector located from the reference voltage space-vector value. After mapping the reference voltage space-vector, the proposed scheme for two level inverter can be applied to find out the switching vectors corresponding to mapped reference voltage space-vector.

The inner subhexagon can be reverse mapped to outer subhexagon by adding the voltage space-vector at the center of corresponding outer subhexagon to the vectors of the inner subhexagon. Generation of the actual switching vectors for three-level inverter is found out by reverse mapping, in which the value of selected subhexagon center's vector is added to mapped switching vectors.

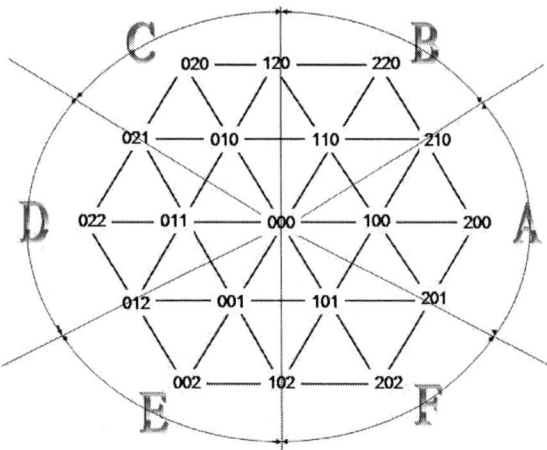

Fig. 1. Space Vectors diagram of a 3-Level Inverter and identification of the subhexagon center.

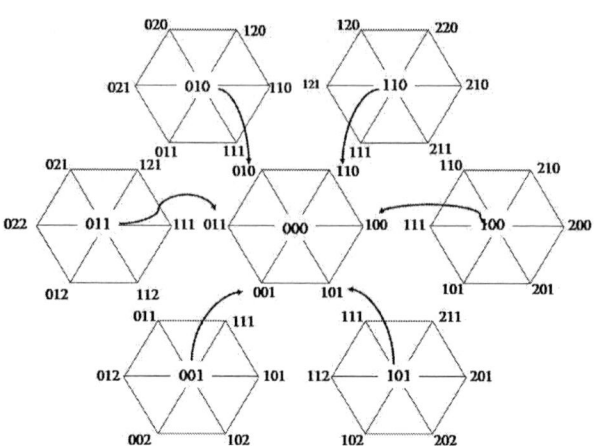

Fig. 2. Mapping of 3-level to 2-level Space-Vector diagram

The four steps involved in the proposed Vector Quantized Space Vector Pulse Density Modulation scheme for three level inverters are (A) Identification of subhexagon center, (B) Mapping of Reference Space Vector Space Vector to inner-subhexagon, (C) Vector Quantized Space Vector Pulse Density Modulation in Two-level and (D) Reverse Mapping to original sbhexagon.

A. Identification of subhexagon center

If the tip of the reference voltage space vector lies within the inner-subhexagon (| Vref | < √3/4 V_{DC}), subhexagon center is 000. If the tip of the reference voltage space-vector is outside the inner-subhexagon, a simple algorithm is used to find the subhexagon center. The regions **A** to **F** as shown in Fig. 1 enclose each outer-subhexagon centers. The location of reference voltage space-vector in regions **A** to **F** is found out by comparing the instantaneous values of three phase control signal.

Fig. 3 illustrates the scheme for finding the subhexagon center which encloses the tip of the reference voltage space-vector from the instantaneous value of three-phase reference sinusoid. Let Va, Vb and Vc represent the instantaneous magnitude of the three-phase reference sinusoid. If the magnitude of the sine wave is positive, it is represented as "1" and if the magnitude is negative, it is represented as "0". During the time interval from $\omega t = 0$ to $\omega t = 60$, Va is positive, Vb is negative and Vc is positive. This corresponds to code vector 101 (+ − +) corresponding to region **A** and subhexagon centre 101. Similarly, during $\omega t = 60$ to $\omega t = 120$, Va is positive, Vb and Vc are negative which implies code vector of 100 (+ − −). That is, if the reference space vector lies within the region **A**, subhexagon centre is 100. Similar procedure can be used to find the remaining code vectors. Hence for regions **B** to **F**, subhexagon centers are 110, 010, 011, 001 and 101 respectively.

B. Mapping of Reference Voltage Vector

Once the subhexagon containing the tip of the reference voltage space vector is identified, the reference voltage space vector is mapped to the two-level space vector diagram [28]–[29]. The mapping can be done by subtracting the vector

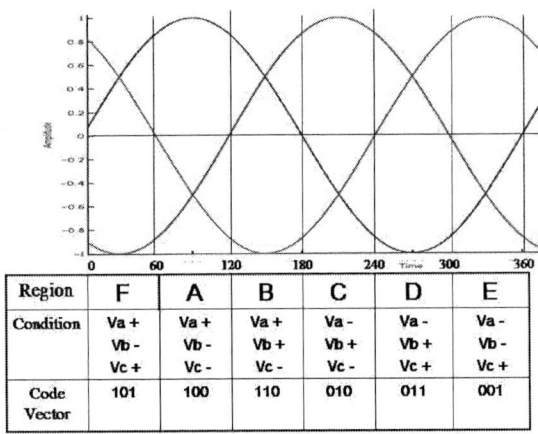

Region	F	A	B	C	D	E
Condition	Va + Vb - Vc +	Va + Vb - Vc -	Va + Vb + Vc -	Va - Vb + Vc -	Va - Vb + Vc +	Va - Vb - Vc +
Code Vector	101	100	110	010	011	001

Fig. 3. Proposed scheme to determine the Subhexagon Center.

value of the selected subhexagon centre, containing the tip of the reference vector, from the original reference voltage space-vector. If (a_i, b_i, c_i) is the instantaneous reference voltage space-vector and (a_c, b_c, c_c) subhexagon centre containing the tip of the reference vector, then mapped reference vector is given by $(a_m, b_m, c_m) = (a_i, b_i, c_i) - (a_c, b_c, c_c)$.

The shifting of outer subhexagons in the space vector diagram of three-level inverter towards zero vector 000 results the mapping of sectors of outer subhexagons to the sectors of inner subhexagon Fig. 2. The mapping is arrived by subtracting the vector value at the centre of the outer-subhexagon from its other vertex vectors.

C. Proposed Vector Quantized Space Vector Pulse Density Modulation in Two-level

After mapping process, the three-level space vector plane is transformed to the two-level space vector plane. The proposed scheme for the Two-level inverter can be applied to find out the switching vectors corresponding to mapped reference vector. The scheme uses Sigma-Delta Modulator with Vector Quantizer. The space-vector diagram of Two-level inverter is divided into seven Voronoi regions for quantizing the mapped reference voltage space-vector.

Fig. 4 represents the proposed modulation scheme. The proposed scheme consists of two sigma-delta modulators each for the resolved m and n components of reference space vector. Each sigma-delta modulator consists of a difference node, a discrete time integrator, a space-vector quantizer and a DAC in the feedback path. The input to the integrator is the difference between the input signal V and the quantized output value S converted back to the predicted analog signal, **Sa**. This difference between the input signal V and the fed back signal **Sa** at the integrator input is equal to the quantization error. This error is summed up in the integrator to produce integrated error vector Ve.

Employing the principles of Vector Quantization (VQ), the integrated error space vector Ve is viewed as a random vector in a two dimensional vector space of two level voltage source inverter. This vector space can be divided into seven Voronoi regions, named **A** to **F** and **O** around each inverter voltage

vectors (Fig. 5). All vectors in a Voronoi region can be quantized to the corresponding inverter voltage vector by the principle of Vector Quantization. The eight inverter voltage vectors can be coded using 3 bits (000 to 111) which will become the code words in the vector quantizer. These code words are actual two level inverter voltage vectors.

Each Voronoi region in the vector space is quantized on to a fixed point as the code vector, which will be centroid of that region. The switching space vectors of inverter V_1 to V_6 and V_0 respectively forms the code vectors of these Voronoi regions **A** to **F** and **O**. The Voronoi region and corresponding output code vector in which sampled reference input vector at any instant can be determined by comparing the instantaneous amplitude of the 3-phase input signal. The algorithm used to find out the subhexagonal center can be applied to find out Voronoi region and code vectors as illustrated in Fig. 3.

The set of available output vectors, V_0 through V_7, is a copy of inverter vectors. The mapping follows the principle of minimum Euclidean distance between the **V** and **S** vectors. So the constructed modulator is equivalent to one selecting at each sampling time, the output vector that minimizes the error vector.

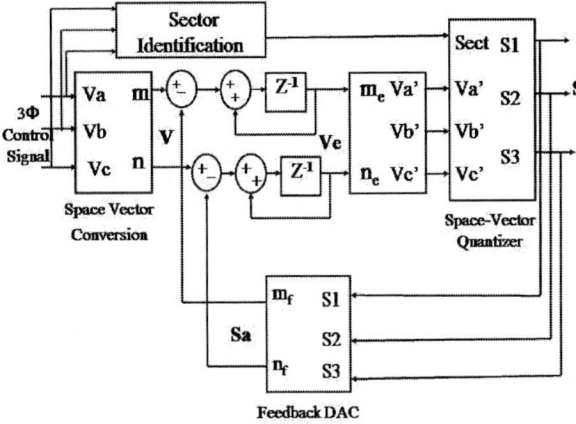

Fig. 4. Proposed Two-level Space Vector Spread Spectrum Modulator

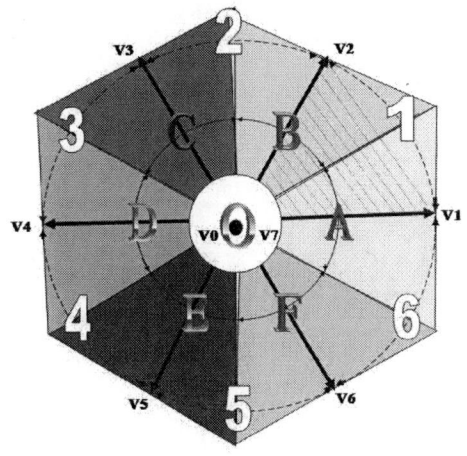

Fig. 5 The Voronoi regions **A** to **G** corresponding to the Vector Quantization regions of 3Φ input control signal.

D. Reverse Mapping

The switching vectors corresponding to mapped two-level reference vector has to be translated back to actual switching vector corresponding to the three-level inverter. The translation can be achieved by reverse mapping the inner-subhexagon to the subhexagon containing the tip of reference space vector. This reverse mapping can be done by adding the vector value at the center of the selected subhexagon to the switching vectors of the two-level inverter [29].

If (a_s, b_s, c_s) is the switching vectors of the two-level inverter and (a_c, b_c, c_c) subhexagon centre containing the tip of the reference vector, then reverse mapped reference vector is given by $\quad (a_s, b_s, c_s) = (a_t, b_t, c_t) + (a_c, b_c, c_c)$

II.   EXPERIMENTAL RESULT AND DISCUSSION

The proposed scheme for three-level inverters can be applied to any generalized inverter configurations like neutral point clamped, H-bridge configuration, capacitor clamped, Cascaded configuration and open end winding configuration. The proposed scheme is experimentally verified for a three-level cascaded inverter configuration by cascading conventional two two-level inverters as shown in Fig. 6 [5]. It is used to drive a 2 HP three phase induction motor with V/f control for different modulation indices covering different speed ranges. The proposed scheme is implemented using the dSPACE DS 1104 RTI platform. The logic for gating pulses for the two inverters in cascaded configuration are derived with Xilinx Virtex-II PRO XC2VP30 FPGA board.

Table-I shows the three-level cascaded inverter configuration switching strategy of the individual inverter top switches. The inverters are switched according to instantaneous two level Pulse Density Modulated (PDM) signal and corresponding subhexagon centre.

The three Pole Voltage waveforms (V_{A2O}, V_{B2O}, V_{C2O}) obtained experimentally are shown in Fig. 7(a) and its time axis expanded view of marked area is shown in Fig. 7(b) for three level operation.

TABLE I.   SWITCHING STRATEGY TO REALISE THREE LEVELS OF THE INDIVIDUAL INVERTERS

Sub Hexagon Center	Modulator signal	Status of Top Switch of Two-Level Inverters		Voltage Level	Switching level
		INV-1	INV-2		
0	0	ON/OFF	OFF	0	0
0	1	OFF	ON	$V_{DC}/2$	1
1	0	OFF	ON	$V_{DC}/2$	1
1	1	ON	ON	V_{DC}	2

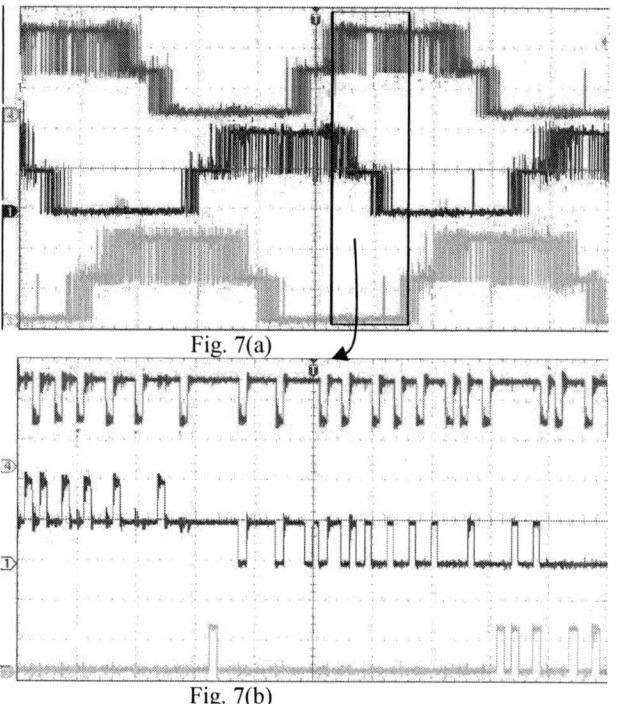

Fig. 7(a)

Fig. 7(b)

Fig. 7(a). Experimental Three Pole Voltage (V_{A2O}, V_{B2O}, V_{C2O}) waveforms for the proposed scheme with Three Level mode of operation (modulation index m= 0.8). Scale: X-axis: 4ms/div; Y-axis : 100 V/div.
Fig. 7(b). Time scale expanded waveforms of the 3 pole voltages of the marked region in Fig. (a). X-axis: 400µs/div; Y-axis : 100 V/div

From time scale expanded waveform of pole voltages shown in Fig. 7(b), it can be noted that the width of each pulse is constant which is equal to the sampling period. But the density of pulses varies depending on the reference space vector there by varying the effective width of pulses. In the proposed scheme switching frequency varies randomly as it is a pulse density modulation scheme. Since the width of each pulse is constant which is equal to the sampling time, minimum pulse width problem associated with PWM does not occur here.

Fig. 8. shows experimental Pole Voltage (V_{AO}), A-Phase voltage (V_{AN}) and A-phase current (I_A) waveforms of induction motor for the proposed scheme with modulation index 0.8 corresponding to three level mode of operation. Fig. 9 and 10 show the experimental pole voltage and phase voltage for modulation indices 0.4 and 1.1 respectively

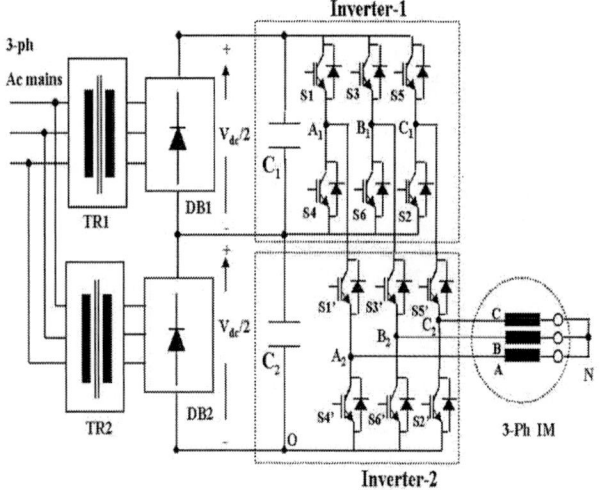

Fig. 6. Three-level inverter configuration by cascading conventional two two-level inverters

978-1-4577-0007-1/11 $26.00 © 2011 IEEE

corresponding to two-level and overmodulation region. In 2-level operation, only bottom inverter-2 is switches and inverter-1 is clamped to zero level.

Fig. 11(a) and (b) shows frequency spectrum of A-phase pole voltage for the proposed scheme and SVPWM for a modulation index 0.8.

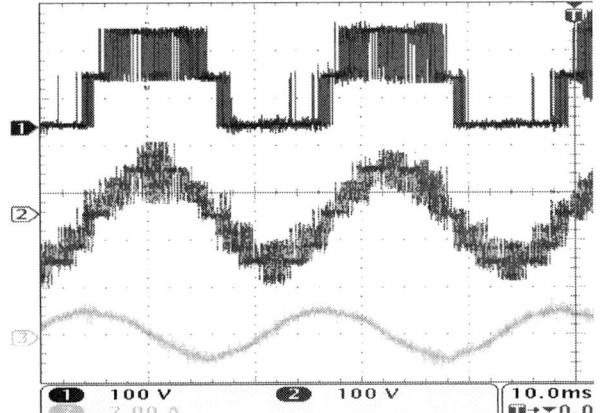

Fig. 8. Experimental Pole Voltage (V_{AO}), A-Phase voltage (V_{AN}) and A-phase current (I_A) for the proposed scheme with Three-level mode of operation (Modulation index 0.8).
Scale: Upper trace (V_{A2O}) and Middle trace (V_{A2N}): Y-axis : 100V/div; Lower trace (I_A) : Y-axis : 2A/div; X-axis: 10ms/div

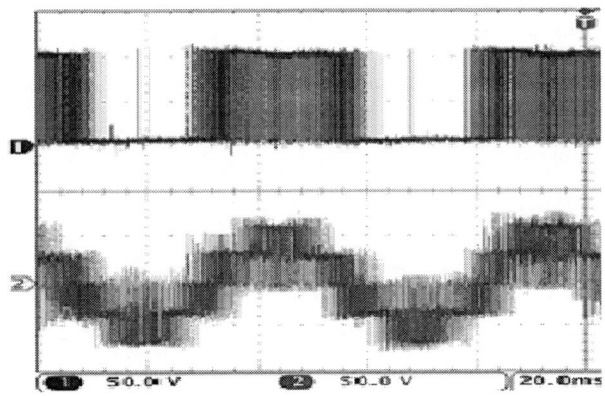

Fig. 9. Upper trace: Pole Voltage (V_{AO}) and Lower trace: Phase voltage (V_{AN}) for proposed scheme with modulation index m=0.4. Scale : X-axis: 20ms/div; Y-axis : 50 V/div

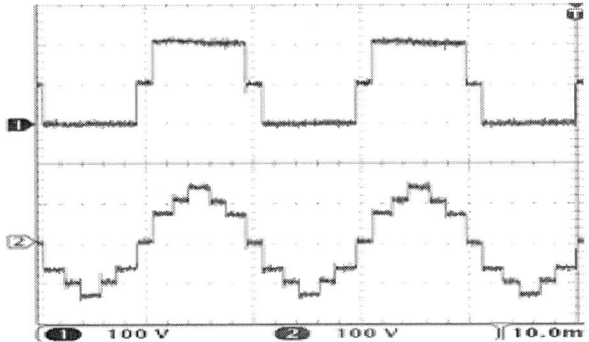

Fig. 10. Upper trace: Pole Voltage (V_{AO}) and Lower trace: Phase voltage (V_{AN}) for proposed scheme with modulation index m=1.1. X-axis: 10ms/div; Y-axis : 100 V/div

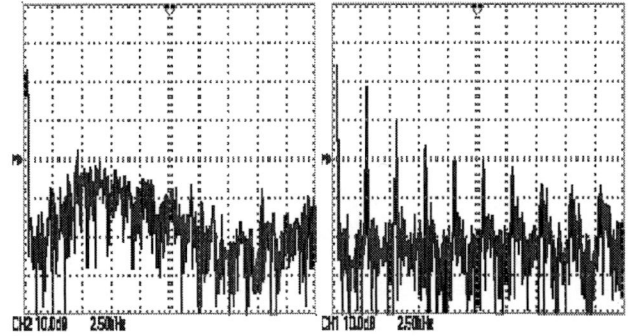

Fig. 11(a). Frequency spectrum of A-phase Pole voltage (V_{AO}) for the proposed scheme with modulation index m=0.8. Scale : Y-axis : 10 dB/div. X-axis : 2.5 KHz/div

Fig. 11(b). Frequency spectrum of A-phase Pole voltage (V_{AO}) for SVPWM with modulation index m=0.8. Scale : Y-axis : 10 dB/div. X-axis : 2.5 KHz/div

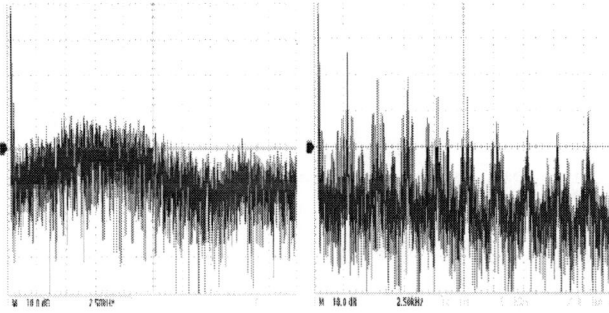

Fig. 12(a). Frequency spectrum of A-phase Pole voltage (V_{AO}) for the proposed scheme with modulation index m=0.4. Scale : Y-axis : 10 dB/div.   X-axis : 2.5 KHz/div

Fig. 12(b). Frequency spectrum of A-phase Pole voltage (V_{AO}) for SVPWM with modulation index m=0.4. Scale : Y-axis : 10 dB/div X-axis : 2.5 KHz/div

Harmonic spectral spreading property of proposed scheme can be noted without any large concentrations of power at discrete frequencies. At low frequency switching harmonics the peaks are 15 to 20 dB less in the proposed scheme compared to conventional SVPWM. Similarly Fig. 12(a) and (b) show the frequency spectrum for the modulation index 0.4 of the proposed scheme and SVPWM. Further it can be seen that noise level is 10 dB below in three-level mode (m = 0.8) compared to two-level mode (m = 0.4). Hence annoying acoustic noise emitted from ac motors operating with a carrier frequency in the audible range may be substantially reduced in an inexpensive manner by using the proposed scheme.

IV. CONCLUSION

A Vector Quantized Space Vector based Spread Spectrum Modulation scheme is proposed and experimentally verified for three phase Three-Level Voltage Source Inverters. The scheme uses first order sigma delta modulator with two dimensional Vector Quantization. The Vector Space of inverter space vector region is divided into vornoi regions to find out switching vectors. A new Space Vector Quantization scheme is proposed to find out voronoi regions to derive switching vectors by comparing the instantaneous amplitude of three phase control input. The scheme does not use look up tables for sector identification and switching vector calculation. Instead of the Cartesian coordinate system, the

978-1-4577-0007-1/11 $26.00 © 2011 IEEE

sixty degree coordinate system is used to represent the space vectors. Since only integer values are involved in sixty degree coordinate system it reduces the computational overheads compared to Cartesian coordinate system. The proposed modulator was implemented for three-level cascaded inverter configuration driving 2-HP induction motor. Experimental results are presented covering different speed ranges such as two-level, three-level and over modulation regions.

In the proposed scheme density of pulses is varied instead of pulse width resulting in random variation in switching frequency. The experimental results show that harmonic spectrum of the proposed scheme disperses the switching harmonics as a wide-spectrum noise. Compared to SVPWM, the switching harmonics peaks are suppressed to the range of 15 to 20 dB. The noise level in three level mode of operation is still lesser compared to two level mode of operation in the scheme. The minimum switching pulse width in the proposed scheme is the sampling time, resulting in avoidance of minimum pulse width problem.

REFERENCES

[1] L. M. Tolbert, F. Z. Peng, and T. G. Habetler, "Multilevel converters for large electric drives," IEEE Trans. Ind. Appl., vol. 35, no. 1, pp. 36–44, Jan./Feb. 1999.

[2] Jose Rodriguez, Jih-Sheng Lai and Fang Zheng Peng, " Multilevel Inverters: A Survey of Topologies, Controls and Applications ", IEEE Transactions on Industrial Electronics, vol.49, No.4, August 2002, pp 724-738.

[3] P. F. Seixas, M. A. Severo Mendes, P. D. Donoso-Garcia, and A. M. N. Lima, "A space vector PWM method for three-level voltage source inverters," in Proc. IEEE APEC, 2000, vol. 1, pp. 549–555.

[4] K. Zhou and D. Wang, "Relationship between space-vector modulation and three-phase carrier-based PWM: A comprehensive analysis," IEEE Trans. Ind. Electron., vol. 49, no. 1, pp. 186–196, Feb. 2002.

[5] W. Yao, H. Hu, and Z. Lu, "Comparisons of space-vector modulation and carrier based modulation of multilevel inverter," IEEE Trans. Power Electron., vol. 23, no. 1, pp. 45–51, Jan. 2008.

[6] Hyo L. Liu, Gyu H. Cho and Sun S. Park, "Optimal PWM Design for High Power Three-Level Inverter Through Comparative Studies," IEEE Trans. Power Electron., vol. 10, no. 1, pp. 38–47, Jan. 1995.

[7] J.W. Gray and F.J. Haydock, "Industrial power quality considerations when installing adjustable speed drive systems," IEEE Trans. Indusry Applications, vol. 32, no. 3, pp. 646–652, May/June 1996.

[8] Samir Kouro, Mariusz Malinowski, K. Gopakumar, Josep Pou, Leopoldo G. Franquelo, BinWu, Jose Rodriguez, Marcelo A. Pérez, and Jose I. Leon, "Recent Advances and Industrial Applications of Multilevel Converters," IEEE Transactions on Industrial Electronics, Volume 57, Issue 8, August 2010, pp. 2553 – 2580.

[9] K. K. Tse, H. S. Chung, S. Y. R. Hui, and H. C. So, "Analysis and spectral characteristics of a spread-spectrum techniques for conducted EMI suppression," IEEE Trans. Power Electron, vol. 15, no. 2, pp. 399–410, Mar. 2000.

[10] J.T. Boys and P.G. Handley, "Spread spectrum switching: low noise modulation technique for PWM inverter drives," IEE Proceedigs-B, Vol. 139, No. 3, May 1992, pp. 252-260.

[11] Andrzej M. Trzynadlowski, Frede Blaabjerg, Stanislaw Legowski, " Random Pulse Width Modulation Techniques for Converter-Fed Drive System – A Review," IEEE Trans. Indusry Applications, vol. 30, no. 5, pp. 1166–1175, Oct. 1994.

[12] Ki-Seon Kim; Young-Gook Jung; Young-Cheol Lim, "A New Hybrid Random PWM Scheme", IEEE Transactions on Power Electronics Volume 24, Issue 1, Jan. 2009

[13] Hyo L. Liu and Gyu H. Cho, " Three-Level Space Vector PWM in Low Index Modulation Region Avoiding Narrow Pulse Problem," IEEE

Transactions on Power Electronics, Vol. 9, No.5, Sep. 1994, pp. 481 – 486 .

[14] Nieznanski J, " Analytical Establishment of the Minimum Distortion Pulse Density Modulation," Electrical Engineering 80(1997), pp 251-258.

[15] Nieznanski J, "Performance Characterization of Vector Sigma- Delta Modulation," Proc. of Conference of the Industrial Electronics Society IECON '98, pp 531-536.

[16] Nieznanski J, Wojewodka A. and Chrzan P.J, "Comparison of Vector Sigma - Delta Modulation and Space- Vector PWM," Conference of the Industrial Electronics Society IECON 2000 Volume 2, pp1322 - 1327.

[17] Atushi Hirota, Satoshi Nagai and Mutsuo Nakaoka, " A Novel Delta-Sigma Modulated Space Vector Modulation Scheme using Scalar Delta-Sigma Modulator," Proc. Power Electronics Specialists Conference, 2003, PESC '03, pp 485 – 489 vol.2.

[18] Atushi Hirota, Satoshi Nagai and Mutsuo Nakaoka, "A Simple Configuration Reducing Noise Level Three-Phase Sinewave Inverter Employing Delta-Sigma Modulation Scheme," Proc. Applied Power Electronics Conference and Exposition, 2006, APEC'06. pp 1485- 1489.

[19] Atushi Hirota, Bishwajit Saha, Sang-Pil Mun and Mutsuo Nakaoka, "An Advanced Simple Configuration Delta-Sigma Modulation Three-Phase Inverter Implementing Space Voltage Vector Approach," Proc. Power Electronics Specialists Conference, 2007, PESC 2007, pp 453 – 457.

[20] Glen Luckjiff Ian, Dobson and Deepak Diwan, "Interpolative Sigma Delta Modulators for High frequency Power Electronics Applications," Proc. of Power Electronics Specialists Conference, PESC '95 , vol.1, pp 444 - 449.

[21] Glen Luckjiff and Ian Dobson, "Hexagonal $\Sigma\Delta$ modulators in Power Electronics," IEEE Transactions on Power Electronics, Vol. 20, No.5 (Sep. 2005), pp1075 – 1083

[22] Glen Luckjiff and Ian Dobson, "Hexagonal Sigma Delta Modulation," IEEE Transactions on Circuits and System I, Vol. 50, No.8 (Sep. 2003), pp991 – 1005.

[23] J. C. Candy and G. C. Temes, "Oversampling methods for A/D and D/A conversion in Oversampling Delta– Sigma Data Converters, Theory, Design and Simulation," New York: IEEE Press, 1992.

[24] Pervez M. Aziz, Henrik V. Sorensen, and Jan Van der Spiegel, "An Overview of Sigma-Delta Converters," IEEE Signal Processing Magazine, Volume 13, Issue 1, September 1996, pp 61-84.

[25] Robert M. Gray, "Vector Quantisation," IEEE ASSP Magazine (April 1984), pp 4-28.

[26] Gersho A.,"On the Structure of Vector Quantisation," IEEE Transactions on Information Theory, IT-28 (March1982), pp157 –166.

[27] Biji Jacob and M.R. Baiju, " Spread Spectrum Scheme for Two-Level Inverters using Space Vector Sigma-Delta Modulation," 5th IET International Power Electronics, Machines and Drives, PEMD 2010.

[28] Jae Hyeong Seo, Chang Ho Choi and Dong Seo Hyun, A New Simplified Space-Vector PWM Method for Three- Level Inverters, IEEE Transactions on Power Electronics, Vol. 16, No.4, July 2001, pp 545–550.

[29] Aneesh Mohammed A.S, Aneesh Gopinath and M.R. Baiju-"A Simple Space Vector PWM Generation Scheme for Any General n-Level Inverter," IEEE Transactions on Industrial Electronics, Volume 56, Issue 5, May 2009, pp. 1649 – 1656.

[30] N. Celanovic, and D. Boroyevich, "A fast space vector modulation algorithm for multilevel three phase converters," IEEE Trans on Industry Applications, Vol.37, No.2, 2001, pp637-641.

[31] Sanmin Wei, Bin Wu, Fahai Li and Congwei Liu, " A General Space Vector PWM Control Algorithm for Multilevel Inverters, " Proc. Applied Power Electronics Conference and Exposition, 2003, APEC'03, pp. 562-568.

Optimization of Operating Parameters in a Unipolar PWM Inverter

Taufik Taufik and Michael McCarty
Electrical Engineering Dept.
Cal Poly State University
San Luis Obispo, California, USA

Makbul Anwari
Electrical Engineering Dept.
Umm Al-Qura University
Saudi Arabia

Anton Satria Prabuwono
Faculty of Information Sci. & Tech.
Universiti Kebangsaan Malaysia
UKM Bangi, Selangor D.E., Malaysia

Abstract–The increase use of renewable dc or battery as a source has further emphasized the importance in efficient design of inverter. This paper discusses a study to determine optimum operating parameters of a commonly used pwm unipolar inverter. Variations on the input voltage level and switching frequency will be investigated to see their impacts on the operation of the inverter as well as harmonics produced. Results through modeling and calculations demonstrate that meticulous selections of design parameters yield inverter's operation with the highest efficiency at the lowest level of output distortions.

Keywords–Inverter; Harmonics; Power Quality

I. INTRODUCTION

Currently in the environmentally conscious, green age, renewable energy resources have become increasingly desirable. A push from traditional power generation has been seen because of its damage to the environment and its depletion of the energy supply. Solar energy, wind power, and even some fuel cells have been implemented more readily due to their pollution-free nature. These resources are constantly being used in order to meet the increasing demand for energy. Also, as demand increases, efficiency must improve as well to minimize the expensive cost of wasted energy.

These modern resources output a DC source which cannot be directly interfaced with the utility grid. As a result, inverters must be implemented in order to use the current infrastructure. Thus, the wide application of grid-connected inverter topologies such as reported in [1], [2], and [3] has been seen.

The efficiency and harmonic distortion of grid-connected inverters is very important because they are directly interfaced with the grid. This means that the inverters immediately affect the efficiency and harmonic distortion of the overall system. Efficiency is also vital because of the added cost of input power per output power as efficiency drops. This lost energy makes the devices less economical and increases their heat output. The heat output also makes the device more expensive because of the need to dissipate this heat.

Harmonic distortion of output voltage and current can lead to voltage distortion reaching a transformer that can't handle harmonics. As a result, k-rated transformers may be necessary despite the additional doubling in cost and increase in weight [4]. This cost can also make the economics of the inverter less efficient. Also, at the

higher frequencies of harmonics, increased eddy currents and hysteresis losses can heat up a transformer. These effects of distortion can also add up and cause detrimental effects to the entire power system [5][6].

Many studies have been conducted on the efficiency and harmonic distortion of inverters. However, all of these studies do not involve the inverter in its simplest form. Inverter efficiency is investigated in a variety of forms. In [7] and [8], high efficiencies are obtained, however a DC-DC converter stage is used. In [9], [10], [11], and [12], the effect of wide input voltage ranges is specifically explored, but once again this is done through the use of a DC-DC converter stage that steps the voltage up or down to the desired DC inverter input. In [13], [14], and [15], soft switching topologies are used to increase efficiency by reducing switching losses. Harmonic distortion studies have also been conducted. In [16] and [17], the harmonic content of the output is investigated, and in [18] and [19] IEEE 519 standards and recommendations for harmonic content are addressed and met. However, in all of these studies a filtering stage is used. Whether investigating efficiencies or harmonics, the basic inverter topology is not used by itself.

One major use of inverters is in photovoltaic applications, especially those of grid-tied. Photovoltaic arrays and their relationship with inverters have also been studied. A photovoltaic module's maximum power point voltage can range anywhere from 23 to 38 V [20]. Modules can be arranged in many different combinations of series and parallel to meet the needs of the converter and load. In [21] several different combinations of photovoltaic arrays in series and parallel are analyzed to determine the maximum possible output power. Traditionally, a central inverter topology is used with respect to configuration where all photovoltaic modules are connected to a main centralized inverter. In [22] and [23] string, multi-string, and team configurations are discussed where modules are connected in parallel and in turn connected to several inverters in parallel. This allows for the use of fewer inverters during lighter irradiation improving efficiency and mismatch. However, all of these studies focus on the voltage and configuration that yields the maximum possible output power of the array and they do not address the desired input voltage of the inverter to give the maximum efficiency of the inversion process.

In this paper, the effect of different input and output variables on a basic unipolar PWM inverter topology is

978-1-4577-0007-1/11 $26.00 © 2011 IEEE

investigated; specifically the efficiency and harmonic content. These changing variables will include the dc input voltage, switching frequency, and output load. The control signals of the inverter can be adjusted for each input voltage and switching frequency to produce the same output voltage at the fundamental frequency.

This study will ultimately culminate in determining the optimum input voltage and switching frequency of an inverter based on its load. This would be a vital resource for engineers designing inverters for the highest possible efficiency and lowest possible harmonic distortion. The switching frequency could be set as a fixed parameter or adjusted on the fly for optimum conditions. The input voltage constraint could be used to appropriately configure photovoltaic modules in any number of series and parallel as well as to appropriately design a DC-DC converter stage to output this desired voltage. Furthermore, since a good portion of the harmonic content can be filtered out, it would be difficult to determine the effect different variable had on the THD after filtering. Therefore, output filtering will not be used. Lastly, since this study could be used to size appropriately DC-DC converters' outputs, a step-up or step-down stage will not be implemented at the input.

The paper begins with a brief introduction on the inverter topology used. Following this, the efficiency is analyzed using Mathcad. Switching losses are based on the PM50CLA060 Intelligent Power Module from Powerex and the efficiency is compared to the manufacturer's loss calculation software. The inverter is then modeled mathematically in Mathcad and the total harmonic distortion (THD) is evaluated.

II. UNIPOLAR PWM INVERTER

Figure 1 shows a standard voltage-sourced, full-bridge inverter using four switches. In PWM inverters the switches operate by comparing a sine wave with a triangle wave, switching the inverter at the high frequency of the triangle wave. This has the disadvantage of complex control circuitry; however, it makes up for this in easily filterable harmonic content.

While bipolar inverters switch between a positive and negative voltage, unipolar inverters switch between a positive, negative, and zero voltage. As the desired sinusoidal output voltage approaches its maximum, the duty cycle of the positive voltage increases. As the sine wave decreases from its peak, the duty cycle decreases until zero, upon which it begins increasing the duty cycle of the negative output voltage. The switching topology used involves one half of the full bridge operating at high frequencies, while the other half operates at the lower output frequency. This is summarized below.

S1 Conducts: $V_{Sine} \geq V_{Tri}$
S2 Conducts: $V_{Sine} > 0$
S3 Conducts: $V_{Sine} < 0$
S4 Conducts: $V_{Sine} < V_{Tri}$

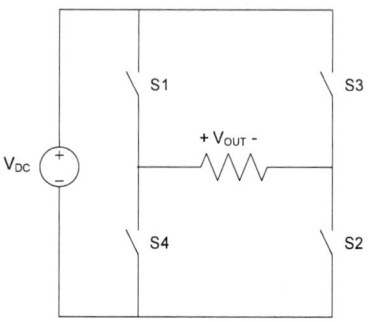

Fig. 1. Full Bridge Inverter

In unipolar PWM inverters the amplitude of the fundamental frequency is linearly proportional to the DC input voltage and the amplitude modulation. The amplitude modulation is the ratio of the sinusoidal control signal amplitude to that of the triangular signal. So in order to maintain a fixed 120 V output at 60 Hz the amplitude modulation must decrease when the DC input voltage increases.

III. MATHEMATICAL ANALYSIS

The inverter parameters used are summarized in Table I. The input voltage, switching frequency, and load are specified as ranges. This is because these values are varied over this range to determine the optimum efficiency and lowest harmonic distortion.

TABLE I
INVERTER PARAMETERS

Input Voltage (V_{dc})	200 - 300
Fundamental Output Voltage (V_{rms})	120
Load (W)	50 - 1000
Switching Frequency (kHz)	2 - 18
Output Frequency (Hz)	60

All calculations will be performed using Mathcad. This reduces user error in making calculations and allows for three dimensional graphing making the viewing of data easier. Trends previously displayed in ten different figures can be conveniently seen in one. Also, Mathcad is more proficient in determining THD when compared with other software. This will be discussed in more detail upon analysis.

First, the efficiency will be calculated. An example case will be shown and compared with the manufacturer's loss simulation software before an overall result is calculated. This particular case will be at 500 W, 250 V_{DC}, and 20 kHz. Although the 20 kHz is outside the range of study, it is used because it is the limit of the IPM.

Let P = 500 W   (1)

$$Let\ V_{rms} = 120\ V \tag{2}$$

$$Let\ pf = 1 \tag{3}$$

$$I_{rms} = \frac{P}{V_{rms}pf} = \frac{500}{120*1} = 4.167\ A \tag{4}$$

$$I_{peak} = \sqrt{2}I_{rms} = \sqrt{2} * 4.167 = 5.893\ A \tag{5}$$

$$I_{AVG}^{IGBT} = \frac{5.893}{2\pi}\int_0^\pi sin(\omega t)d(\omega t) = \frac{5.893}{2\pi}[-cos(\pi) + cos(0)] = 1.876A \tag{6}$$

Taking this value as the collector current, relevant information taken from the Powerex PM50CLA060 IPM datasheet [24] is summarized in Table II.

TABLE II.
RELEVANT DATASHEET INFORMATION

Collector-Emitter Voltage (V_{CE})	0.625 V
FWD Forward Voltage (V_F)	0.75 V
Turn On Energy (E_{ON})	0.13 mJ/Pulse
Turn Off Energy (E_{OFF})	0.03 mJ/Pulse

The following equations made available in [25] were used to calculate the conduction and switching losses at 20 kHz. A modulation index of M = 170 V / 250 V = 0.68 was used.

$$P_C^{IGBT} = \frac{V_{CE}I_{peak}}{2\pi}\left[1 + \frac{M\pi}{4}pf\right] + \frac{r_{ce}I_{rms}^2}{2\pi}\left[\frac{\pi}{4} + M\left(\frac{2pf}{3}\right)\right] = 0.9\ W \tag{7}$$

$$P_S^{IGBT} = \frac{1}{\pi}f_s\left(E_{on} + E_{off}\right)\frac{V_{DC}}{V_{ref}}\frac{I_{peak}}{I_{ref}} = 2.667\ W \tag{8}$$

$$P_C^{Diode} = \frac{V_f I_{peak}}{2\pi}\left(1 - \frac{M\pi}{4}pf\right) = 0.328\ W \tag{9}$$

$$P_S^{Diode} = \frac{1}{\pi}f_s E_{rec}\frac{V_{DC}}{V_{ref}}\frac{I_{peak}}{I_{ref}} = 0 \tag{10}$$

$$P_{Loss/IGBT} = P_C^{IGBT} + P_S^{IGBT} + P_C^{Diode} + P_S^{Diode} = 3.9W \tag{11}$$

For the given conditions, Mitsubishi's Melcosim Loss Simulator was used to simulate the losses of the exact IPM used. The results can be seen in Figure 2. A total loss per IGBT was simulated as 3.9 W for a percent difference of 0.15%.

The loss per IGBT can be multiplied by four to get the total IGBT losses. This is then added to the output power to determine the input power and Figure 3 shows the inverter efficiency variation with respect to input voltage and switching frequency. A slight decrease in efficiency can be seen as the input voltage is increased, however a more obvious decrease in efficiency can be seen as switching frequency is varied. Figure 4 shows a more familiar trend--the efficiency variation with respect to output power and switching frequency for a fixed 200 V input. The output power is varied from 50 to 1000 W. The familiar asymptotic curve of efficiency can be seen

while efficiency still varies linearly with switching frequency.

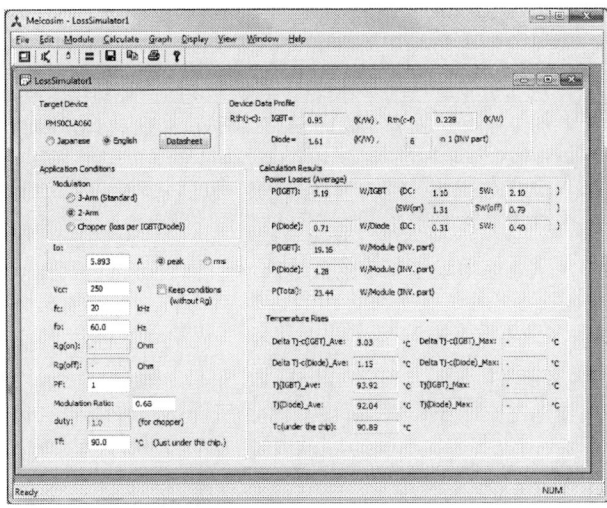

Fig. 2. Melcosim Loss Simulator for Calculated Conditions

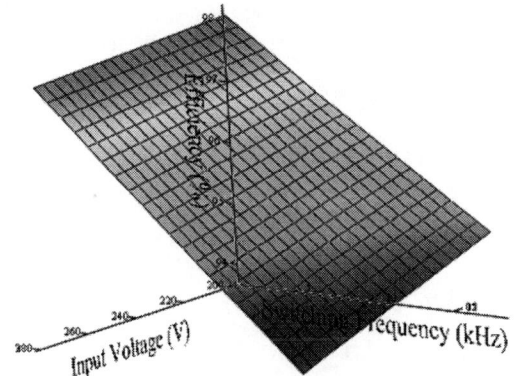

Fig. 3. Efficiency vs. Input Voltage & Switching Frequency at 500W

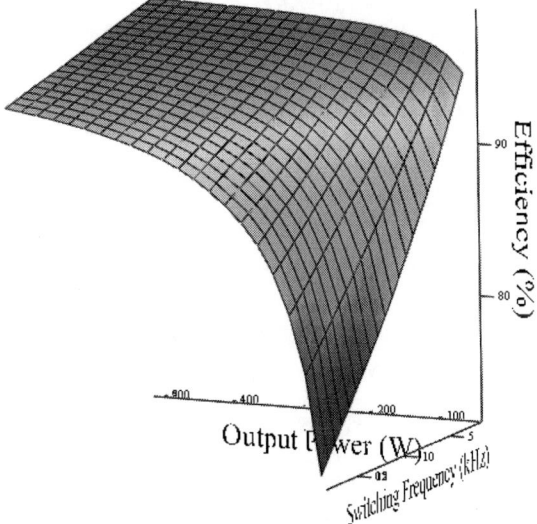

Fig. 4. Efficiency vs. Output Power & Switching Frequency at 200Vin

Total harmonic distortion can be calculated by determining the total rms voltage as well as the rms voltage of the fundamental component. However, this is not the way standard circuit simulation software determines THD. Circuit simulation software normally determines the first 50 harmonics and calculates THD as the ratio of the 2nd through 50th rms voltage with that of the fundamental component. This is rather limiting because, at 60 Hz, the highest harmonic frequency that can be measured is 3 kHz. Since the inverter is switching at frequencies up to 18 kHz, measuring only up to the 50th harmonic would neglect harmonics occurring at and around multiples of the switching frequency.

In this study no output filtering is performed and therefore it would seem that the total harmonic distortion is calculable by hand. However, due to the somewhat complicated switching, it is not as simple and would be far too complex to perform these calculations by hand. Mathcad serves as an intermediary between hand calculations and circuit simulation software. It allows for the true total harmonic distortion to be calculated in its pure mathematical form. The remainder of the section will show how an inverter was built in mathcad in order to perform the complex integrals necessary to find the THD. This begins with the comparison of a triangle wave oscillating at the switching frequency with a sine wave oscillating at 60 Hz. A Unipolar inverter switching topology will be used to generate switching signals. These can then be used to determine the output voltage.

$$V_1(t) = \begin{vmatrix} 1 \ if \ V_{sin}(t) > V_{tri}(t) \\ 0 \ otherwise \end{vmatrix} \tag{12}$$

$$V_2(t) = \begin{vmatrix} 1 \ if \ -V_{sin}(t) > 0 \\ 0 \ otherwise \end{vmatrix} \tag{13}$$

$$V_2(t) = \begin{vmatrix} 1 \ if \ -V_{sin}(t) > 0 \\ 0 \ otherwise \end{vmatrix} \tag{14}$$

$$V_4(t) = \begin{vmatrix} 1 \ if \ V_{sin}(t) < V_{tri}(t) \\ 0 \ otherwise \end{vmatrix} \tag{15}$$

$$V_{out}(t) = \begin{vmatrix} V_{DC} \ if \ V_1(t)^\wedge V_2(t) \\ -V_{DC} \ if \ V_3(t)^\wedge V_4(t) \\ 0 \ otherwise \end{vmatrix} \tag{16}$$

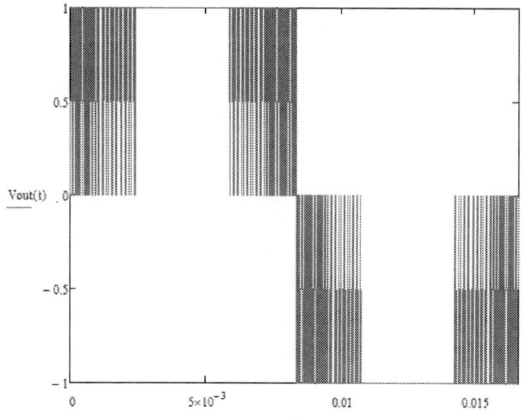

Fig. 5. Output Waveform for MathCAD THD Calculation

Figure 5 shows the output waveform generated by the equations above. The amplitude of the sine wave can

be adjusted to produce the desired output voltage at the fundamental frequency. This is necessary in order to produce 120 Vrms at 60 Hz for varying switching frequencies and input voltages.

Notice in Figure 5 how rapid switching occurs in the beginning. The pulse widths gradually increase until there is 100% duty cycle at the sine wave's peak. From here, the pulse widths decrease to almost zero upon which the voltage goes negative. Now the complicated integrals seen below can be performed in order to determine the THD.

$$V_{rms} = V_{dc}\sqrt{f_{sin}\int_0^{\frac{1}{f_{sin}}} V_{out}(t)^2 dt} \tag{17}$$

$$a_1 = 2f_{sin}\int_0^{\frac{1}{f_{sin}}} V_{out}(t)\cos(2\pi f_{sin}t)\,dt \tag{18}$$

$$b_1 = 2f_{sin}\int_0^{\frac{1}{f_{sin}}} V_{out}(t)\sin(2\pi f_{sin}t)\,dt \tag{19}$$

$$V_{1,rms} = V_{dc}\sqrt{\frac{a_1{}^2 + b_1{}^2}{2}} \tag{20}$$

$$THD = 100\% * \frac{\sqrt{V_{rms}{}^2 - V_{1,rms}{}^2}}{V_{1,rms}} \tag{21}$$

Through an iterative process, all switching frequency and input voltage cases were performed. For each case, the amplitude of the sine wave was adjusted to produce 120 Vrms at 60 Hz. Figure 6 shows how the THD varies. As the switching frequency varies, the THD remains relatively constant. This is because the total rms output voltage actually remains relatively constant for all switching frequencies while the fundamental component is fixed at 120 V. This is most likely due to the total on time of each cycle being the same regardless of switching frequency. However, the THD does increase as the voltage increases. This is to be expected. As the input voltage gets further away from the desired peak output voltage of 170 V, more switching has to occur at the peak of the output wave. This causes the output voltage to have more harmonic content away from 60 Hz increasing the THD. At 270 V and 280 V the THD seems to begin to level off from its linear increase. This is because at this input voltage the sine wave could not be adjusted to a small enough value to produce 120 Vrms at 60 Hz. So rather than being 120 V, it actually varies from 121 to 126 V causing the THD to be less than expected. It is worth mentioning that the THD is not dependent on the load. Since this output voltage remains the same regardless of loading, the THD will follow.

It was mentioned previously that circuit simulation software can noticeably under measure THD. The voltages of all of the harmonics for a particular case were calculated in Mathcad. Figure 7 shows the frequency spectrum of the output voltage for 2 kHz switching at 210 V. A 120 V peak can be seen at the fundamental frequency followed by peaks at multiples of 33 of the switching frequency. Simulation software would only see the first two peaks and ignore the four that follow. For higher switching frequencies, all peaks except the first would be ignored completely causing an inaccurate THD reading.

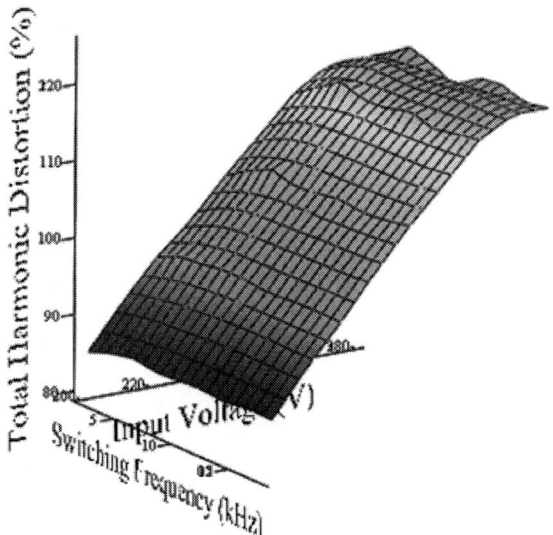

Fig. 6. Total Harmonic Distortion vs. Switching Frequency and Vin

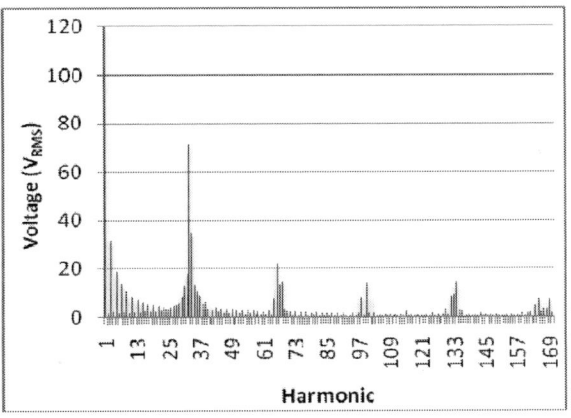

Fig. 7. Frequency Spectrum of Output Voltage

IV. CONCLUSION

The results of this study suggest particular steps that should be taken in designing the most optimum inverter and photovoltaic system. In order to maintain the highest efficiency, the lowest switching frequency should be used as well as a DC input voltage that is just above the desired peak of the output voltage. This is the case regardless of load. With respect to harmonic distortion, similar conditions apply. The lowest possible input voltage should also be used with the best case being that of the peak of the desired output voltage. Without filtering, switching frequency is not dependent on THD and any value could be used. With filtering, higher switching frequencies are most likely preferred. This is a result of higher frequencies being easier to filter because of the harmonic content being further away from the fundamental frequency as well as the smaller size of components. Also, higher frequencies make low pass filtering easier. High pass filtering techniques are more susceptible to noise. However, there exists a trade-off

here between efficiency and harmonic filtering. If a higher frequency is chosen because of filtering ease, the efficiency will suffer. The importance of each of these should be evaluated before a decision is made.

Not only has this study suggested optimal inverter design parameters, but it has pointed out a deficiency in circuit simulation software. This same deficiency is most likely to occur with power quality analyzers in real world situations. When high frequency harmonics are present caution should be taken when trusting the reading of these devices.

ACKNOWLEDGMENT

The authors would like to thanks Faculty of Information Science and Technology, Universiti Kebangsaan Malaysia for financial support.

REFERENCES

[1]. Yaosuo Xue; Liuchen Chang; Pinggang Song, "Recent developments in topologies of single-phase buck-boost inverters for small distributed power generators: an overview," *Power Electronics and Motion Control Conference, 2004. IPEMC 2004. The 4th International* , vol.3, no., pp. 1118- 1123 Vol.3, 14-16 Aug. 2004.

[2]. Kjaer, S.B.; Pedersen, J.K.; Blaabjerg, F.; "Power inverter topologies for photovoltaic modules-a review," *Industry Applications Conference, 2002. 37th IAS Annual Meeting. Conference Record of the* , vol.2, no., pp. 782- 788 vol.2, 2002.

[3]. Yaosuo Xue; Liuchen Chang; Sren Baekhj Kjaer; Bordonau, J.; Shimizu, T.; "Topologies of single-phase inverters for small distributed power generators: an overview," *Power Electronics, IEEE Transactions on* , vol.19, no.5, pp. 1305- 1314, Sept. 2004.

[4]. B. W. Kennedy, *Energy Efficient Transformers*, McGraw-Hill Professional, 1997.

[5]. R. Barnes, "Harmonics in power systems", *Power Engineering Journal*, Volume 3, Issue 1, pp 11 – 15, January 1989.

[6]. W. A. Maslowski, "Harmonics in power systems", *Proc. IEEE 1992 Annual Textile, Fiber and Film Industry Technical Conference*, pp. 11/1 – 1110, 1993.

[7]. Edelmoser, K.H.; Himmelstoss, F.A.; "Analysis of a new high-efficiency DC-to-AC inverter," *Power Electronics, IEEE Transactions on* , vol.14, no.3, pp.454-460, May 1999.

[8]. Meksarik, V.; Masri, S.; Taib, S.; Hadzer, C.M.; "Development of high efficiency boost converter for photovoltaic application," *Power and Energy Conference, 2004. PECon 2004. Proceedings. National*, vol., no., pp. 153- 157, 29-30 Nov. 2004.

[9]. Chien-Ming Wang; "A novel single-stage full-bridge buck-boost inverter," *Applied Power Electronics Conference and Exposition, 2003. APEC '03. Eighteenth Annual IEEE*, vol.1, no., pp. 51-57 vol.1, 9-13 Feb. 2003.

[10]. Uematsu, T.; Tanaka, K.; Takayanagi, Y.; Kawasaki, H.; Ninomiya, T.; "Utility interactive inverter controllable for a wide range of DC input voltage," *Power Conversion Conference, 2002. PCC Osaka 2002. Proceedings of the* , vol.2, no., pp.498-503 vol.2, 2002.

[11]. Zargari, N.R.; Ziogas, P.D.; Joos, G.; "A two-switch high-performance current regulated DC/AC converter module," *Industry Applications, IEEE Transactions on*, vol.31, no.3, pp.583-589, May/Jun 1995

[12]. Edelmoser, K.H.; Himmelstoss, F.A.; "Improved 1kW solar inverter with wide input voltage range," *Signals, Circuits and Systems, 2003. SCS 2003. International Symposium on*, vol.1, no., pp. 201- 204 vol.1, 10-11 July 2003.

[13]. Shireen, W.; Kulkarni, R.A.; "A soft switching inverter module with modified DC-link circuit for high frequency DC-AC power conversion, "*Applied Power Electronics Conference and Exposition, 2003. APEC '03. Eighteenth Annual IEEE*, vol.1, no., pp. 507- 511 vol.1, 9-13 Feb. 2003.

[14]. Guan-Chyun Hsieh; Chun-Hung Lin; Jyh-Ming Li; Yu-Chang Hsu; "A study of series-resonant DC/AC inverter," *Power*

978-1-4577-0007-1/11 $26.00 © 2011 IEEE

Electronics Specialists Conference, 1995. PESC '95 Record., 26th Annual IEEE , vol.1, no., pp.493-499 vol.1, 18-22 Jun 1995.

[15]. Hui, S.Y.R.; Gogani, E.S.; Jian Zhang; "Analysis of a quasi-resonant circuit for soft-switched inverters," *Power Electronics, IEEE Transactions on* , vol.11, no.1, pp.106-114, Jan 1996.

[16]. Maswood, A.I.; Yusop, A.K.; Rahman, M.A., "A novel suppressed-link rectifier-inverter topology with near unity power factor," *Power Electronics, IEEE Transactions on* , vol.17, no.5, pp. 692-700, Sep 2002.

[17]. Zue, A.O.; Chandra, A.; "Simulation and stability analysis of a 100 kW grid connected LCL photovoltaic inverter for industry," *Power Engineering Society General Meeting, 2006. IEEE*, vol., no., pp.6.

[18]. Naik, R.; Mohan, N.; "A novel grid interface for photovoltaic, wind-electric, and fuel-cell systems with a controllable power factor of operation," *Applied Power Electronics   Conference and Exposition, 1995. APEC '95. Conference Proceedings 1995., Tenth Annual*, vol., no.0, pp.995-998 vol.2, 5-9 Mar 1995.

[19]. Oliva, A.R.; Balda, J.C.; "A PV dispersed generator: a power quality analysis within the IEEE 519," *Power Delivery, IEEE Transactions on*, vol.18, no.2, pp. 525-530, April 2003.

[20]. Kjaer, S.B.; Pedersen, J.K.; Blaabjerg, F.; "A review of single-phase grid-connected inverters for photovoltaic modules," *Industry Applications, IEEE Transactions on*, vol. 41, no.5, pp. 1292- 1306, Sept-Oct. 2005.

[21]. Tria, L.A.R.; Escoto, M.T.; Odulio, C.M.F.; "Photovoltaic array reconfiguration for maximum power transfer," *TENCON 2009 - 2009 IEEE Region 10 Conference*, vol., no., pp.1-6, 23-26 Jan. 2009.

[22]. Imhoff, J.; Pinheiro, J.R.; Russi, J.L.; Brum, D.; Gules, R.; Hey, H.L.; "DC-DC converters in a multi-string configuration for stand-alone photovoltaic system," *Power Electronics Specialists Conf., 2008. PESC 2008. IEEE* , vol., no., pp.2806-2812, 15-19 June 2008.

[23]. Myrzik, J.M.A.; Calais, M.; "String and module integrated inverters for single-phase grid connected photovoltaic systems - a review," *Power Tech Conference Proceedings, 2003 IEEE Bologna* , vol.2, no., pp. 8 pp. Vol.2, 23-26 June 2003.

[24]. "PM50CLA060 Datasheet." Powerex, Inc.

[25]. Franke, W.-T.; Mohr, M.; Fuchs, F.W., "Comparison of a Z-source inverter and a   voltage source inverter linked with a DC/DC-boost-converter for wind turbines concerning their efficiency and installed semiconductor power," *Power Electronics Specialists Conference, 2008. PESC 2008. IEEE*, vol., no., pp.1814-1820, 15-19 June 2008.

A New Multi Carrier Based PWM for Multilevel Converter

Bahr Eldin S. Mohammed and K.S.Rama Rao
Department of Electrical and Electronic Engineering
Universiti Teknologi PETRONAS,
Bandar Seri Iskandar, 31750 Tronoh, Perak, Malaysia

Abstract- **This paper addresses a new multi-carrier modulation technique called wave shift multi-carrier modulation (WSHM), which used to control the cascade multilevel converter (CMC). The proposed switching technique generates lower voltage total harmonic distortion (THD) in comparison with conventional multi-carrier based pulse width modulation (PWM) schemes, phase-shift (PSHM) and level-shift (LSHM) modulations. To compare the performance of proposed method with PSHM and LSHM techniques, a simulation circuit of seven and nine levels for CMC is designed and simulated. To complete the comparison between the proposed method and PSHM, LSHM the dynamic behavior of dynamic voltage restorer (DVR) based seven-level CMC for power quality improvement is investigated. Digital simulations are carried out using PSCAD/ EMTDC to validate the performance of CMC.**

I. INTRODUCTION

Multilevel converters (MLC) are emerging as a new breed of power converter options for power system applications. Recent advances in power switching devices enabled the suitability of MLCs for high voltage and high power applications because they are connecting several devices in series without the need of component matching. The general structure of the MLC is to synthesize a sinusoidal voltage by several levels of voltages, typically obtained from capacitor voltage sources. Three types of capacitor voltage synthesis based multilevel converters are reported [1] as follows:

1. Diode-Clamped Multilevel Converter (DCMC).
2. Flying-Capacitor Multilevel Converter (FCMC).
3. Cascaded Multilevel Converters (CMC)

Compared DCMC and FCMC converters, a CMC as shown in Fig.1 is easy to design and assemble because of the uniform circuit structure of the converter units and modularized circuit layout. Easy packaging is also possible in CMC topology as each level has the same structure, and there are no extra clamping diodes or voltage-balancing capacitors, which are required in the DCMC and the FCMC. The number of output voltage levels can then be easily adjusted by changing the number of full-bridge converters. The CMC synthesizes a desired voltage from several independent sources of DC voltages, which may be obtained from batteries, fuel cells or solar cells [2]. In general, the output voltage of CMC is controlled as follows [3]:

i. By controlling the pulse width of i. the output voltage by fundamental frequency switching (FFS) method or PWM technique while keeping the magnitude of dc voltage fixed.

ii. By controlling the magnitude of dc voltage while keeping the pulse width of output voltage fixed or by controlling the modulation index in case of PWM.

iii. By controlling both the magnitude of dc voltage and the pulse width of the output voltage.

The basis of selecting the method of controlling the converter output voltage is dependent on the ability to control the THD. In this paper, the second method will be adopted to control the inverter output voltage. The objective of achieving minimum THD is based on two switching methods, the fundamental frequency switching (FFS) and multi-carrier-based PWM technique.

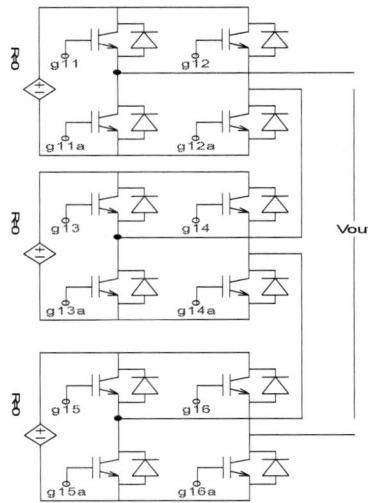

Fig.1. One phase of seven-level CMC

FFS modulation can be easily implemented for the CMC due to its unique structure. All switching angles can be calculated off-line and then stored in a look-up table for digital implementation. Compared with the carrier-based PWM schemes, FFS features low switching losses since all the IGBT switches operate at fundamental frequency. As the expressions for switching angles are nonlinear and transcendental, deriving a valid solution over the full range of amplitude modulation index (ma) is not always possible. Thus the switching angles should be calculated to minimize the magnitude of those harmonics that cannot be eliminated [4].Various PMW techniques applied to the multilevel converters are discussed in [5]-[7]. The PWM techniques can be classified into two categories: the triangle intersection technique and the direct

978-1-4577-0007-1/11 $26.00 © 2011 IEEE

digital technique (space vector modulation). With the development of digital technology, the space vector modulation is widely used, due to not only relatively easy hardware implementation, but also its features of good dc link voltage utilization and low current ripple. But this method has a very significant drawback that if the voltage level is more than five, the control algorithm becomes too complex to implement [3]. Thus it is reasonable to adapt in this paper the triangle intersection techniques in the high level application.

II. PHASE-SHIFT MULTI-CARRIER MODULATION (PSHM)

CMC with m voltage levels requires (m − 1) triangular carriers. In the phase-shifted multi-carrier modulation, all the triangular carriers have the same frequency and the same peak-to-peak amplitude, but there is a phase shift between any two adjacent carrier waves, given by:

$$\varphi_{g_h} = \frac{360^0}{(m-1)} \qquad (1)$$

The gate signals are generated by comparing the modulating wave with the carrier wave. Fig. 2 shows the principle of PSHM for one phase of seven levels CMC presented in Fig.1 where six triangular wave carriers are required with a 60° phase displacement between any two adjacent carriers. The advantage of PSHM is that the switching frequency and conduction period is same for all devices and rotating of switching patterns is not required [4].

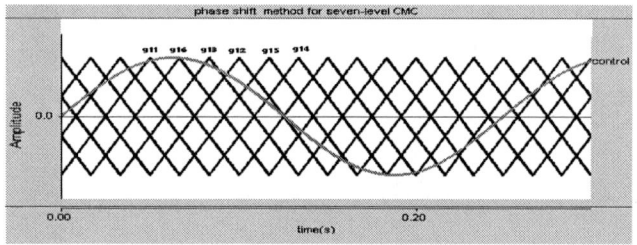

Fig.2. PSHM for seven-level CMC

Figs. 3 to 8 shows the simulated voltage waveforms and their harmonic content of three-phase seven and nine levels CMC using PSHM under the condition of modulation frequency, f_m= 50 Hz, carrier frequency, f_{cr} = 900 Hz, amplitude modulation index, m_a = 0.9 and frequency modulation index, m_f = 18. The line voltage harmonic spectrum shown in Figs. 5 and 8 is based on 50 Hz base frequency and the THD considered for the first 63 harmonics.

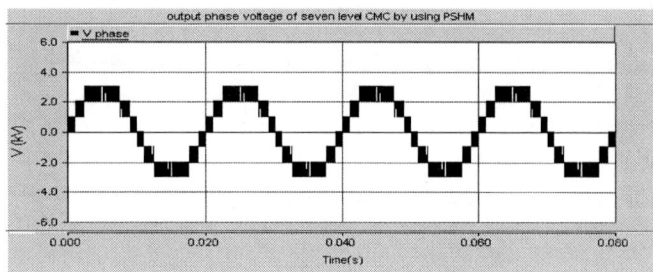

Fig. 3. Output phase voltage of seven-level CMC

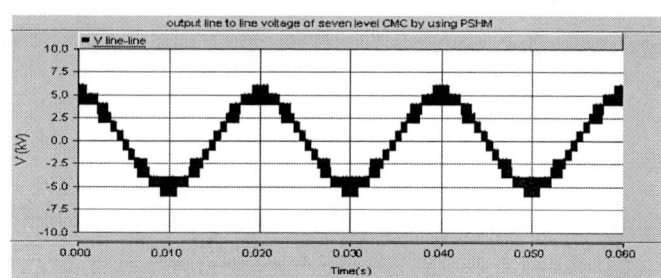

Fig. 4. Output line-to-line voltage of seven-level CMC

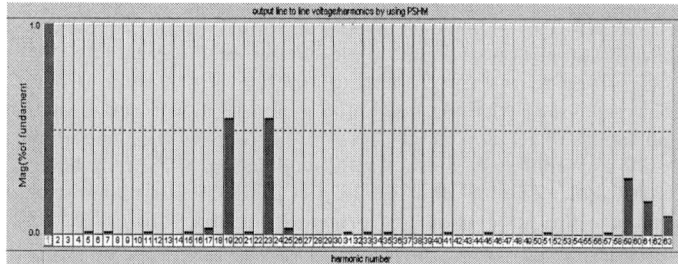

Fig. 5. Seven-level CMC output line voltage harmonics (THD = 16.2264 %)

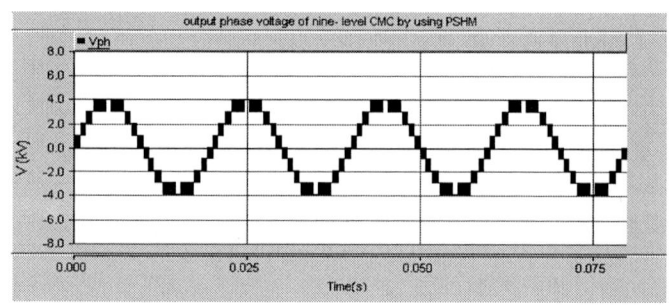

Fig.6. Output phase voltage of nine-level CMC

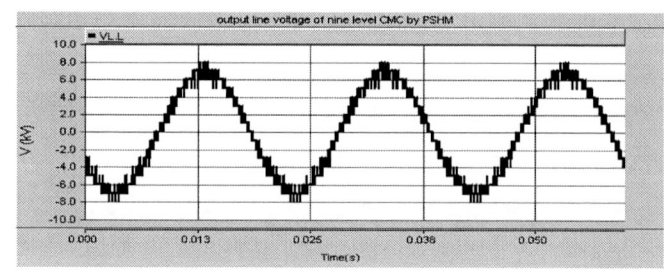

Fig. 7. Output line to line voltage of nine-level CMC

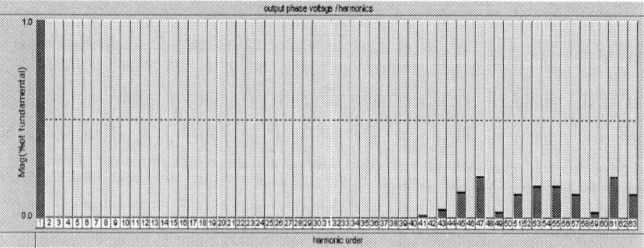

Fig. 8. Nine-level CMC output line voltage harmonics (THD = 10.0737 %)

III. LEVEL SHIFT MULTI-CARRIER MODULATION (LSHM)

For m-level CMC using level-shifted multicarrier modulation scheme, (m – 1) triangular carriers are required, all having the same frequency and amplitude. The (m – 1) triangular carriers are vertically disposed such that the bands they occupy are contiguous. The amplitude modulation index is defined as:

$$m_a = \frac{V_m}{V_{cr}(m-1)} \qquad (2)$$

Where V_m is the peak amplitude of the modulating wave and V_{cr} is the peak amplitude of each carrier wave. There are three schemes for level shift multi-carrier modulation listed as follows:

(i) In-phase disposition (IPD), where all carriers are in phase.
(ii) Alternative phase opposite disposition (APOD), where all carriers are alternatively in opposite disposition.
(iii) Phase opposite disposition (POD), where all carriers above the zero reference are in phase but in opposition with those below the zero reference.

In this paper only IPD modulation scheme is addressed as it provides the best harmonic profile of all three-level shift multi-carrier modulation schemes [4]. Fig. 9 shows the principle of the IPD modulation for one phase of seven-level CMC reported in Fig.1.

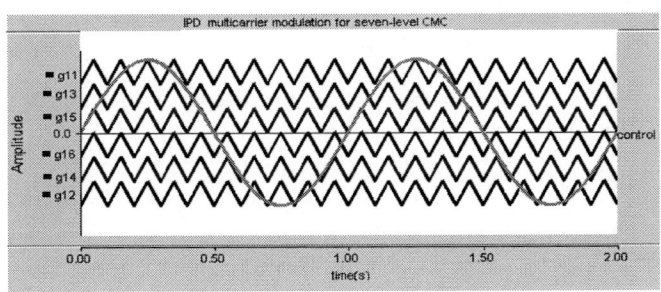

Fig. 9. LSHM for seven-level CMC

A. Simulation of Three Phase Seven and Nine- Levels CMC by Using LSHM

Fig.10 to 15 shows the simulated voltage waveforms and their harmonic content of three phase seven and nine levels CMC using LSHM under the same conditions of PSHM simulation.

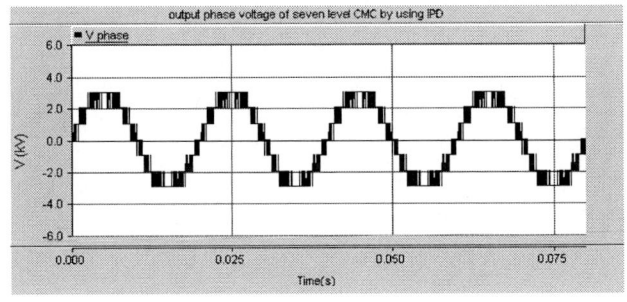

Fig. 10. Output phase voltage of seven-level CMC

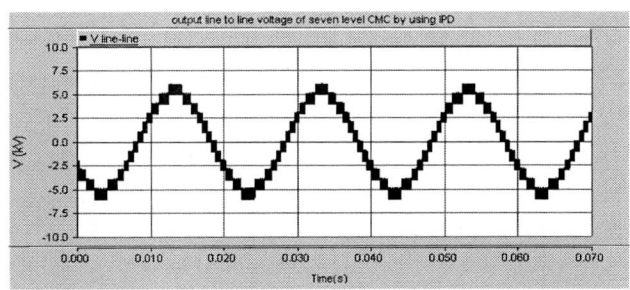

Fig.11. Output line to line voltage of seven-level CMC

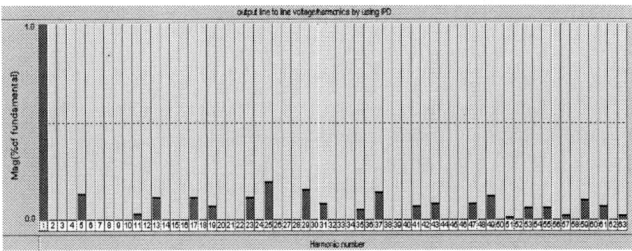

Fig. 12. Seven-level CMC output line voltage harmonics (THD = 14.5639 %)

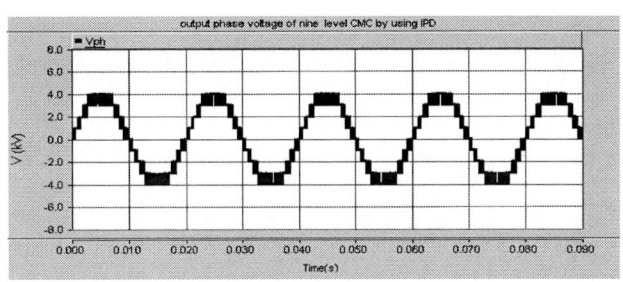

Fig.13. Output phase voltage of nine-level CMC

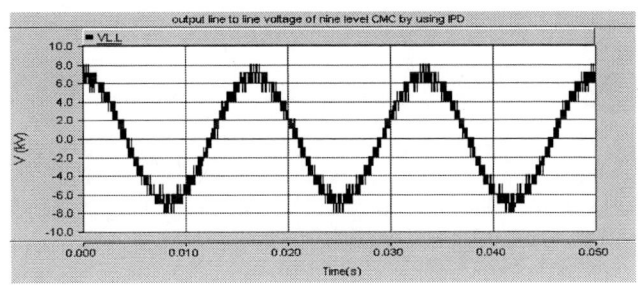

Fig.14. Output line to line voltage of nine-level CMC

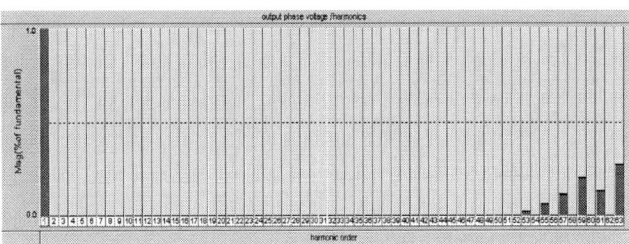

Fig. 15. Nine-level CMC output line voltage harmonics (THD = 8.4603 %)

IV. PROPOSED METHOD

The proposed modulation technique is a combination of phase shift multi-carrier and level-shifted multi-carrier modulation (in-phase disposition (IPD)) schemes which overcomes the problem of rotating of switching pattern of level-shifted multi-carrier modulation and small phase displacement at phase voltage of CMC. For m level CMC in the proposed method, (m – 1) triangular carriers are required. In the carrier wave all the triangles have the same frequency, same peak to peak amplitude and are vertically disposed, but there is a phase shift between any two disposed carrier waves as in (3).

$$\phi_{sh} = \frac{360^0}{4(m-1)} \qquad (3)$$

The amplitude modulation index is defined as in (4)

$$m_a = \frac{V_m}{V_{cr}(m-1)} \qquad (4)$$

Fig.16 shows the principle of the proposed method for one phase of seven-level CMC reported in Fig.1.

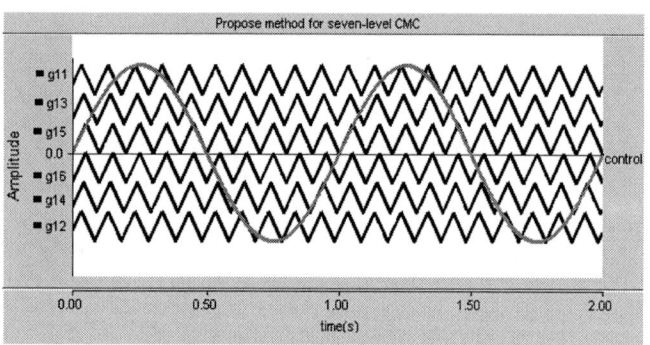

Fig. 16. Proposed method for seven-level CMC

A. Simulation of Three Phase Seven and Nine- Levels CMC by Using Proposed Method

Figs.17 to 22 shows the simulated voltage waveforms and their harmonic content of three-phase seven and nine-levels CMC using the proposed method under under the same conditions of PSHM simulation.

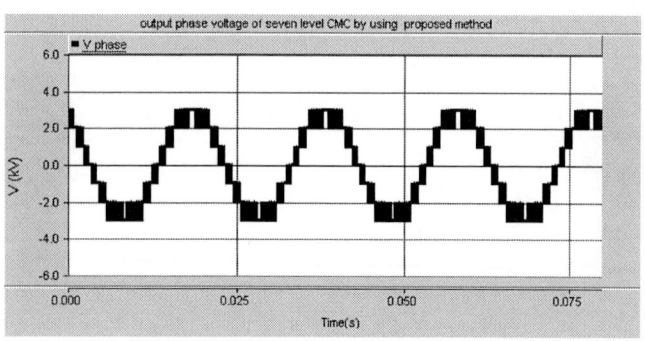

Fig. 17. Output phase voltage of seven-level CMC

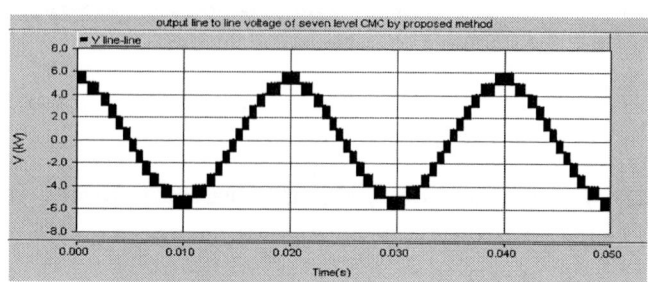

Fig.18. Output line to line voltage of seven-level CMC

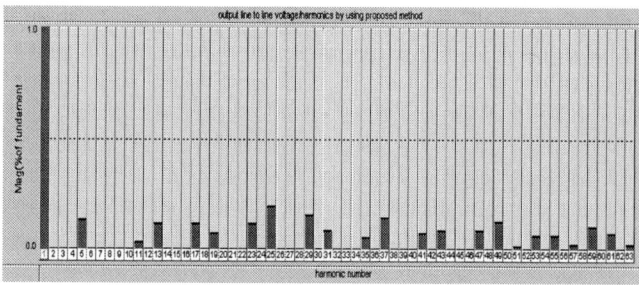

Fig. 19. Seven-level CMC output line voltage harmonics (THD = 13.512 %)

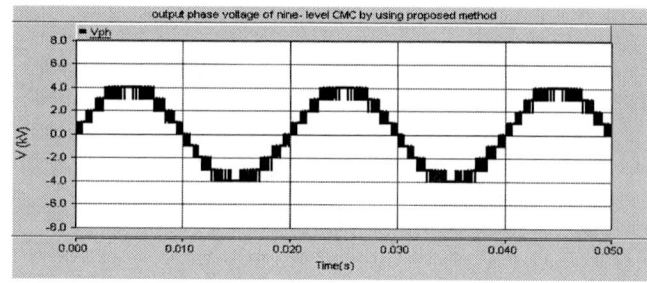

Fig.20. Output phase voltage of nine-level CMC

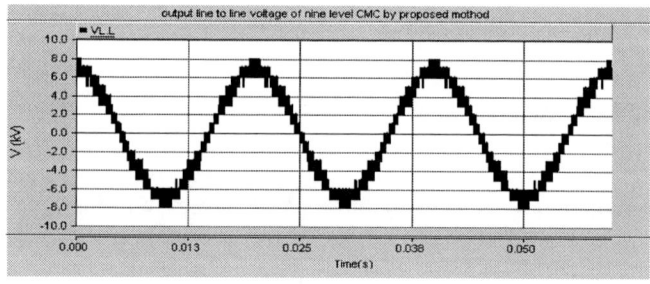

Fig.21. Output line to line voltage of nine-level CMC

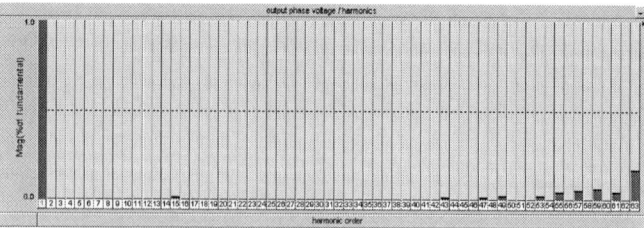

Fig. 22. Nine-level CMC output line voltage harmonics (THD = 8.2846 %)

978-1-4577-0007-1/11 $26.00 © 2011 IEEE

V.   DYNAMIC VOLTAGE RESTORER (DVR)

The power electronic converter based series compensator that can protect critical loads from all supply side disturbances other than outages is called a DVR. The DVR is capable of generating or absorbing independently controllable real and reactive power at its ac output terminals. The main focus in this paper is to compare the performance of the proposed switching technique with multi-carrier-based PWM schemes, phase-shift and level-shifted modulation. Based on the pervious analysis of seven and nine level CMC, a seven level CMC is selected as a candidate design for DVR. This paper analyzed the dynamic behavior of the DVR and the controlling system based on the proposed method as well as phase-shift and level-shifted modulation techniques.

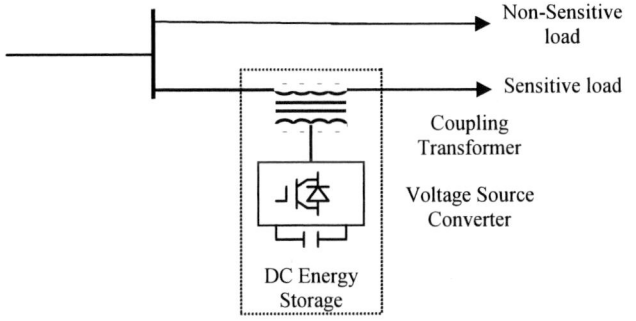

Fig. 23. Schematic representation of the DVR

The DVR injects a set of three-phase ac output voltages in series and in synchronism with the distribution feeder voltages. The amplitude and phase angle of injected voltages are

variable there by allowing control of the real and reactive power exchange between the DVR and the distribution system. The reactive power exchange between the DVR and the distribution system is internally generated by the DVR without ac passive reactive components. The DVR can protect the

sensitive load by injecting voltage of controllable amplitude, phase and frequency into the distribution feeder via a series injection transformer. Thus DVR can only supply partial power to the load during very large variations (sag or swells) in source voltage [8]-[12]. The real power exchanged at the DVR Output ac terminals are provided by the DVR input dc terminals or by an external energy source or energy storage system. Fig. 23 shows a typical DVR connected in series with 11 kV distribution feeder that supplies a sensitive load.

A test system comprising of 13.0 kV, 100 MVA, 3-phase transmission line, represented by Thevenin equivalent, feeding into the primary side of a three-winding transformer is selected as shown in Fig. 24. Two (R-L) loads are connected to the tertiary windings via two two-winding transformers. The DVR is connected in series with a sensitive load 2 through a coupling transformer, with a leakage reactance of 10 %. A unity transformer turn ratio is used for the DVR coupling transformer with no booster capabilities. A DVR based CMC is connected to the 11 kV tertiary winding to provide instantaneous voltage support at the load point. Two different simulation studies are carried out using PSCAD/EMTDC.

In the first case the system is tested without the DVR for a three-phase fault via a fault resistance of 0.36Ω. The voltage sag is observed from 0.3 s to 0.6 s as shown in Figs. 25.

The second simulation is carried out using the same scenario as above but now with the DVR based seven-level CMC in operation in three different modes as follows: (i) DVR control system based on PHSM. (ii) DVR control system based on LHSM. (iii) DVR control system based on proposed technique. The simulation results are presented as shown in Figs. 26 to 28. The DVR based seven level CMC is simulated to be in operation only for the duration of the fault of 0.3 s and the total simulation period is observed to be 0.8 s. As the DVR is in operation, the voltage sag is mitigated almost completely at the end of 0.6 s. The sag mitigation is performed with smooth, stable and rapid DVR controls based on the proposed method. Acceptable overshoots are observed when the DVR comes in and out of operation.

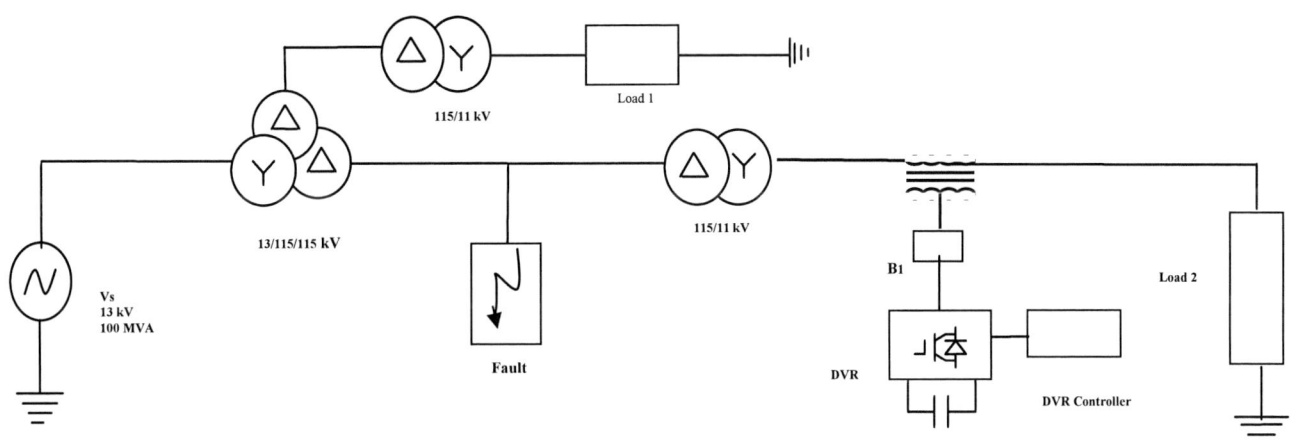

Fig.24. Schematic diagram of test system including DVR

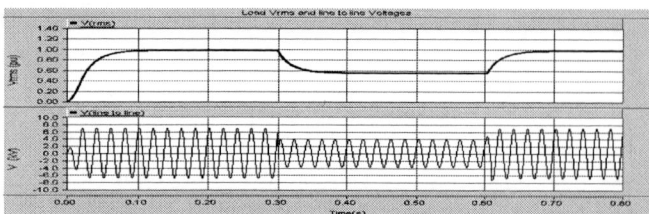

Fig. 25. Load voltage during three-phase fault without DVR

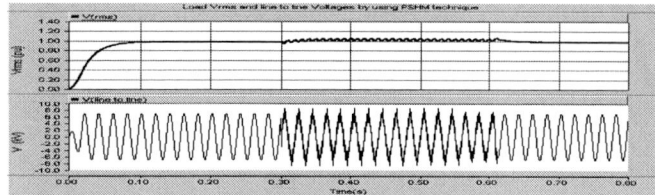

Fig. 26. Load voltage during three-phase fault with DVR based on PSHM

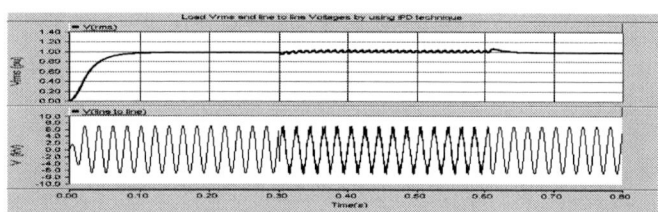

Fig. 27. Load voltage during three-phase fault with DVR based on LSHM-IPD

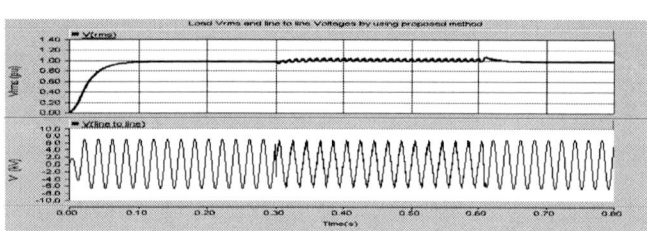

Fig. 28. Load voltage during three-phase fault with DVR based on the proposed method

VI. COMPARSION BETWEEN WSHM AND PSHM,LSHM SCHEMES

The line voltage THD profile of nine-level CMC with the WSHM method and the other two PSHM and LSHM methods under different values of m_f are shown in Figs. 29(a) and 29(b) based on 50 Hz base frequency and the THD considered for 255 harmonics. It can be seen that the THD produced by WSHM method lower than by PSHM and LSHM techniques.

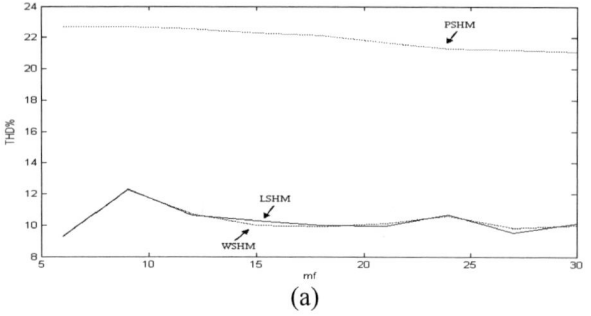

(a)

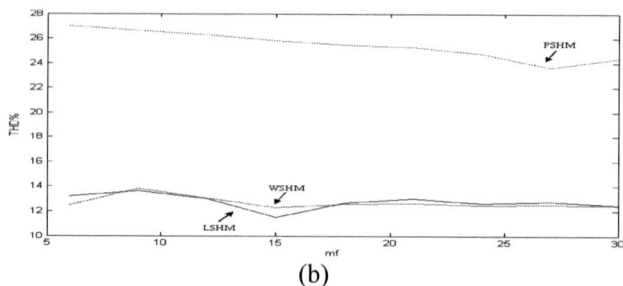

(b)

Fig.29 THD profile of line voltage produced by the nine-level CMC (a) $m_a = 1$, (b) $m_a = 0.85$

CONCLUSIONS

This paper presents simulation of seven and nine levels CMC and one of the custom power equipment namely, DVR based seven-level CMC in a distribution system. Three types of multicarrier based PMW techniques were considered to control the output voltage of CMC. Among those three modulation techniques, it has been found and proved that the proposed method is better than others in terms of THD reduction. The simulation results have demonstrated excellent voltage regulation capabilities of the seven levels CMC based DVR using the proposed new multi-carrier based PWM technique.

REFERENCES

[1]. Jih-Sheng Lai, Fang Zhen g Peng, "Multilevel converters a new breed of power converters," Industry Application Conference, Thirtieth IAS Annual Meeting, Conference Record of the IEEE, pp.2348-2356, August 2002

[2]. Muhammad. H. Rashid, Power Electronics Circuits, Devices and Applications, Third Edition, Person Prentice Hall, pp.40-6430, 2004.

[3]. Husam. K. Al. H, "Investigation of a cascade multilevel inverter as advanced static compensator," Department of electrical engineering and computer engineering, University of Manitoba, Canada, August 2002.

[4]. Bin, Wu, High Power Converters and AC Drives, Jon Willy & Sons. Inc, Hoboken, New Jersey, pp.119-142, 2006.

[5]. John .N. Chiasson, L. M. Tolbert, K.J McKenzie, Zhong Du , "Control of a multilevel converter using resultant theory," IEEE Transactions on Control Systems Technology, Vol. 2, No.3, pp .345-354, May 2003

[6]. R.Bensraj, S. P. Natarajan and V. Padmathilagam, "Multi-carrier trapezoidal PWM strategies based on control freedom degree for msmi," RPN Journal of Engineering and applied Sciences, Vol. 5, No.5, May 2010.

[7]. P.G. Song, E. Y. Guan, L. Zhao, S. P. Liu, "Hybrid electrical vehicles with multilevel cascaded converter using genetic algorithm," IEEE Conference on Industrial Electronics and Applications, pp.1-6, May 2006.

[8]. Arindam Ghosh , Gerad Ledwich, Power Quality Enhancement Using Custom Power Devices, Kluwer's Power Electronics and Power System Series Editor, M. A. Pai, Kluwer, Academic Publishers, PP.241-376,2002.

[9]. Ashwin Kumar Schoo, T. Thyagarajan, "Modeling of facts and custom power devices in distribution network to improve power quality," Third International Conference on Power Systems, Kharagpur, India, pp.1-7, December 27-29, 2009.

[10]. Olimpo Anaya, Lara and E. Acha, "Modeling and analysis of custom power systems by PSCAD/ETMDC," IEEE Transaction on Power Delivery, Vol.17, No.1, pp.266-272, January 2002.

[11]. Hojat Hatami, Farhad Shahnia, Afshin Pashaei, S. H. Hosseini, "Investigation on D-STATCOM and DVR operation for voltages control in distribution networks with a new control strategy," IEEE Power Tech, pp.2207-2212, June 2008.

[12]. P. Vasudevanaidu, M. Tech, Y. Narendra Kumar, "A new simple modeling and analysis of custom power controllers," Third International Conference on Power Systems, Kharagpur, India, pp.1-6, December 2009.

978-1-4577-0007-1/11 $26.00 © 2011 IEEE

Parallel Configuration in Energy Management Control for the Fuel cell-Battery-Ultracapacitor Hybrid Vehicles

Jenn Hwa Wong[1], N.R.N.Idris[1] and Makbul Anwari[2],

[1]Department of Energy Conversion,
Faculty of Electrical Engineering,
University Teknologi Malaysia,
81300 Skudai, Johor,
Malaysia.

[2] Department of Electrical Engineering,
University of Umm Al-Qura,
Makkah 21955,
Saudi Arabia.

Abstract- **This paper proposes a parallel energy-sharing configuration of energy management control for fuel cell hybrid vehicles (FCHVs) application. The hybrid energy source consists of fuel cells (FCs) and energy storage units (ESUs) made up of battery pack and ultracapacitor (UC) module. The aim of the control is to regulate the DC link voltage, which is connected to the traction DC motor via the h-bridge converter. Each source is connected to the DC bus/load using parallel active topology. A total of six control loops are constructed in a supervisory system in order to regulate the DC bus voltage, control of current flow and at the same time, to monitor the state of charge (SOC) of the energy storage devices. The effectiveness of the proposed parallel energy-sharing control system is discussed and analyzed and then verified by experiments.**

Keywords: Fuel cell, battery, ultracapacitor, hybrid source, fuel cell hybrid vehicle and energy management control.

I. INTRODUCTION

Application of FCs in electric vehicle is one of the promising solutions to provide high-energy efficient, quiet operation and less pollutant drive train. In contrast to the battery powered electric vehicle, FCHV offers a longer driving range for as long as the fuel supply is available on board. However, there are few problems associated with using FC alone to power the vehicle, such as a relatively short lifespan, poor dynamic response, long start-up time, high cost, and inability to capture of regenerative energy during braking mode [1-3]. Moreover, peak power demands from FC leads to fuel starvation phenomenon and is hazardous to its lifespan. Therefore, hybridization of FC with ESU is necessary in order to overcome the above mention problems. It is also shown [2] that a proper sizing of ESU in FCHV could greatly reduce the size and cost of the vehicle itself.

The ESU could consist of battery modules, UC modules or a combination of both (dual ESUs). There have been several studies carried out to investigate on the hybridization of the FCHV. From the comparative study performed in [2,4,5], results showed that the hybridization of FC-battery-UC vehicle leads to a more practical solution, higher fuel economy and increases battery lifetime. Of course, these

aspects also depend on the hybridization structures, control algorithm adopted and energy management control in the hybrid system.

This paper focuses on the energy management and power-flow control. From the literatures, most of the proposed energy management strategies are of series energy-sharing configurations [1,2,6]. In a series configuration, the power flow of energy sources is controlled in such a way that the energy dense sources are used to deliver the power to the power dense source, and the power dense source is then used to regulate the DC bus and to respond to the peak power demand.

This paper proposes a parallel energy-sharing configuration in controlling the power-flow from multiple energy sources to power an electric vehicle propulsion system. The hybrid source is composed of a FC generator, battery units and UC modules. Through parallel configuration of power-flow, the DC bus is regulated by all energy sources simultaneously but with different contributions depending on the characteristics of the sources and defined by the control laws.

Arrangement of the paper is as follows: Section II discusses on the proposed energy management system while Section III discusses on the proposed control strategy. The results are presented in Section IV together with discussions. Lastly, the conclusion is presented in Section V.

II. ENERGY MANAGEMENT SYSTEM

Energy management is one of the most important factors in determining the efficiency, dynamic performance as well as reliability of a FCHV. This is true especially with the utilization of combined ESUs (battery and UC). In order to efficiently utilize these sources and to avoid component failure, the proposed algorithm in this paper is developed based on the characteristic of the vehicle load components, FC, UC and battery. These are discussed as follows.

- *FCHV load components* can be categorized into two types: constant load and transient load. Constant load consists of based load (e.g. on-board electric load and air conditioning), rolling resistance, aerodynamic drag and gravitational pull on an inclined road. These loads can be assumed constant and they should be supplied by the FC. On the other hand, transient load is associated with the power needed during acceleration, deceleration or braking. These loads cause power transient and should be supplied or absorbed by the energy buffer or storage units.

- *Fuel cell (FC)* shows a slow transient response and has a relatively high internal resistance. In addition, FC system has the disadvantage of slow start-up. However, FC is able to supply power continuously for as long as the reactants are available. Hence, it could function as a power generator in the hybrid source by constantly supplying the average or steady state power. The power flow during this mode is as shown in Fig. 1(a). Depending on the speed of vehicle and the state-of-charge (SOC) of the ESUs, the FC can also be used to charge ESUs when they have low energy content.

- Ultracapacitor (*UC*) has a high power density and hence able to provide a large amount of power within a relatively short period. Moreover, UC is a robust device, has long lifecycle, require low maintenance and low internal resistance. Consequently, UC can be used as the energy buffer during the peak power demand. The possible power flow during the transient stage is as shown in Fig. 1(b) where the UC is responsible in proving the main source of power during acceleration and absorb most of the braking energy. Nevertheless, UC is known to have a relatively low energy density and fast self-discharge characteristics. Vehicle may face start-up problem once the energy content in the UC is depleted due to self-discharged process. Under this condition, one cannot rely on the FC to power the vehicle, supply the initial acceleration power and charge on the UC at the same time because of its slow start-up time. Therefore, during the start-up stage, power must generally comes from other source of energy, i.e. battery.

- *Battery* has an advantage of high specific energy but relatively low in specific power. The power response is faster than FC, but slower than UC. Furthermore, battery has a limited lifespan (300-2000 cycles) depending on several factors such as: types of the battery, depth of discharge cycles, discharge rate, cell operating temperature, charging regime, number of overcharge and others. Hence, to optimize the lifespan of battery, it is necessary that battery current slope be limited in order to reduce the peak transient stress, hence, the peak power response generally comes from the UC. As discussed early, depending on the SOC of the UC, the main power during start-up might come mostly from the battery as depicted in Fig. 1(c).

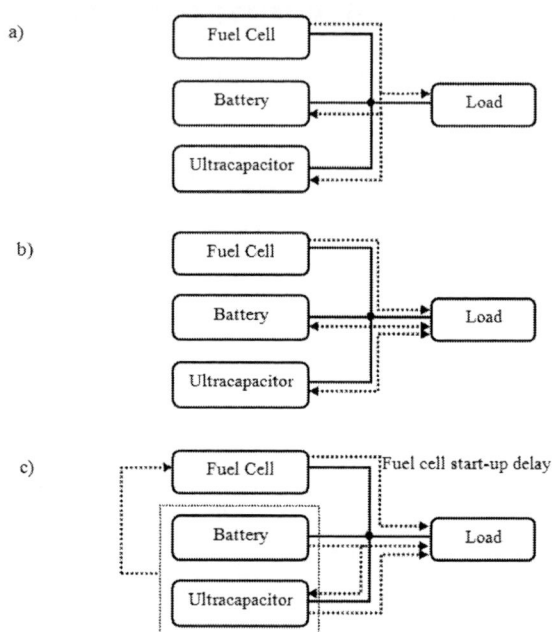

Fig. 1 Mode of power flow in the hybrid system: a) during steady state, b) during transient, c) during start-up

III.   PROPOSED ENERGY CONTROL SYSTEM

A parallel energy-sharing configuration is proposed to control the power flow between the DC bus, energy source (i.e. FC) and ESUs (i.e. UC and battery). The aim of the control is to provide a regulated DC bus voltage, which is connected to the converter of the motor drive system and vehicle accessories. The regulated DC voltage is obtained by controlling the power flow between FC, ESU and battery using the power electronic converters as depicted in Fig. 2. In the parallel active connection, the FC and ESUs are connected to the DC bus via DC/DC converters. The FC is connected to the DC bus using a single quadrant boost converter that also at the same time blocks the power flow back to the FC. On the other hand, the battery and UC are connected to the DC bus via half-bridge converters that allow bi-directional energy flow.

The structure of the proposed parallel energy-sharing management control is as shown in Fig. 3. It contains a total of six control loops (all using PI-controllers): a DC bus voltage control loop, three inner current control loops, an UC voltage control loop and a battery charging control loop. The DC bus voltage control loop is used to regulate the DC bus voltage, which generates a current reference to the three inner current control loops. The three inner current loops are used to control the currents of the FC, battery and UC respectively. In order to limit the slope of the reference current for the battery and FC within their safe values, low pass filters with time constants τ_1 and τ_2 are used for the two current references respectively. The current reference of the UC is obtained by subtracting the reference current generated by the DC voltage loop with the output current from the battery and FC; this is to ensure that only the UC current reference contains the demanding peak transient elements of the load current reference. To enable the battery to operate in a narrow charge-discharge cycle, the battery current reference is subtracted with the FC output current. By doing so, the peak power demand from the FC is avoided and at the same time only the FC will supply the continuous steady state power.

In the proposed control strategy, the UC is used mainly for two reasons: to provide the peak power requirement during acceleration and to absorb the vehicle kinetic energy during regenerative braking. It is therefore important to ensure that the UC is always ready to provide the peak power as well as to absorb the braking power. For this reason, the SOC of the UC is made dependent on the vehicle speed such that the available space of energy storage in the UC is proportional to the vehicle kinetic energy. For instance, if the vehicle is moving fast (i.e. large kinetic energy), more room is made available in the UC for regenerative braking and vice versa. Thus the UC voltage is given by (1).

$$V_{UC}(v) \leq \sqrt{V_{UC,max}^2 - \frac{M}{C_{UC}}v^2} \tag{1}$$

In (1), V_{UC} is the terminal voltage of the UC, $V_{UC,max}$ is the allowable maximum voltage of the UC, M is the mass of the vehicle, v is the speed of vehicle and C_{UC} is the capacitance of the UC. To ensure that the UC always has adequate energy from the battery and FC for vehicle accelerations, UC charging command (I_{UC-C}) is added to the battery and FC current references. Conversely, UC need to be discharged to provide sufficient volume for the vehicle kinetic energy during regenerative braking. This can be realized by summing up the UC discharge command (I_{UC-D}) to the UC current control loop. To avoid battery being charged by UC, the UC discharge command is limited to the load current demand.

In the proposed energy-sharing control system, battery is only charged by the FC and is controlled through the battery charging control loop. A simple charging method is implemented to charge on the battery, which is based on constant current-constant voltage (CCCV) method. The initial SOC of the battery (lead-acid) can be obtained based on its open-circuit terminal voltage [7]. The battery charging command ($I_{SOC\ Batt}$) is generated from the battery charging control loop and added to the FC current reference.

Based on the discussion above, the reference signals for each control loop are summarized as below:

$$V_{bus\ ref} = constant \tag{2}$$

$$I_{UC\ ref} = I_{load} - (I_{FC\ feedback} + I_{Batt\ feedback}) + I_{UC-D} \tag{3}$$

$$I_{Batt\ ref} = I_{load} + I_{UC-C} - I_{FC\ feedback} \tag{4}$$

$$I_{FC\ ref} = I_{load} + I_{Batt-C} + I_{UC-C} \tag{5}$$

where $V_{bus\ ref}$ is the dc bus voltage, $I_{UC\ ref}$, $I_{Batt\ ref}$ and $I_{FC\ ref}$ are the current loop reference signals for the UC, battery and FC respectively. I_{load} is the load current demand, I_{UC-C} and I_{UC-D} is the UC charge and discharge signal that are generated from the UC voltage control loop. $I_{FC\ feedback}$ and $I_{Batt\ feedback}$ are the current feedback signals for the FC and battery respectively and I_{Batt-C} is the battery charging command.

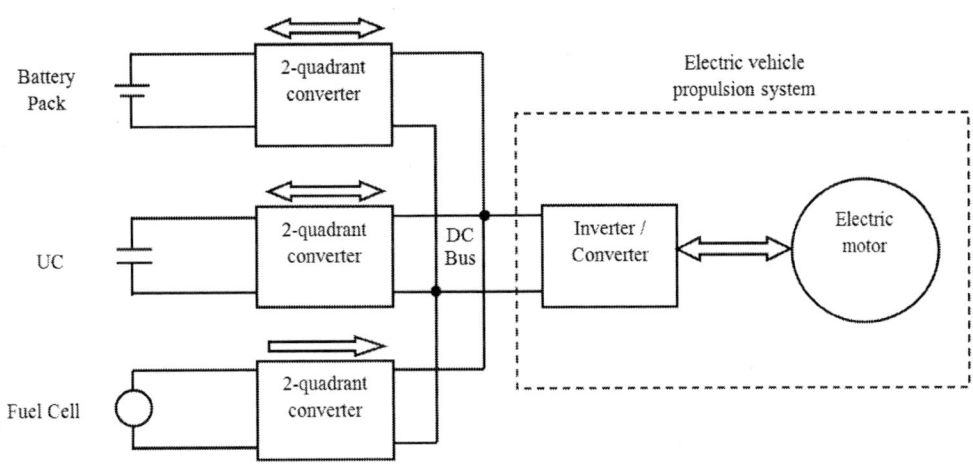

Fig. 2   Proposed parallel energy sharing system

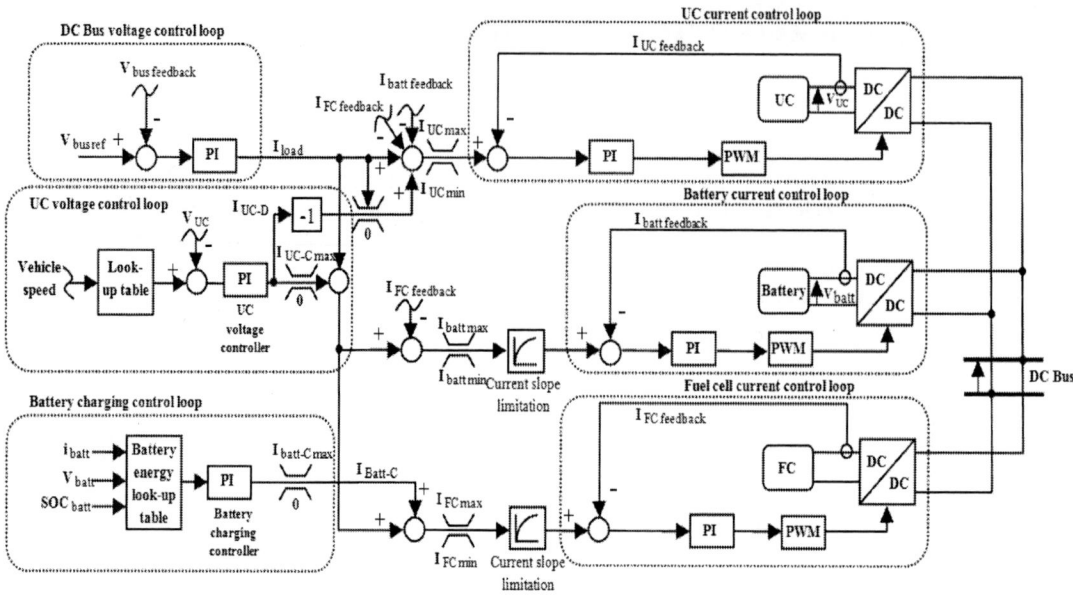

Fig. 3   Proposed parallel energy sharing control for the FC-battery-UC hybrid system

IV.   RESULTS AND DISCUSSIONS

Simulations as well as experiments are carried out to verify the viability of the proposed method. The simulations are carried out using of MATLAB/Simulink simulation package. For the experimental set-up, the battery pack is made up of 4 series connected 12V , 45AH calcium-calcium battery. The UC used in the experiment is a 165F, 48V BMOD0165 EO48 BO1 BOOSTCAP from Maxwell. Due to unavailability of the FC at the moment, its behavior is emulated using a HP6675A power supply with an output voltage at 48V. The control algorithm is implemented using dSPACE DS1104 controller board with an overall sampling period of 100μs. The DC bus voltage is regulated at 80V where a shunt DC motor rated at 0.25hp 120V 3000r.p.m. is connected to it via a H-bridge converter to represent the vehicle propulsion system. Initially, the battery is fully charged and the SOC of UC is set at 87.5% (V_{UC} =42 V). An ideal start-stop drive cycle is applied to the proposed hybrid system. Some key waveforms are recorded during the motor acceleration, deceleration and steady state speed as shown in Fig. 4. The DC motor accelerates at t=10s from stand still to steady state speed of 1300 rpm and then decelerates at t=310s to halt.

As the DC motor started to accelerate, the following conditions are observed:

- During motor acceleration, the UC module supplies most of the peak power demand followed by the battery and FC within a limited current slope.
- The UC power provides the fastest dynamics; the battery power is second followed by the FC.

- Synchronously, the UC power, after a sharp discharge during motor acceleration, decreases slowly to a constant discharge current at around 1A.
- Based on the speed of the DC motor, the UC voltage falls to it reference voltage to make room for regenerative energy from the DC motor.
- The steady-state load current is approximately 2A, which is totally supplied by the FC alone. The battery is back to idle state after the transient response.

When the DC motor starts to decelerate, it can be observed that the regenerative braking energy from the DC motor regenerates back to the dc bus. One can observe that:

- The sharp braking power is recuperated by the UC and at the same time, the UC is also getting charged from the FC and battery.
- The FC and battery supplies power to charge the UC. The battery goes to idle condition and the UC is charged up only by FC.
- When the UC module is charged up to its reference voltage, the FC power slowly reduces to zero.
- When the UC is fully charged, the battery will capture the excess energy.

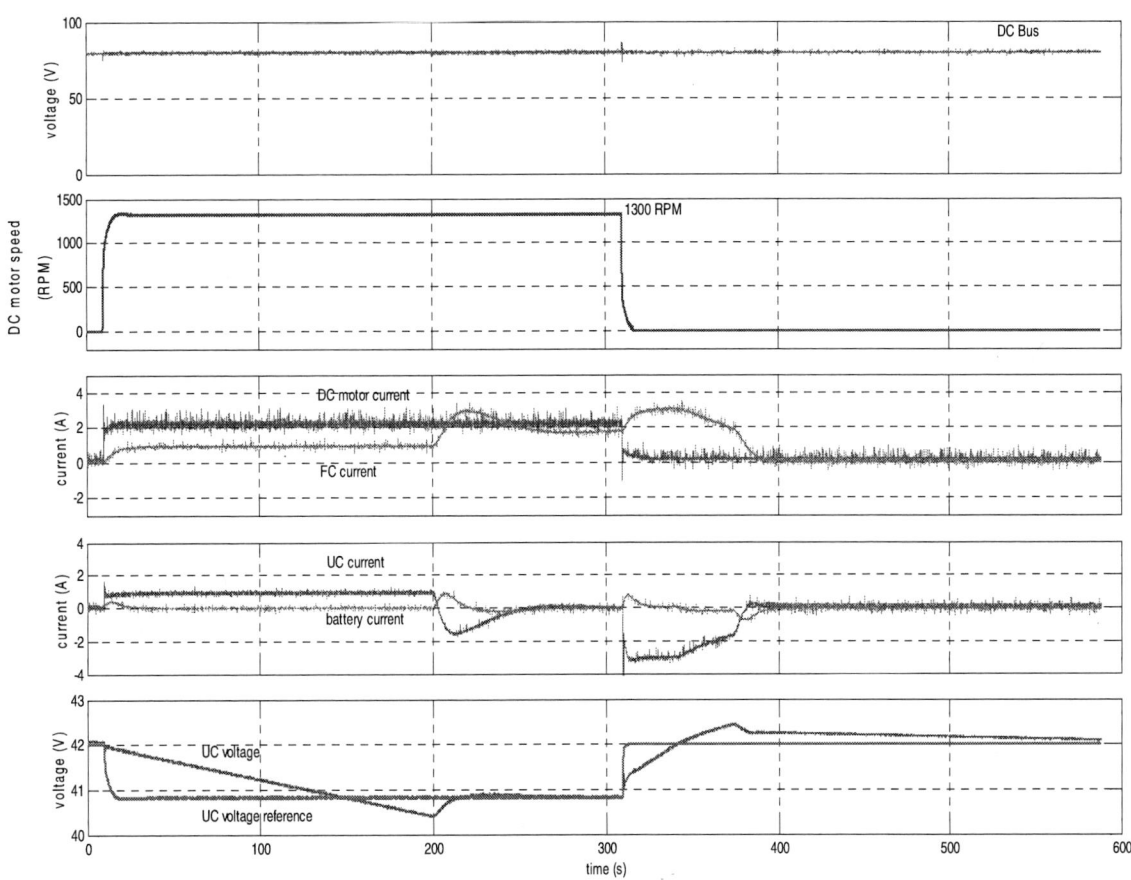

Fig. 4 An ideal start-stop drive cycle response for the proposed hybrid system

During the steady state conditions, one can observe the following:

i. Steady state speed
- During the steady state speed, the UC is discharged accordingly to the speed of motor to make room for regenerative power.
- However, if the UC voltage is lower than its reference voltage, the battery followed by the FC will charge the UC.
- Once the UC reached its reference voltage, the FC supplies all of the constant load power. At this time, the battery and UC are in the idle condition.

ii. Stand still condition
- At the stand still (zero speed) condition, the FC and battery charge the UC up to its reference voltage.
- The FC, battery and UC are in idle condition when there are no load demands.

V. CONCLUSIONS

This paper proposes and discusses on the design and control structure of a FCHV using a parallel energy-sharing configuration. The hybrid source in this paper consists of FC generator, battery and UC units. Through the proposed energy control system, the peak power stress on the FC and battery is avoided. The UC is used to provide peak power demand and to recuperate most of the braking power at the same time. Voltage of the UC is controlled accordingly, depending on the vehicle's speed in order to ensure sufficient energy for vehicle acceleration and also adequate space to capture the kinetic energy during braking. The proposed method does not guarantee perfect results in all situations, but provides a satisfactory energy management method in control the overall FCHV system. The validity of the proposed energy control scheme is supported by experimental results.

ACKNOWLEDGEMENT

The authors would like to thank the Ministry of Higher Education (MoHE) of the Malaysian government for providing the funding (under FRGS) to conduct the research.

REFERENCES

[1] Phatiphat Thounthong, Stephane Rael, Bernard Davat, "Energy management of fuel cell/ battery/ supercapacitor hybrid power source for vehicle applications", *Journal of Power Sources* Vol. 193, No. 1, pp. 376-385, 2009.

[2] Wenzhong Gao, "Performance Comparison of a Fuel Cell-Battery Hybrid Powertrain and a Fuel Cell-Ultracapacitor Hybrid Powertrain", *IEEE Transaction On Vehicular Technology*, Vol. 54, No.3, pp. 143-150, May 2005.

[3] Jennifer Bauman and Mehrdad Kazerani, "A Comparative Study of Fuel Cell-Battery, Fuel cell-Ultracapacitor, and Fuel Cell-Battery-Ultracapacitor Vehicles", *IEEE Transaction On Vehicular Technology*, Vol. 57, No. 2, pp. 760-769, March 2008.

[4] N. Schofield, H. T. Yap and C. M. Bingham, "Hybrid Energy Sources for Electric and Fuel cell Vehicle Propulsion", in *IEEE 2005 Vehicle Power and Propulsion Conference*, 2005, pp. 522-529.

[5] E. Schaltz, A. Khaligh and P. O. Rasmussen, "Investigation of Battery/ Ultracapacitor Energy Storage Rating for a Fuel Cell Hybrid Electric Vehicle", in *IEEE 2008 Vehicle Power and Propulsion Conference*, 2008, pp.1-6.

[6] Phatiphat Thounthong, Stephane Real and Bernard Davat, "Control Strategy of Fuel cell/ Supercapacitors Hybrid Power Sources for Electric Vehicle", *Journal of Power Sources*, Vol. 158, No. 1, pp. 806-814, July 2006.

[7] A. A. Ferreira, J. A. Pomilio, G. Spiazzi and L. de Araujo Silva, "Energy Management Fuzzy Logic Supervisory for Electric Vehicle Power Supplies System", *IEEE Transactions on Power Electronics*, Vol. 23, No. 1, pp. 107-115, Jan 2008.

978-1-4577-0007-1/11 $26.00 © 2011 IEEE

Performance Comparison of SVPWM and Hysteresis Current Control for Dual Motor Drives

Jurifa Mat Lazi[1], Zulkifilie Ibrahim, Marizan Sulaiman, Irma Wani Jamaludin, Musa Yusuf Lada

Faculty of Electrical Engineering (FKE),Universiti Teknikal Malaysia Melaka (UTeM),

Hang Tuah Jaya, 76100 Durian Tunggal, Melaka, Malaysia
jurifa@utem.edu.my[1]

Abstract- **Dual motor drives fed by single inverter are purposely designed to reduced sizes and cost with respect to single motor drives fed by single inverter. This paper presents the speed responses behavior of Dual Permanent Magnet Synchronous Motor (PMSM) driven base on two different PWM control schemes, which are Space Vector Pulse Width Modulation (SVPWM) and Hysteresis Current Controller. These two techniques are compared for wide range of speed and for variation of load. MATLAB/Simulink has been chosen as the simulation tools. The comparison between SVPWM controller and Hysteresis Current Controller for Dual PMSM fed by single inverter is presented. Both techniques show satisfactory speed regulation for a wide range of speed either with load, no load or variation of load.**

Keywords— **Dual Motor Drives, Single Inverter, SVPWM, Hysteresis Current control**

I. INTRODUCTION

In many applications, one motor is controlled by one converter. These systems are called SMSC, single machine single converter system [1]. Multi machine systems (MMS) are more and more used for industry today. Those systems allow to extend the field of high power applications or to increase their flexibility, mechanical simplicity and safety operating. However, it includes a lot of power switches which are large in size, costly and bulky. The high cost and large size need of the inverter make such dual inverter, dual motor drive configurations economically less competitive. Therefore, the need for dual motor drives fed by single inverter is rising consequently to reduce sizes and cost with respect to the single motor drives, either in industrial or in traction application.

But, the reduction number of power electronics switches and other components will results the paralleling of the drives systems. If the load torque for each motor is still the same, there is no speed changes will be encountered because every motor will have the same behavior [2] . On the other hand, a variation of load on one of the motors will create perturbations on the electrical part and perhaps a malfunctioning of the system. For this type of disturbance, a control drive is needed to compensate the disturbance in order to make the system back to its origin. After several reading, an average technique has been selected to overcome the loss of adhere of the motor. The technique is average of the mean of phase current.

Generally MMSC can be divided into two main categories, which are master-slave and mean control system [3]. In Master-Slave scheme, one motor which is selected as the master is directly control. The motor with the highest load is set as the master motor and the other one is slave motor, which has the same applied voltage, same electric pulsation and also the same speed [4]. Then the behavior of slave motor will be ignored. In some conditions, the performance of slave motors may not acceptable [3].

Whereas in mean control, there are several techniques have been applied . One of them is average of current [5], [6]. In this scheme, the control system is basically similar to that of a single machine. And the machine internal parameter such as flux, do not show desirable behavior. The second technique is averaging over the parameters of the equivalent circuit at steady state [7]. But in the case of motor parameters are not similar, some of the results, may not be not acceptable. The other techniques are the averaging the voltage space vector [3], [8]. Through this technique, for each motor, a single motor controller was applied, and then a reference voltage is obtained for each machine. In view of the fact that an inverter can only provide one reference voltage vector, thus a vector average is taken over the motor reference voltages, and the result is generated by the inverter. Besides, there are other technique such as mean and differential torque [5], [6] and the Optimum torque over current ratio [9].

II. MODULATION TECHNIQUE

To drive a motor, a modulation technique needs to be used in order to generate pulses with certain rules and goals through supplying DC voltage for the inverter. Basically, modulation technique can be classified into two types, which are voltage control and current control [10]. Voltage control modulation can be divided into three types of Pulse Width Modulation (PWM). The first type is six step PWM, the second is

978-1-4577-0007-1/11 $26.00 © 2011 IEEE

sinusoidal PWM and the third is Space Vector PWM also known as SVPWM. Besides, for current control, there are two techniques used, which are hysteresis current controller and delta modulation [7].

The pulse width modulation (PWM) makes the inverter output the waveforms which are made up of many pulses with certain rules and goals through supplying DC voltage for the inverter. Since it is the task for DC/AC switching mode to produce a sinusoidal AC output voltage, therefore, to control the flux linkage and frequency with ease, PWM is the essence in adjusting the speed drive systems. Among many forms of PWM, the SPWM and SVPWM are the most common form [6], the former is more familiar and the latter becomes mature promptly especially in the middle and high power systems.

A. SVPWM (Space Vector Pulse Width Modulation)

Space vector PWM (SVPWM) refers to a special technique of determining the switching sequence of the inverter power switches for obtaining variable output voltage which is defined spatially. Compared with the former three phase sinusoidal modulation (SPWM) method, SVPWM has advantages of lower current harmonics and possible higher modulation index. This technique has a wide linear modulation range without using distorted modulation and it also guarantees that only one switch changes at any time.

Figure 1 shows the basic configuration of SVPWM for Dual PMSM drives. Firstly, the speed controller estimates the torque through current, i_q^*, then, two current controller will convert the i_q^* and i_d^* signals to voltages (V_q^*, V_d^*). These voltages are then transformed to α-β model by inverse Park's transformation. Then, using inverse Clark's equation, V_α^* and V_β^*, are converted to three phase voltages, V_a, V_b, V_c. These signal are used to generate SVPWM pulses before they can be used to drive a PMSM. Since it is dual motor, the average of d-q voltages ($V_{d,A}^*$, $V_{q,A}^*$, $V_{d,B}^*$, $V_{q,B}^*$) for both motors are calculated in order to be transform to three-phase voltages. As a feedback, three phase current output, need to be transform back to DQ model by using Clark's and Park's equations. Then, the actual i_d and i_q will be compared with the references current before proceed again by the current controller.

TABLE I
TRANSFORMATION SUMMARY FOR COORDINATE SYSTEM IN FIELD ORIENTED CONTROL

Park's equation : a,b,c → α,β	Inverse Park's equation : d,q → α,β
$i_\alpha = i_a$ $i_\beta = \dfrac{1}{\sqrt{3}} . i_a + \dfrac{2}{\sqrt{3}} i_b$ $i_a + i_b + i_c = 0$	$i_\alpha = i_d . \cos(\theta) - i_q . \sin(\theta)$ $i_\beta = i_d . \sin(\theta) + i_q . \cos(\theta)$
Clark's equation : α,β → d,q	Inverse Clark's : α,β → a,b,c
$i_d = i_\alpha . \cos(\theta) + i_\beta . \sin(\theta)$ $i_q = -i_\alpha . \sin(\theta) + i_\beta . \cos(\theta)$	$i_a = i_\alpha$ $i_b = -\dfrac{1}{2} . i_\alpha + \dfrac{\sqrt{3}}{2} . i_\beta$ $i_c = -\dfrac{1}{2} . i_\alpha - \dfrac{\sqrt{3}}{2} i_\beta$

B. Hysteresis Current Controller

The basic structure of the dual PMSM drives with hysteresis current control in the stationary reference frame and with PI speed controller is shown in Figure 2. Three independent hysteresis current controllers in the three phase a,b,c reference frame are applied in this scheme. In high performance servo drives, hysteresis current controllers are used to ensure that the actual currents flowing into the motor are as close as possible to the current references.

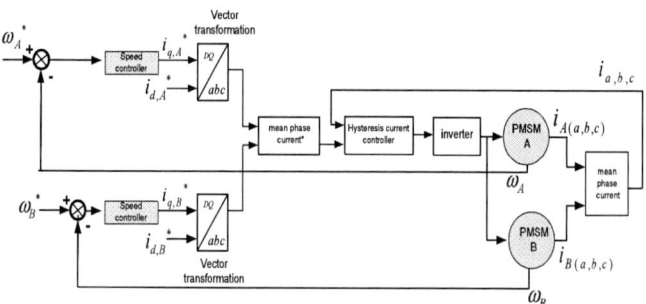

Fig. 2. Hysteresis Current Control for Dual PMSM configuration

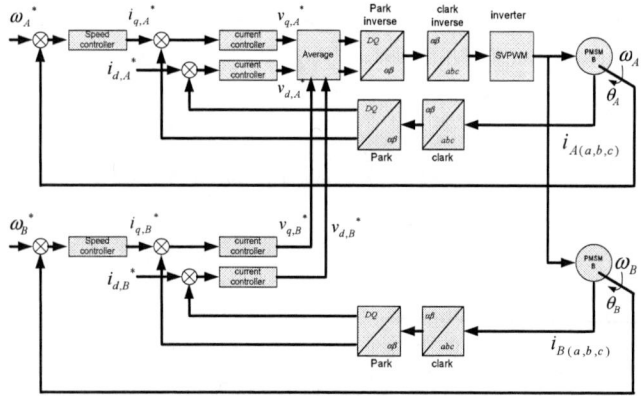

Fig. 1. SVPWM for Dual PMSM drives configuration

978-1-4577-0007-1/11 $26.00 © 2011 IEEE

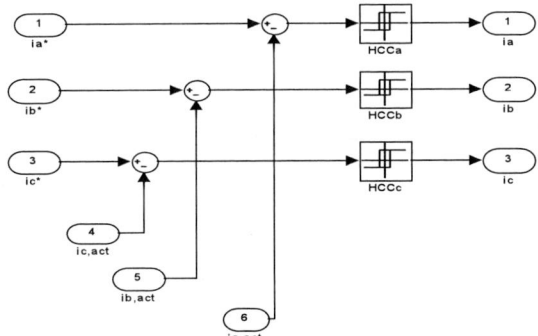

Fig.3. Hysteresis Current Control

Figure 3 shows the block diagram for hysteresis controller in order to produce the output signal. The actual phase currents (i_a, i_b, i_c) are compared with reference phase current (i_a*, i_b*, i_c*) using three independent comparator in hysteresis controller. The logic condition for six inverter switches is chosen by the output of the comparator [1].

When the phase "a" current is smaller than (i*-Δi), where Δi is the hysteresis band, the output of the comparator is "1", the "a" phase will be connected with the positive track of DC link. In contrast, if the phase "a" current is bigger than (i*- Δi), the output of the comparator will become "0", and the "a" phase will connected to the negative track of DC bus. A similar procedure exists in the other legs. The reason that this is called a hysteresis controller is that the leg voltage switches to keep the phase current within the hysteresis band. The phase currents are, therefore, approximately sinusoidal in steady state.

The smaller the hysteresis band, the more closely do the phase currents represent sine wave. Small hysteresis band, however, imply a high switching frequency, which is a practical limitation of the power device. Increased switching frequency also implies increased inverter losses.

III. MATHEMATICAL MODEL

The simulated machine is a smooth air gap PMSM without any damping circuits in the rotor. The rotor field is constant and created by permanent magnets and the e.m.f are considered as sinusoidal. The simplified electric equations for motor "A" can be presented as below [5]:

$$v_A = Ri_A + L\frac{di_A}{dt} + jp\omega_{r,A}\psi_{r,A} \tag{1}$$

$$T_A - T_{L,A} = J\frac{d\omega_{r,A}}{dt}$$
$$with \ T_A = \frac{3}{2}p\Im m\{i_A\psi_{r,A}\} \tag{2}$$

$$\omega_{r,A} = \frac{d\theta_A}{dt} \tag{3}$$

Where;

ω_r : Motor Angular velocity,
Ψ_r : Rotor flux,
T : Electrical torque,
T_L : Load torque,
J : Moment of Inertia.
θ : Instantaneous angular position

The model of the motor "B" can be derived from (1) to (3) by changing the subscript "A" to "B".

With the assumptions, motor "A" and motor "B" are equal in all parameters but have different loads. The space vectors of the rotor fluxes, $\psi_{r,A}$ and $\psi_{r,B}$ are equal in magnitude and its instantaneous position θ_A and θ_B respectively in the stationary frame. Consider a rotating reference frame d,q whose direct axis "d" is along the direction of $(\psi_{r,A}+\psi_{r,B})/2$ and its instantaneous angular position is $\theta=(\theta_A+\theta_B)/2$. Based on this reference, the electromagnetic torque of the motors "A" and "B" can be expressed as:

$$T_A = \frac{3}{2}p.\psi_{r,A}.i_{qA} \tag{4}$$

$$T_B = \frac{3}{2}p\psi_{r,B}.i_{qB} \tag{5}$$

And the average of the current and torque are as follows:

$$i_\Sigma = \frac{i_A+i_B}{2} \tag{6}$$

$$T_\Sigma = \frac{T_A+T_B}{2} \tag{7}$$

IV. SIMULATION RESULTS

Averaging of Dual PMSM drives simulation has been done by using MATLAB/Simulink with referring the control strategy shown in Figure 1 and Figure 2. Quantities observed are speed responses for both motors for variation of speed and different loads.

The first technique simulated is depicted in Figure 1. This simulation uses the average of phase current as the input for the hysteresis current controller. In contrast with SVPWM

978-1-4577-0007-1/11 $26.00 © 2011 IEEE

technique, Hysteresis uses actual current feedback from the motor to be compared with reference current, while in SVPWM, no current feedback needed. This method ensures that the actual currents flowing into the motor as close as possible to the current references.

The second control technique analyzed in this paper is synthesized by the control diagram of Figure 2. This control strategy is based on the generation of two different sets of reference currents $(i_{d,A}*, i_{q,A}*, i_{d,B}*, i_{q,B}*)$. These current are converted to three-phase current $(i_{a,b,c})$ by using vector transformation. Average of these three-phase reference currents are calculated before compared with actual phase current through hysteresis current controller. The current controllers are able to evaluate the expected voltage of motor "A" and "B".

TABLE II
SPECIFICATIONS OF MOTOR

No	Motor Specifications	Value
1	Rated Torque	8 Nm
2	Rated Speed	209 rad/s
3	Inertia	0.0006329 kgm^2
4	Resistance	0.9585 Ω
5	Inductance	0.00525 H
6	Magnet Flux	0.1827 Vs
7	DC link Voltage	300 V

The relevant parameters of the motors are listed in TABLE II. Both motors have the same specifications and applied for both SVPWM and Hysteresis Current Control Technique. The transient responses of the drives for SVPWM and Hysteresis Current Controller are shown in Figure 4 to Figure 7.

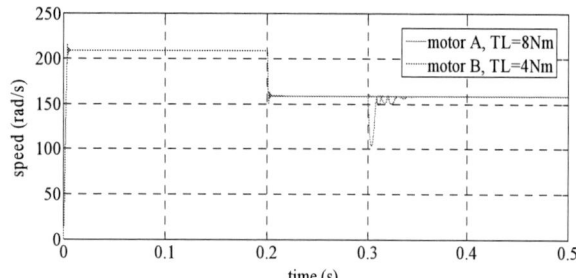

(a)   Speed response for motor A and motor B (SVPWM)

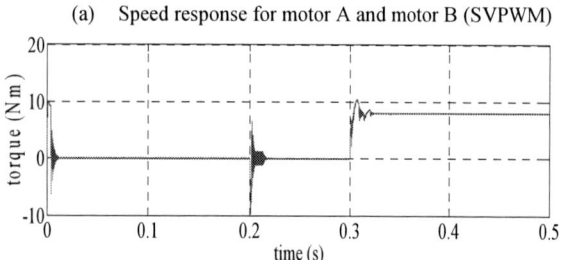

(b)   Torque response for motor A (SVPWM)

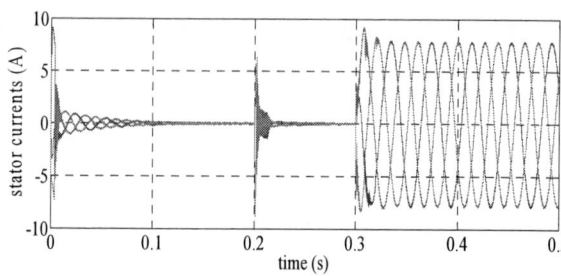

(c)   Stator current for motor A (SVPWM)

Fig.4. Transient responses for SVPWM method, (a). Speed response for motor A and motor B, (b) Torque response for motor A. (c) Stator current for motor A

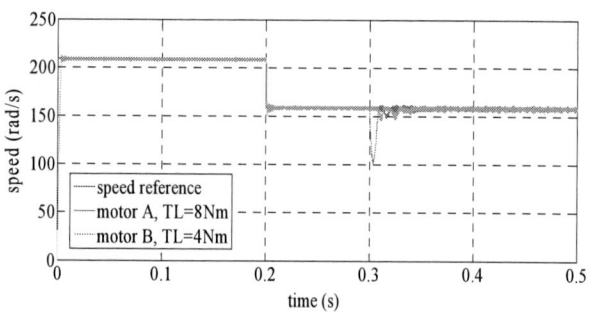

(a)   Speed response for motor A and motor B (Hysteresis Current Controller)

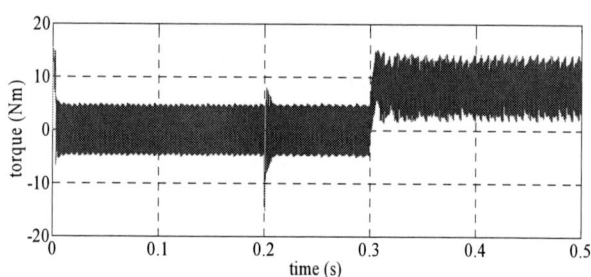

(b)   Torque response for motor A (Hysteresis Current Controller)

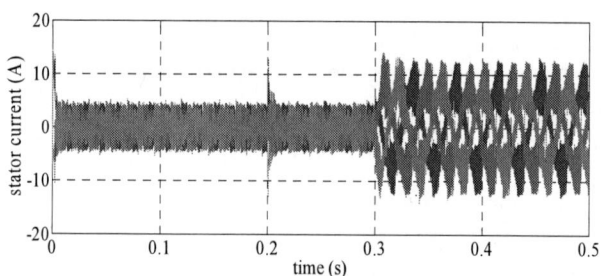

(c)   Stator current for motor A (Hysteresis Current Controller)

Fig. 5. Transient responses for Hysteresis Current Control method (a). Speed response for motor A and motor B, (b). Torque response for motor A, (c). Stator current for motor A

Fig.4. and Fig.5. show the transient response during start-up at t=0s, speed reduction about 50Nm (209rad/s to 159 rad/s) at t=0.25s and load variation at t=0.3s for SVPWM and Hysteresis Current Control method respectively. From t=0s until t=0.3s, the systems are running without load. At t=0.3s, motor "A"s are given 8Nm which is the rated torque, and motor "B"s are given 4Nm, which is half of the rated torque. The systems show that both techniques follow well the speed reference command. Although, both techniques show a little undershoot during load variation at t=0.3s, they manage to get to their reference speed after several millisecond. From these figures also, it can be observed that the output with SVPWM technique has low torque ripple and low current distortion as well as persistent for the rest of the system.

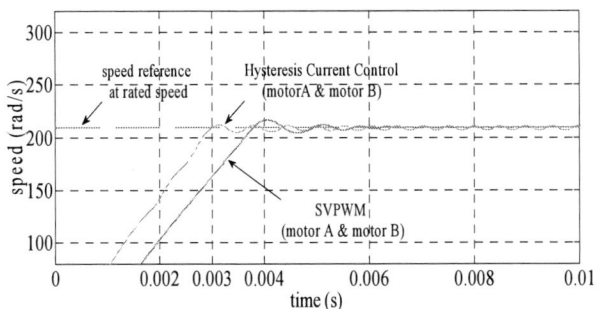

(a). Speed responses during start-up. Both motors are running without load

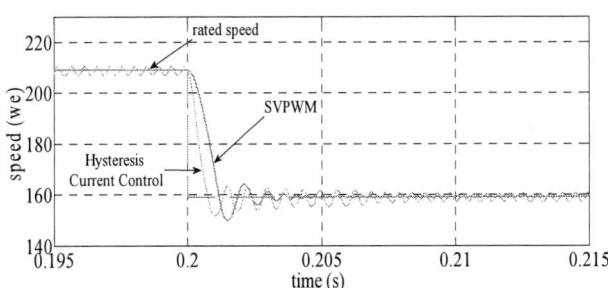

(b). Speed responses during reduction of speed about 50Nm

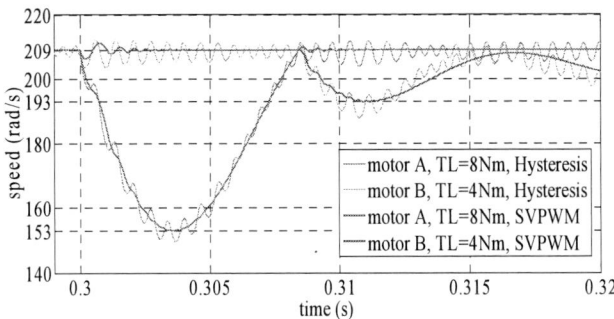

(c). Speed responses during load variation at rated speed (TL= torque load)

Fig.6. Comparison of Speed responses for motor A and motor using SVPWM and Hysteresis Current Control method – Close-up view.

Fig.6s, show the close-up view for both motor A and motor B, using both method, SVPWM and Hysteresis Current Control at rated speed. Fig.6(a). proves that, although Hysteresis Current Control method gives more ripple, it achieve steady state faster than SVPWM method. Their settling time are 0.3ms and 0.38ms respectively. Fig. 6(b) shows, the behavior of Hysteresis method achieve the steady state faster than SVPWM similar when the reduction of speed is applied to the system at t=0.2s. Speed responses for both motors and both techniques during load variation at t=0.3s are depicted on Figure 6(c). Both methods for motor A, which given more load, TL=8Nm) experience the same undershoot which is about 44 rad/s (209rad/s to 153 rad/s). But Hysteresis Current Control speed response shows more ripple compared to SVPWM method. For motor B, which given less load (TL=4Nm), the reduction speed due to load disturbance is about 16Nm (209rad/s to 193rad/s). This figure also prove that, the motor with higher load will experience higher undershoot, and the lighter load will get better performance.

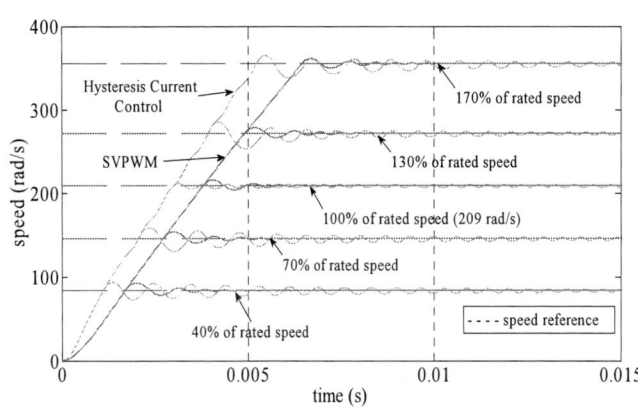

Fig.7. Speed responses for motor A and B using SVPWM and Hysteresis Current Controller during start-up

Fig. 7. shows the speed behavior for both SVPWM and Hysteresis Current Control methods during start-up. Obviously, that Hysteresis Current Controller method reaches the steady state faster than SVPWM method for all cases, but as can be observed, SVPWM produced fewer ripples than Hysteresis method. Also noted that, the optimize controller is at rated speed condition, where the magnitude of the ripple is lower that other cases.

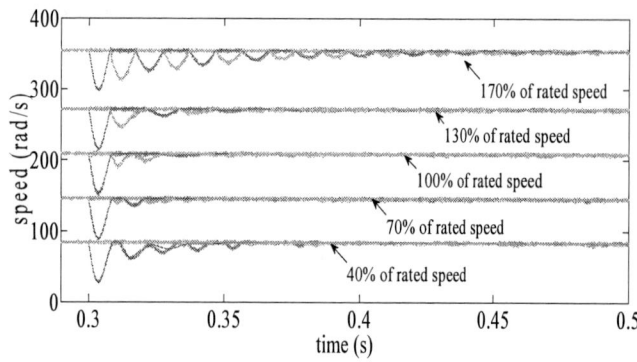

Fig. 8. Speed responses for motor A and B by using SVPWM and Hysteresis Current Controller during load variation

Fig.8. illustrate the behavior of motor A and motor B using SVPWM and Hysteresis Current Control during load variation at t=0.3s. Motor "A"s are given 8Nm torque load and motor "B"s are given 4Nm torque load. Both methods show good responses due to wide range of speed. As can be seen, both methods are stable from 50% to 170% of rated speed. But the oscillation for rated speed condition is having fewer undershoot compared to lower and upper of rated speed condition. The worst case is at 170% of rated speed, where the oscillation last for more than 0.1 second. And the minimum speed that can be applied is at 40% of rated speed. Below than that, the response shows the unacceptable behavior.

V. CONCLUSION

The comparison between SVPWM controller and Hysteresis Current Controller for Dual PMSM fed by single inverter is presented. Both techniques show acceptable speed regulation for a wide range of speed either with load, no load or variation of load. For overall, output with SVPWM has low torque ripple and low current distortion compared to Hysteresis Current Controller. In other way, Hysteresis Current Control technique can reach the speed demand faster than Space Vector Pulse Width Modulation (SVPWM) technique, but it has more oscillation before it comes to steady state. For variation of load testing, both controller techniques manage to stabilize the system within the acceptable duration. Both techniques also show that they can be applied from range of 40% to 170% of rated speed.

REFERENCES

[1] Z. Li and S. Fengchun, "Torque control of dual induction motors independent drive for tracked vehicle," in *2008 10th Intl. Conf. on Control, Automation, Robotics and Vision*, 2009, pp. 68-72.

[2] Y. He, *et al.*, "A comparative study of space vector PWM strategy for dual three-phase permanent-magnet synchronous motor drives," in *Applied Power Electronics Conference and Exposition (APEC), 2010 Twenty-Fifth: Annual IEEE.*, 2010, pp. 915-919.

[3] H. Mohktari and A. Alizadeh, "A new multi-machine control system based on Direct Torque Control algorithm," in *The 7th International Conference on Power Electronics, EXCO 2007*, 2008, pp. 1103-1108.

[4] D. Bidart, *et al.*, "Mono inverter dual parallel PMSM-structure and control strategy," in *34th Industrial Electronics Annual Conference , 2008, IECON 2008*, 2009, pp. 268-273.

[5] M. Acampa, *et al.*, "Optimized control technique of single inverter dual motor AC-brushless drives," in *Universities Power Engineering Conference, 2008, UPEC 2008*, 2008, pp. 1-6.

[6] M. Acampa, *et al.*, "Predictive control technique of single inverter dual motor AC-brushless drives," in *Proceeding of the 2008 International Conference on Electrical MAchines*, 2008, pp. 1-6.

[7] P. Kelecy and R. Lorenz, "Control methodology for single stator, dual-rotor induction motor drives for electric vehicles," in *Power Electronics Specialist Conference, 1995, PESC'95*, 1995, pp. 572-578.

[8] J. Wang, *et al.*, "Comparative Study of Vector Control Schemes for Parallel-Connected Induction Motors," in *power Electronics Specialist Conference, 2005, PESC 05, IEEE 36th*, 2006, pp. 1264-1270.

[9] D. I. A.Del Pizzo, I.Spina, "Optimum Torque/Current control of dual PMSM single VSI Drive," *15th International Symposium on Power Electronics-Ee 2009, Novi Sad, Republic of Srbia*, October 28th-30th ,2009 2009.

[10] C. Mitsui, *et al.*, "Efficiency study of adjustable speed drive with dual motor connection," in *Power Engineering Conference,2007, IPEC 2007*, 2007, pp. 555-558.

A bridgeless Cuk PFC converter

M. R. Sahid; A. H. M. Yatim: N. D. Muhammad

Faculty of Electrical Engineering, Universiti Teknologi Malaysia,
81310 UTM Skudai, Johor, Malaysia
Email: rodhi@fke.utm.my

Abstract- **A bridgeless Power Factor Correction (PFC) circuit based on Cuk converter is proposed in this paper. The operation during each sub-interval modes of the converter operated in Discontinuous Conduction Mode (DCM) is discussed. The small-signal and large signal models are presented using Current Injected Equivalent Circuit Approach (CIECA). PLECS/Simulink is used to verify the capability of the proposed converter to regulate the output voltage while the input current regulation is inherent. This converter is capable to operate in universal input voltage condition.**

I. INTRODUCTION

Since the past several decades, the power factor correction (PFC) circuit is no more a strange issue among switch-mode power supply (SMPS) designer due to its capability to draw energy effectively from the mains. Besides improving the power factor, PFC circuits are also designed to give the highest efficiency by reducing the number of components conducted during its operation.

Recently, the bridgeless PFC circuit has gained its interest and popularity as part of the topology used in designing SMPS [1]. The first bridgeless PFC circuit proposed in 1983 [2] by D.M. Mitchell based on Boost converter, has paved the way for higher efficiency PFC converter. This can be achieved by reducing the number of components conducted, i.e. diode rectifier. As depicted in Fig. 1, while the conventional PFC converter would have at least two diode rectifiers conducted all the time, the bridgeless PFC only require one diode rectifier conducted, especially during turn ON condition.

A substantial work on bridgeless Boost PFC has been carried out since the last few years. Most of the works are focused on improving the efficiency of the converter by either reducing the energy loss within each component or by introducing the soft-switching techniques [3]. However, one of the drawbacks in Boost converter is the DC output voltage level, which is always higher than the input voltage. For a universal input voltage PFC converter, the output voltage is around 380V DC or more, while most electronic applications always operate at lower value namely 5V to 50V DC. Introducing another DC-DC converter that step down the 380V DC to 5V DC would sacrifice much of the converter efficiency. Thus, considering a PFC converter that capable to give lower voltage at the output would be a great alternative. Besides Boost, several bridgeless PFC topologies such as Buck [4], Buck-Boost [5], SEPIC [6] and Cuk [7] have been proposed.

In this paper, a new bridgeless PFC circuit based on Cuk converter is proposed. One of the advantages inherent in each

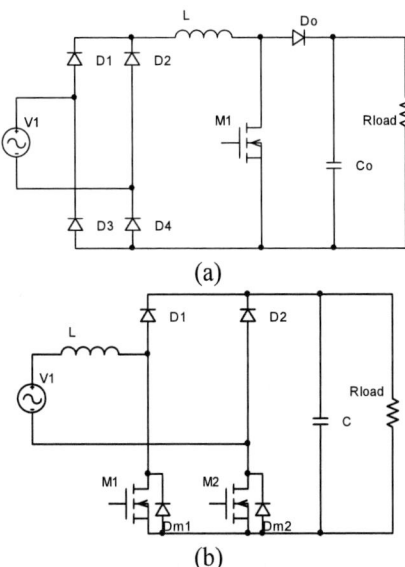

Fig. 1. Schematic of (a) the conventional Boost PFC and, (b) bridgeless Boost PFC converter.

Cuk converter is its high quality input and output current. The location of the inductors at the input and output port of the Cuk converter is the main reason to justify that these two currents would never be turned OFF abruptly.

II. OPERATIONS AND ANALYSIS OF THE PROPOSED CONVERTER

The proposed bridgeless Cuk PFC converter schematic with operation during positive and negative half-line period is shown in Fig. 2(a), 2(b) and 2(c) respectively. The number of components conducted, i.e. input diode, during each half-line period is less compared to the conventional Cuk PFC converter [7] and even less than the normal bridgeless Cuk PFC [8]. For instance, the normal bridgeless and conventional Cuk should have at least two input diode conducted either during MOSFET turned OFF period for normal bridgeless or conducted all the time for conventional PFC. On the other hand, as shown in Fig. 2(b) and (c), the fully-bridgeless Cuk converter only has one input diode conducted all the time which is either Ds2 or Ds1. However, the drawback is the overall number of components used to develop this converter is more compared to the other two Cuk PFC topologies mentioned earlier. It is due to two set of Cuk converters exist during each half-line period.

978-1-4577-0007-1/11 $26.00 © 2011 IEEE   81

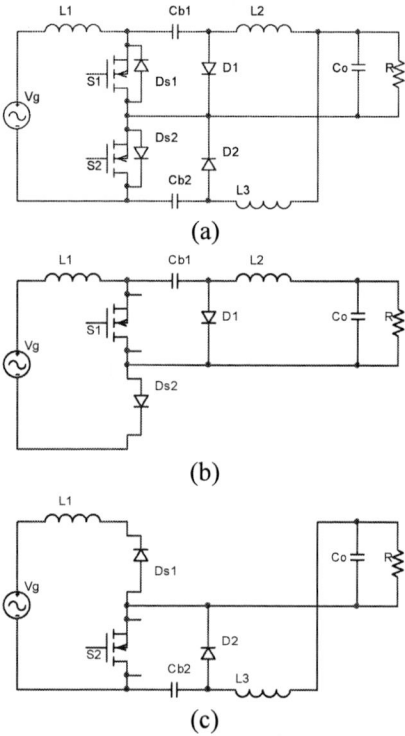

(a)

(b)

(c)

Fig. 2. Schematic of (a) the proposed bridgeless Cuk PFC with operation during, (b) positive and, (c) negative half-line period.

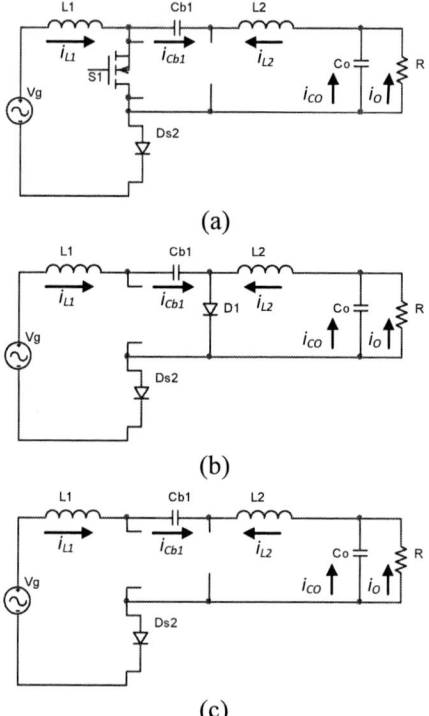

(a)

(b)

(c)

Fig. 3. Equivalent circuit during (a) d_1T_S, (b) d_2T_S, and (c) d_3T_S.

In this paper, the analysis during positive half line period is discussed while the same model can be used for negative half line period. Since the discontinuous conduction mode (DCM) operation is our concern, three subinterval modes exist within each switching period namely, d_1T_S, d_2T_S and d_3T_S. The equivalent circuit for these three modes is depicted in Fig 3(a), (b) and (c) while the key waveforms are shown in Fig. 4. Throughout the analysis to obtain the high frequency model, the input voltage, $v_g(t)$ is assumed to be constant within each switching period. Then, for the low frequency model, this assumption is no longer valid and the sinusoidal waveform will replace the value of $v_g(t)$ to get more accurate model for the purpose of PFC model [9].

To further ease the analysis, the Current Injected Equivalent Circuit Approach (CIECA) [10,11] is used to obtain the small-signal model of the proposed converter. In CIECA, the average input and output current injected to/from the circuit within each switching period, T_S, is analyzed. In this converter, the input current is i_{L1} and the output current is i_{L2} with the average value are represented as

$$i_{L1-AVG-Ts} = \frac{d_1^2 T_S v_g (v_g + v_O)}{2L_1 v_O}$$

(1)

and

$$i_{L2-AVG-Ts} = \frac{d_1^2 T_S v_g (v_g + v_O)}{2L_2 v_0}$$

(2)

From these two high frequency models, by defining the input voltage as $v_g = v_m |\sin(\omega t)|$, the low frequency models are obtained by integrating the high frequency model up to half line period, $T_L = \pi$ rad, such that,

$$
\begin{aligned}
i_{L1-AVG-TL} &= \frac{1}{\pi} \int_0^\pi \frac{d_1^2 T_S v_g (v_g + v_O)}{2L_1 v_O} d\omega t \\
&= \frac{v_m d_1^2 T_S}{v_O L_1} \left(\frac{v_m}{4} + \frac{v_O}{\pi} \right)
\end{aligned}
$$

(3)

and

$$
\begin{aligned}
i_{L2-AVG-TL} &= \frac{1}{\pi} \int_0^\pi \frac{d_1^2 T_S v_g (v_g + v_O)}{2L_2 v_0} d\omega t \\
&= \frac{v_m d_1^2 T_S}{v_O L_2} \left(\frac{v_m}{4} + \frac{v_O}{\pi} \right)
\end{aligned}
$$

(4)

With small perturbation introduce to i_{L1}, i_{L2}, d_1, v_m and v_O, the small signal models are,

$$\hat{i}_{L1} = \frac{1}{r_1} \hat{v}_m + j_1 \hat{d}_1 - g_1 \hat{v}_O$$

(5)

and

$$\hat{i}_{L2} = g_2 \hat{v}_m + j_2 \hat{d}_1 - \frac{1}{r_2} \hat{v}_O$$

(6)

where

978-1-4577-0007-1/11 $26.00 © 2011 IEEE

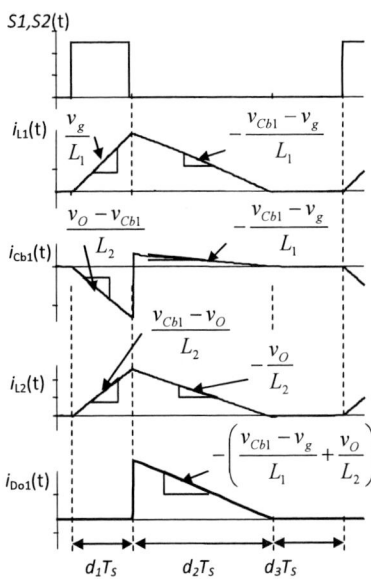

Fig. 4. Key waveforms for the proposed converter in DCM.

$$\frac{1}{r_1} = \frac{D_1^2 T_S}{2\pi L_1 V_O}\left(V_m \pi + 2V_O\right),$$

$$j_1 = \frac{D_1 T_S V_m}{2\pi L_1 V_O}\left(V_m \pi + 4V_O\right),$$

$$g_1 = \frac{T_S D_1^2 V_m^2}{4L_1 V_O^2},$$

$$g_2 = \frac{D_1^2 T_S}{2\pi L_2 V_O}\left(V_m \pi + 2V_O\right),$$

$$j_2 = \frac{D_1 T_S V_m}{2\pi L_2 V_O}\left(V_m \pi + 4V_O\right),$$

$$\frac{1}{r_2} = \frac{D_1^2 T_S V_m^2}{4L_2 V_O^2}.$$

The small signal equivalent circuit is shown in Fig 5. The input-to-output and control-to-output transfer functions are

$$\left.\frac{\hat{v}_O}{\hat{v}_m}\right|_{\hat{d}_1 = 0} = \frac{g_2}{\left(sC_O + \dfrac{1}{R} + \dfrac{1}{r_2}\right)} \quad (7)$$

and

$$\left.\frac{\hat{v}_O}{\hat{d}_1}\right|_{\hat{v}_m = 0} = \frac{j_2}{\left(sC_O + \dfrac{1}{R} + \dfrac{1}{r_2}\right)} \quad (8)$$

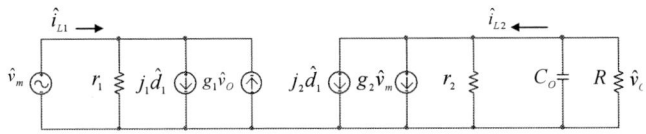

Fig. 5. Small-signal equivalent circuit.

III. DESIGN AND SIMULATION RESULTS

The small-signal and large-signal model for proposed fully-bridgeless Cuk converter is designed based on the parameters shown in TABLE I. It can be seen that the proposed converter will accept universal input voltage values which ranging from 115Vrms to 240Vrms. As shown in Fig. 6, the converter control-to-output transfer function as presented in Equation (8) is plotted using MATLAB to see the variation of the bode-plot gain and phase with universal input voltage values. It seems that by increasing the input voltage, the gain of the control-to-output bode-plot will increase as well. However, the phase-plot is very much similar when the input voltage is increased.

TABLE I
Circuit parameters

Parameters	Values
Input voltage, Vg	$115 - 240$Vrms
Line frequency, f_L	50Hz
Switching frequency, f_S	50kHz
Input inductor, L1	150uH
Intermediate inductors, L2 & L3	75uH
Bulk capacitors, C1 & C2	1uF
Output capacitor, Co	1.41mF
Output voltage, Vo	50V DC
Output power, Po	100W

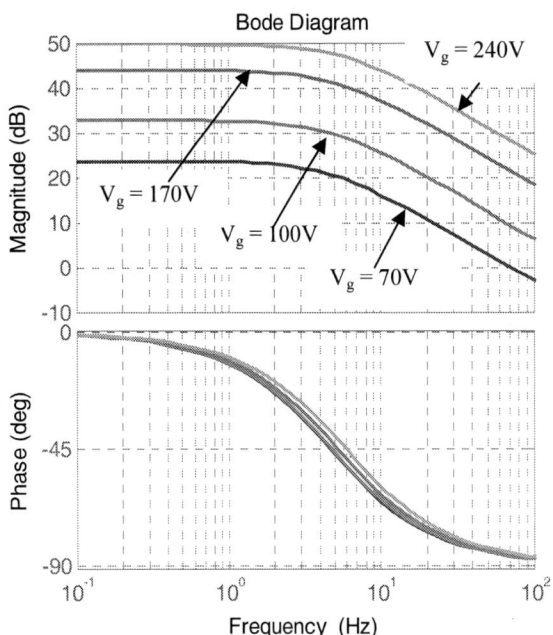

Fig. 6. Gain and phase plot variation with input voltage at 100V, 140V, 240V and 340V rms.

Then, the PLECS/Simulink is used to verify the small-signal model derived earlier. Using the PLECS AC Model, the bode-plot of the converter is plotted using the real switch circuit. As depicted in Fig. 7, the bode-plot is compared with the derived transfer function obtained from CIECA as presented in Equation (8). For V_g = 240V rms, the phase-plot between these two methods are similar while for the gain-plot, a gain gap as small as 2 dB is noticed.

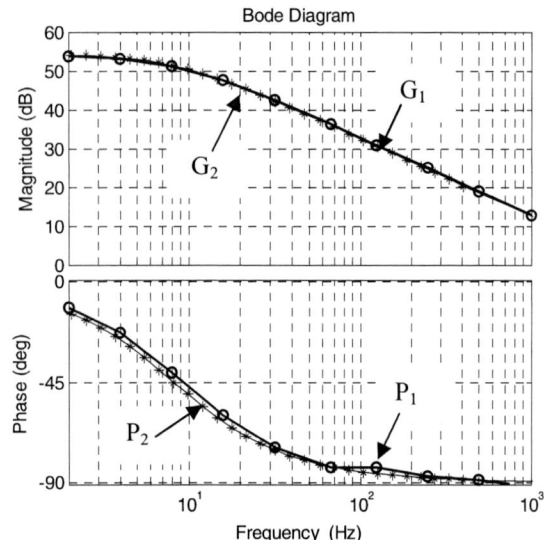

Fig. 7. Control-to-output bode-plot obtained from PLECS/Simulink(∅) (G_1&P_1) and from CIECA model (*) (G_2&P_2).

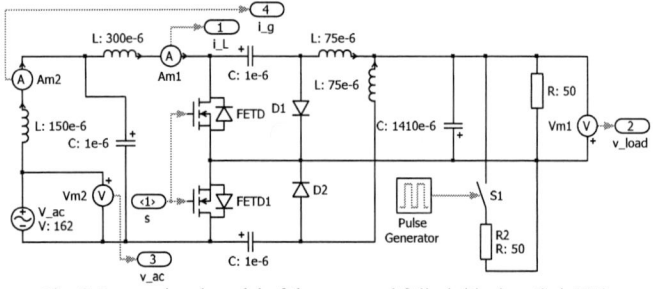

Fig. 8. Large-signal model of the proposed fully-bridgeless Cuk PFC converter.

Using simple PI controller, the converter closed-loop system is designed and simulated by utilizing the voltage follower method. The voltage loop bandwidth is designed at 20 Hz with 82° phase margin. The PLECS/Simulink is again used to develop the large signal model as shown in Fig. 8.

The inductor current, output (load) voltage and input voltage waveforms for V_g = 115 Vrms and V_g = 240 Vrms are shown in Fig. 9 and Fig. 10 respectively. Suppose the output voltage of a Cuk converter is negative, but the waveform oppositely shows a positive output voltage at 50 VDC. As can be seen from Fig. 8, this is due to the connection of the output voltage measurement which is in the reverse polarity.

From Fig. 9 and Fig. 10, the PI controller is able to regulate the output voltage to 50 VDC when the load change from 100 W to 50 W at t = 0.2 s. At t = 0.4 s, the load is change back to 100 W and the output voltage is successfully regulated to 50

VDC. Similar response can be observed when the input voltage is set at 240 Vrms.

Fig. 11 shows the filtered input current when LC input filter is connected with L and C is designed at 150 uH and 1 uF. With the input filter the current total harmonic distortions (THD_i) is measured at 7 % and 14 % for 115 Vrms and 240 Vrms input voltage respectively. At full load, the power factor for 115 Vrms is 0.988 and for 240 Vrms is 0.904.

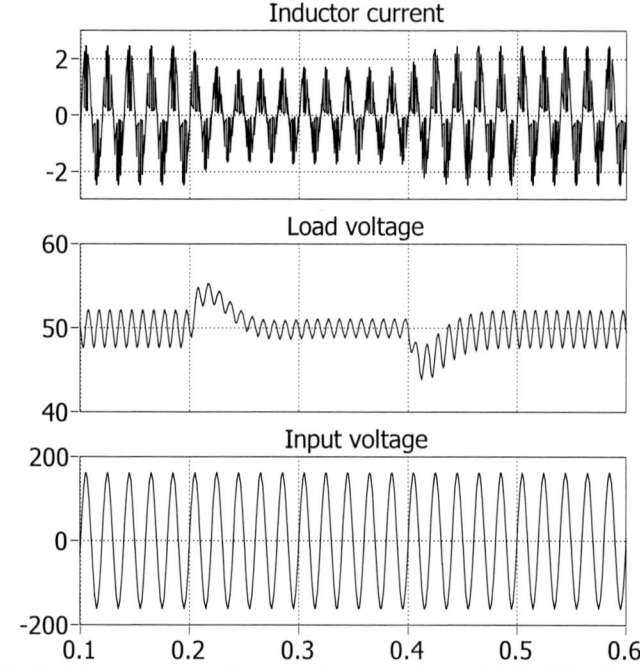

Fig. 9. Inductor current and output voltage waveforms with load disturbance of 100W-50W-100W at 0.2s and 0.4s for 115Vrms input voltage.

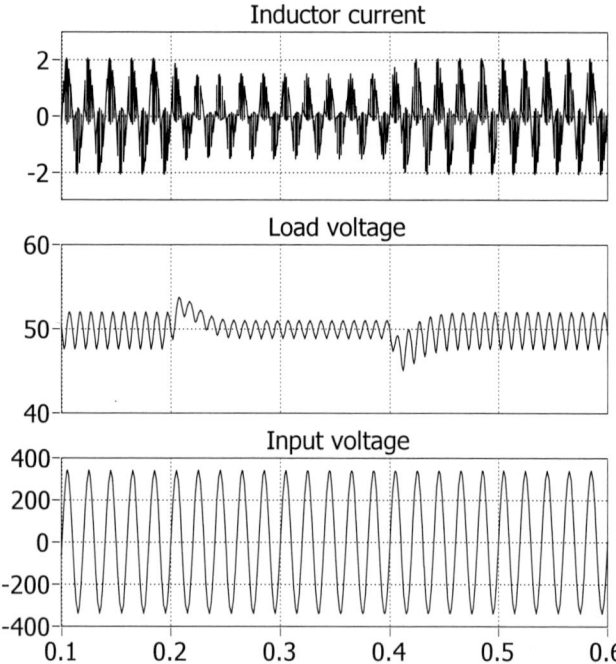

Fig. 10. Inductor current and output voltage waveforms with load disturbance of 100W-50W-100W at 0.2s and 0.4s for 240Vrms input voltage.

978-1-4577-0007-1/11 $26.00 © 2011 IEEE

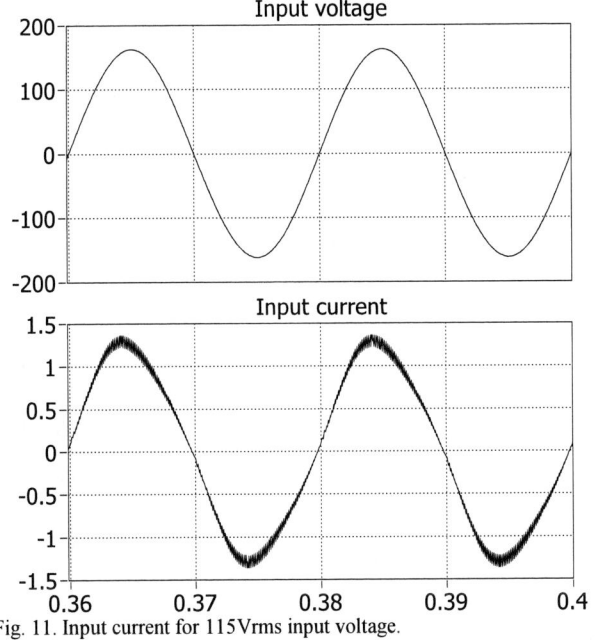

Fig. 11. Input current for 115Vrms input voltage.

IV.   CONCLUSIONS

The small-signal and large-signal model for the proposed bridgeless Cuk PFC converter has been designed and simulated using Matlab/Simulink/PLECS simulator. The small-signal model has been derived using CIECA method and verified using PLECS AC model. Both models show reasonable similarity for wide range of operating frequency. The closed-loop PI controller successfully regulates the output voltage with step-load change up to half the nominal load value.

ACKNOWLEDGMENT

The authors would like to thank to MOHE (Ministry of Higher Education), Malaysia for the financial support.

REFERENCES

[1] Hancock, J. M, "Bridgeless PFC boost low-line efficiency," Power Electronics Technology magazine, February 2008.

[2] D.M. Mitchell, "AC-DC Converter having an improved power factor",U.S. Patent 4,412,277, Oct. 25, 1983.

[3] Woo-Young Choi; Jung-Min Kwon; Eung-Ho Kim; Jong-Jae Lee; Bong-Hwan Kwon; , "Bridgeless Boost Rectifier With Low Conduction Losses and Reduced Diode Reverse-Recovery Problems," Industrial Electronics, IEEE Transactions on , vol.54, no.2, pp.769-780, April 2007

[4] Yungtaek Jang; Jovanovic, M.M.; , "Bridgeless buck PFC rectifier," Applied Power Electronics Conference and Exposition (APEC), 2010 Twenty-Fifth Annual IEEE , vol., no., pp.23-29, 21-25 Feb. 2010.

[5] Wang Wei; Liu Hongpeng; Jiang Shigong; Xu Dianguo; , "A novel bridgeless buck-boost PFC converter," Power Electronics Specialists Conference, 2008. PESC 2008. IEEE , vol., no., pp.1304-1308, 15-19 June 2008.

[6] Ismail, E.H.; , "Bridgeless SEPIC Rectifier With Unity Power Factor and Reduced Conduction Losses," Industrial Electronics, IEEE Transactions on , vol.56, no.4, pp.1147-1157, April 2009.

[7] Brkovic, M.; Cuk, S.; , "Input current shaper using Cuk converter," Telecommunications Energy Conference, 1992. INTELEC '92., 14th International , vol., no., pp.532-539, 4-8 Oct 1992

[8] Sabzali, A.J.; Ismail, E.H.; Al-Saffar, M.A.; Fardoun, A.A.; , "A new bridgeless PFC Sepic and Cuk rectifiers with low conduction and switching losses," Power Electronics and Drive Systems, 2009. PEDS 2009. International Conference on , vol., no., pp.550-556, 2-5 Nov. 2009.

[9] Lin, J.-L., Yang, S.-P. and Lin, P.-W., Small-signal analysis and controller design for an isolated zeta converter with high power factor correction. Electric Power Systems Research, Sept. 2005. 76(1-3): p. 67-76.

[10] Chetty, P.R.K.; "Current Injected Equivalent Circuit Approach to Modeling Switching DC-DC Converters", IEEE Transactions on Aerospace and Electronic Systems, vol. AES-17, no. 6, pp.802 – 808, Nov. 1981

[11] Chetty, P. R. K.; "Current Injected Equivalent Circuit Approach to Modeling of Switching DC-DC Converters in Discontinuous Inductor Conduction Mode", IEEE Transactions on Industrial Electronics, vol. IE-29, np. 3, pp. 230 – 234, Aug. 1982

Development of a Doctor Following Mobile Robot With Mono-vision Based Marker Detection

Mohd Nazri Abu Bakar
School of Mechatronic Engineering
Universiti Malaysia Perlis
Perlis, Malaysia
Email: nazribakar@unimap.edu.my

R. Nagarajan
School of Mechatronic Engineering
Universiti Malaysia Perlis
Perlis, Malaysia
Email: nagarajan@unimap.edu.my

Abd Rahman Mohd Saad
Engineering Centre
Universiti Malaysia Perlis
Perlis, Malaysia
Email: abd.rahman@unimap.edu.my

Abstract—This work presents a simple marker-based person detection method for a mobile robot that can follow a person in indoor environments. Expensive laser range finders or RFID which can provide very accurate range measurements have been used by several researchers for person detection. Recently, vision based approaches have been popular for person detection within a group of multiple people using stereo cameras. In this paper of proposed implementation, an inexpensive single camera has been used to acquire video frames to detect a specific target person and determine its position. A new detection method using color and shape based marker technique has been advanced in this work. The experimental results show that the proposed algorithm can detect a target person under various conditions such as marker features and lighting conditions.

Index Terms—Following robot; Mono-vision; Object tracking; Marker-based detection.

I. INTRODUCTION

The ability to follow a specific person is one of the most important problems for service robots. In recent years, a number of research efforts on person detection and tracking have been documented using vision based methods. Vision based approach is economical but very challenging for robust real time human detection and tracking due to loss of information from captured image caused by non-rigid or complex object motion, uncontrolled lighting conditions and object occlusions [1]. However, many of them use only a fixed camera and some of them propose the method for detection of multiple people but not a specific person [2]. Stereo cameras are widely used for person detection and tracking by mobile robots [1], [3], [4], [5]. The image information provided from cameras such as color and texture is very useful for person identification. Feature based detection of a person is the most popular method. Tsalatsanis et al. [6] applied region growing algorithm to identify the target person by using color information from wearing cloth. Fritsch et al. [7] have used skin color for face detection based on the color segmentation of skin regions. In recent years, shape based technique has also been developed [4], [8] for person detection. However, these detection methods mentioned above are based on stereo vision system mounted on the robot and most of them only focused for multi people tracking.

In this paper, we propose a novel system for specific person detection and tracking using single camera for mobile robot which can be used in hospital environment as doctor assistant. Two red circular patches on doctor's dress are used as vision marker for target person identification. This system performs of background subtraction, target person recognition, position of target person range finding and tracking. This paper is organized as follows: Section II describes briefly the specifications and system configuration of robot. Section III presents a method to detect and track a specific person using shape and color based marker technique. Section IV introduces an approach of motion control to keep the robot follows the target. In section V, we show some of the experimental results in various situations and environmental conditions. Finally, conclusions are derived in section VI.

II. SPECIFIC TARGET PERSON FOLLOWING ROBOT

A. Specifications

The specific target person following robot is developed which having a specially designed platform so that it can go into the limited space between hospital beds. The platform of robot is approximately 1m in height, 0.4m in width, and 25kg in weight. The robot is mounted with two independently driven wheels so that it is able to move at a speed of up to 1.2 m/s and two additional wheels on the rear are used for keeping the robot stability. A single Webcam camera (Logitech C600) is mounted on the top of robot at a height of 1m from ground to capture the image for the target detection and tracking system. The Ultrasonic Range Finder (Maxbotix LV-EZ1) is used for obstacle avoidance system. The robot is powered by Lithium-ion polymer (LiPo) batteries and its operation time is about one hour with full batteries.

B. System Configuration

The development of configuration system in our robot is mainly divided into two function parts: Vision Part for detection and tracking the marker and Motion Part to control the mobile robot follows the Doctor smoothly. This configuration is shown in Figure 1. A single webcam camera captured images and send to Vision Part through an USB port. Various algorithms for image processing of the monocular video are indicated in Vision Part. In searching and localization of the marker, our system has two Modules: Detection Module and

978-1-4577-0007-1/11 $26.00 © 2011 IEEE

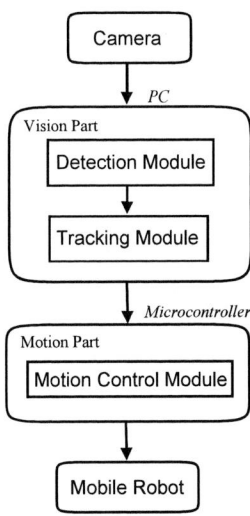

Fig. 1. Overview of the System Configuration

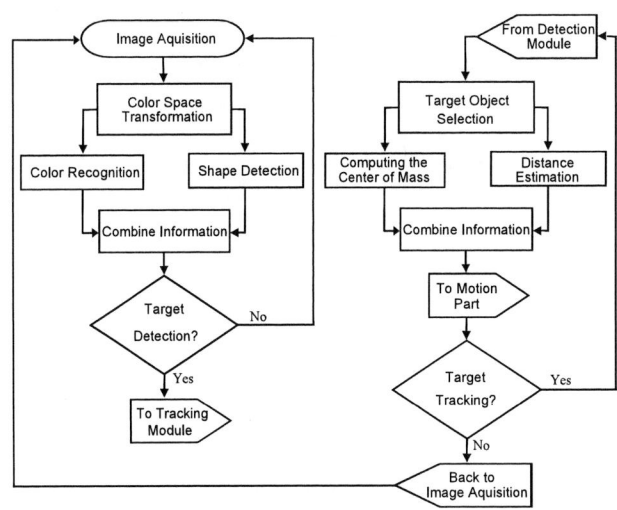

Fig. 2. Algorithm of Marker Detection Module and Tracking Module.

Tracking Module. In Marker Detection Module as shown in Figure 2, the characteristics of marker including color and shape information are needed. A hybridization of method combining color features and edge information of marker is proposed to distinguish the marker from other noise objects. The background subtraction technique is performed to detect the foreground object. Once the marker is detected, the system enters to Tracking Module. In the Tracking Module as shown in Figure 2, the position within the image is determined and the range of the marker from camera is computed. The information from the Tracking Module (distance and direction) is sent to Motion Part through an USB port. At the Motion Part, two independent drive motors are controlled to move the mobile robot appropriately towards the target. A notebook PC (X2 Dual-Core, 2.0GHz) performs the processes including Marker Detection Module and Tracking Module. The Motion Control Module employs a 89V51RD2 microcontroller from Phillips as processor.

III. HYBRIDIZATION METHOD FOR SPECIFIC PERSON DETECTION AND TRACKING

The most popular approaches within computer vision for target detection are usually to select the most suitable feature parameters of a specified part of the input image. This single approach may not be easier to detect the target (marker) because of uncontrolled lighting conditions, due to the same color of the target and that of other objects, temporary occlusions and the difficulties imposed by target tracking. To overcome these situations, we propose a novel detection and tracking approach based on a hybridization of method. This hybridization method consists of a combination of color-based and shape-based detections in order to the track marker which may have a non-uniform color and texture distribution due to variations in lighting. As shown in Figure 3, the marker consists two circular red color stickers diameter 28mm each

and separated by a distance of 40mm between centers. The marker is then printed on the dress/coat of the target person.

A. Color-based Marker Detection

In the first step, the captured 640x480 pixel color input image is filtered with a 5x5 Gaussian filter to reduce speckles of high frequency noise. Then the input image is converted from RGB to HSV color space by the OpenCV [9]. The HSV color space is chosen for its ability to accommodate the variable lighting conditions. In the second step, the image segmentation is proposed to extract the regions of interest in order to remove all colors except the color of marker (red). There are many kinds of techniques in literature for image segmentation [10]. But the most popular technique is by using Color Threshold module [11]. The Color Threshold module is widely used to remove parts of image that fall within a specified color range for foreground detection. The Color Threshold module will check every element of the input image to allow only the pixels above a certain range of value to be realized based on the selected value of threshold. In this image segmentation, only the hue and saturation channel are used. Two range threshold values, H_{low} and H_{high}, for hue channel based segmentation and S_{low} and S_{high} for saturation channel based segmentation are defined. The hue channel threshold is used to segment color of the image, where H_{low} will be close to 180 and H_{high} will be close to 0 for red color detection. The saturation channel threshold will force the value of saturation to have more red color and reduce white light for a perfect red color detection. As a result, the threshold module will generate a monochrome image (mask) where the foreground color is white and the background color is black. Monochrome image means that each pixel is stored as a single bit (0 or 1) and it is important for limiting the computational speed of image processing. The characteristic function of detected foreground in monochrome image is defined as:

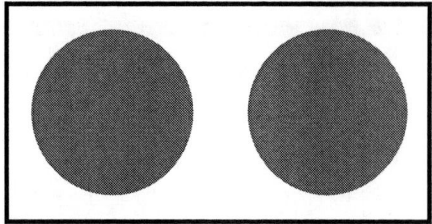

Fig. 3. The Marker.

$$I_M(i,j) \begin{cases} = 1 \text{ for Foreground} \\ = 0 \text{ for Background} \end{cases} \qquad (1)$$

where $I_M(i,j)$ is the monochrome image with pixel position (i,j).

B. Shape-based Marker Detection

Marker-based tracking is not reliable if only the color features method is used. In dynamic environment, especially, there are same color objects which can cause tracking of target to fail. In other words, further identification of the target marker is necessary. Shape-based detection is often used as a further feature for object tracking. Various edge detection methods have been applied for detecting the shape in images. One of the most common image processing used is Canny edge detector. The Canny edge detection algorithm is basically to detect edges in a very robust manner of the grayscale image. The circular patterns within an image can be finding after the process of shapes recognition. In this stage, the Circular Hough Transform (CHT) [12] is utilized to find two red circles of marker. The CHT can be described as a transformation of a set of feature points in the image space into a set of accumulated votes in the parameter space[13]. The equation of circle can be described as:

$$(x - x_0)^2 + (y - y_0)^2 = r^2 \qquad (2)$$

where (x_0, y_0) are the center of the circle in the x-axis and y-axis respectivliy and r is the radius of circle. To determine a circle, it is necessary to accumulate votes in the three-dimensional parameter space (x_0, y_0, r) that can be described with the parametric equations:

$$x = x_0 + r\cos\theta \qquad (3)$$

$$y = y_0 + r\sin\theta \qquad (4)$$

Once the circles are found, the marker recognition is implemented. Sometimes, more false circles may be detected. To identify the correct circles, several aspects should be considered. An algorithm for the marker identification with three criteria is proposed. These criteria consist of radius of circle r_i, distance between two centers of circles as Dx_M in

x-axis and distance between two centers of circles as Dy_M in y-axis. Their values are bounded as given by the inequalities:

$$r_{min} \leq r_M \leq r_{max} \qquad (5)$$

$$Dx_{min} \leq Dx_M \leq Dx_{max} \qquad (6)$$

$$|Dy_M| \leq Dy_{max} \qquad (7)$$

where
r_{min} : Minimum value of radius;
r_{max} : Maximum value of radius;
r_M : Radius of circle;
Dx_{min}: Minimum distance between two centers of circles in x-axis;
Dx_{max}: Maximum distance between two centers of circles in x-axis;
Dx_M : Distance between two centers of circles in x-axis;
Dy_M : Distance between two centers of circles in y-axis;
Dy_{max}: Maximum absolute distance between two centers of circles;
It is visualized that r_M and Dx_M are positive; Dy_M can be of positive or negative values.

C. Marker-Based Tracking Control

The most popular tracking methods with mobile robot are mainly relying on the distance from robot to the targeted object and its direction. These methods [3], [14] measure the distance from the target object to robot by using laser which only able to track the target continuously when there are no obstacles in the tracking paths. Hyukseong Kwon et al. [5] calculates the distance from the robot to target object by using two uncalibrated independently moving cameras for tracking the target. Stereo is also popular for tracking such as [6], [8], [15] but this method has very complexity computation. In our system, we propose a simple method by using only a single camera for tracking the marker. This method mainly depends on the changing distance in pixels between two centers of circles in marker. The distance between two centers can be measured by using Euclidean Distance method, as defined by:

$$Dx_M = x_2 - x_1 < 0 \qquad (8)$$

$$Dy_M = |y_2 - y_1| \qquad (9)$$

where circle 1 is at (x_1, y_1) and circle 2 is at (x_2, y_2). Circle 2 is assumed to be far away from the origin compared to circle 1. The circles within the marker are placed in x-axis only. Then $y_1 = y_2$, $Dy_M = 0$.

D. Distance and Position Estimation

The direction of the movement robot depends on the positional information of the moving marker as an input from the Detection Module. This position is determined by calculating the center of mass (X_M, Y_M) of the marker as:

$$X_M = x_1 + \frac{Dx_M}{2} \qquad (10)$$

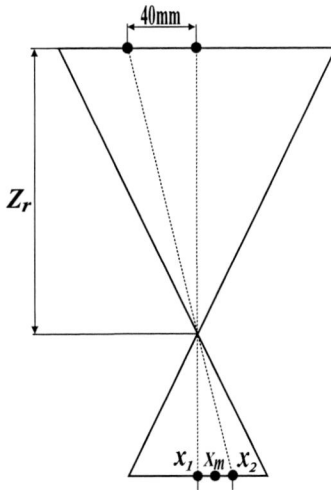

Fig. 4. The Mono-vision view of the robot .

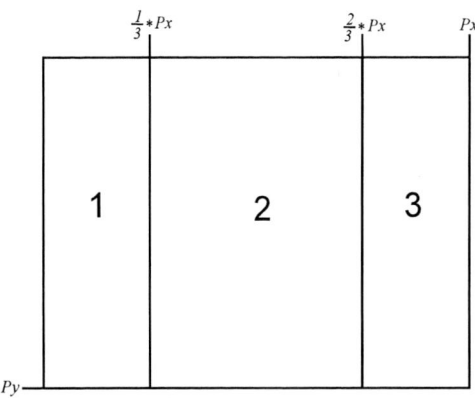

Fig. 5. Image segmentation for mobile robot motion control. Px = 640 and Py = 480.

$$Y_M = y_1 + \frac{Y x_M}{2} \qquad (11)$$

The distance between two centers of circle is necessary to calculate the distance of marker from robot. The proportional relationship method is proposed to calculate the distance of marker from mobile robot (also referred as the range) as follows:

$$Z_r = \frac{C}{X_M} \qquad (12)$$

where Z_r is distance from robot to marker and C is measured in the linear portion of "Range Vs Distance between Circles" plot. Figure 4 shows the view of Mono-vision in horizontal (X-Axis), where x_1 and x_2 are the pixel coordinates of circles, X_M is the pixel coordinate center of mass of marker onto the image plane, Dx_M is the pixel distance between two circles in x-axis and Z_r is the distance from robot to marker in mm.

IV. MOTION CONTROL FOR THE PERSON FOLLOWING ROBOT

A. Motion Control Module

Once the position on the image plane is estimated and the distance of marker from robot is calculated, then the system enters to the Motion Control Module in Motion Part. The position of marker (center of mass) is to be at the center of the segmented image 2 of Figure 5. If the position of marker is not at the center of segmented image 2, then the system will set the position which makes the robot turns in its vertical axis appropriately to bring the center of mass at the center of segmented image 2. Then the range is determined and the robot is made to move toward the target person. The Figure 6 shows the algorithm of Motion Control Module.

V. EXPERIMENTS

The marker-based person tracking algorithm is tested on a mobile robot where a Webcam camera (Logitech C600)

is mounted on the top of robot at a height of 86 cm from ground. Two different experiments have been conducted in an indoor lab environment. The first experiment demonstrates the effort of the detection algorithm that distinguishes the marker from the background, even if other objects with similar colors appear in the image scene. This experiment also shows the ability of detection algorithm to keep track of the marker in various lighting conditions. The second experiment demonstrates the accuracy of distance calculated by this approach. The first experiment is conducted with varying light intensities as observed in lab environment. In addition, no other disturbing factors are included in the environment. The experiment is repeated several times to check the appearance of edge detecting circles at constant lighting conditions. Such an effort indicates the detection of marker to reach 97% accuracy. One of the images and its processed output are given in Figure 7. The same experiment is repeated with various lighting conditions , in a cloud environment and with disturbing factors of object of same color of the marker. The percentage of detection of marker comes to 92% which is still considered high and adequate to be used in the situation of robot following the doctor. Figure 8 to figure 10 illustrate the various situation of disturbing factors included in the experiment.

The second experiment is performed to validate the effectiveness of the algorithm to calculate the distance between the robot and the target. The distance estimation is very importance for robot movement towards the target. The tolerance of accuracy for distance estimation is fixed within the bound as 3%. That is, when outside the bound, it is not counted. Each measurement is repeated 100 times to ensure the accuracy of range finding. The brightness is varied within range from nominal value. Table 1 describes the result of accuracy of distance test. From Table 1, the test results showed that our range measurement is adequate to estimate the distance of target from robot for human detection and tracking.

VI. CONCLUSION

This paper has described a method of detection and tracking a specific person based on colored marker by using a

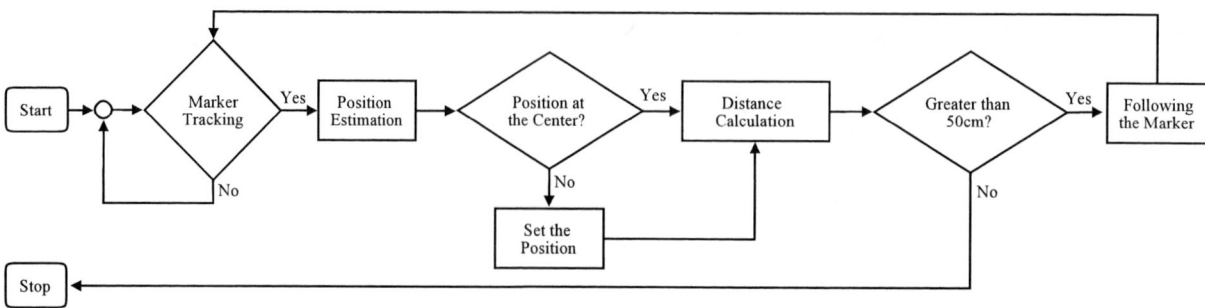

Fig. 6.   The algorithm of Motion Control Module.

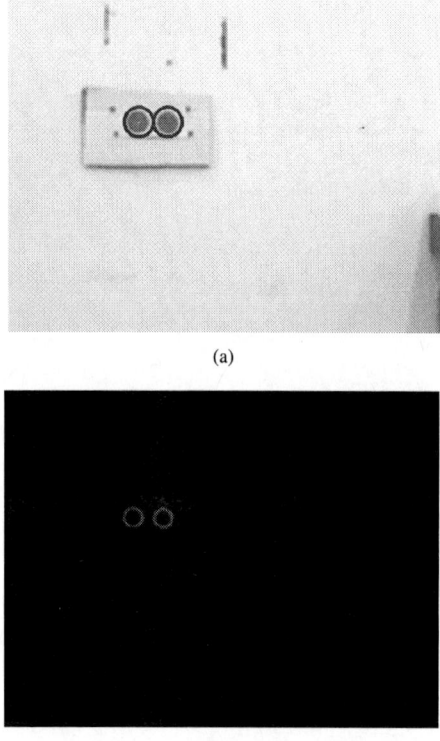

(a)

(b)

Fig. 7. The detection of marker without disturbing factors; (a) original image in RGB; (b) Edge detected monochrome image after color transformation.

TABLE I
THE ACCCURACY OF DISTANCE MEASUREMENT.

Distance (cm)	Accuracy (%)
60	88.6
70	91.1
80	92.4
90	94.9
100	96.2
110	97.5
120	98.7
150	96.2
170	94.9
200	84.8

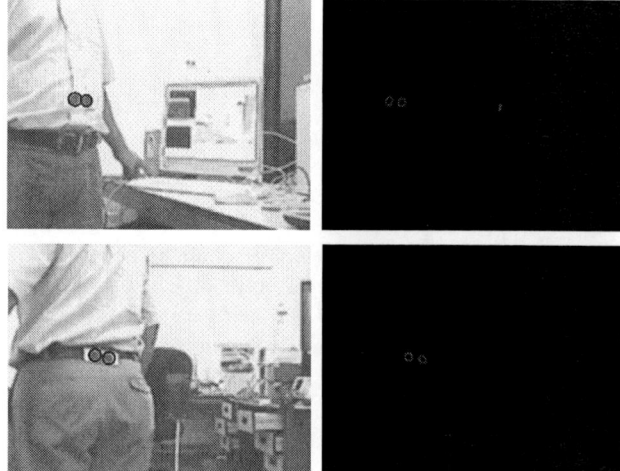

Fig. 8.   Various poses of marker positions.

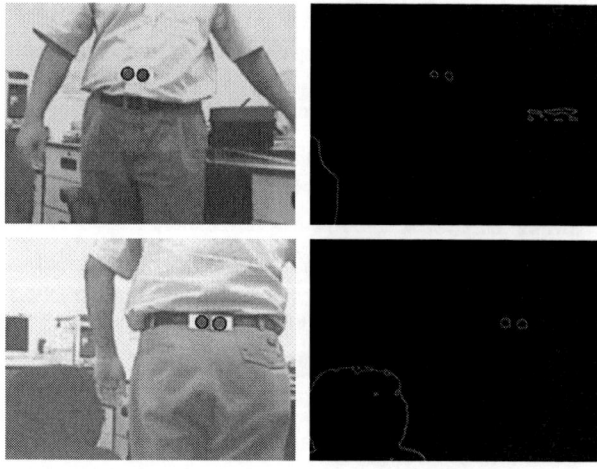

Fig. 9.   The result of marker detection when disturbance due to similar colors.

single camera. The proposed approach uses a hybridization of methods combining color features and edge information to distinguish a marker from other noise object. The test results present the effectiveness and the robustness of the proposed

978-1-4577-0007-1/11 $26.00 © 2011 IEEE

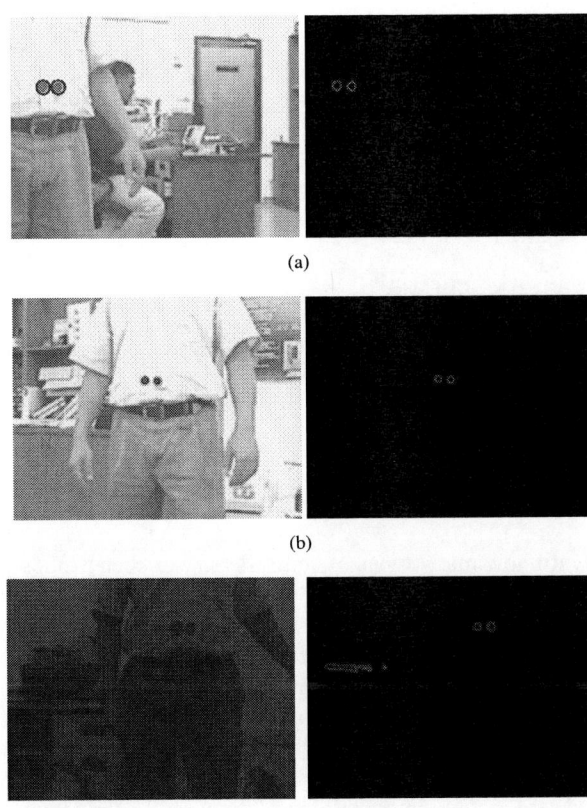

(a)

(b)

(c)

Fig. 10. Detection of marker in various lighting condition.

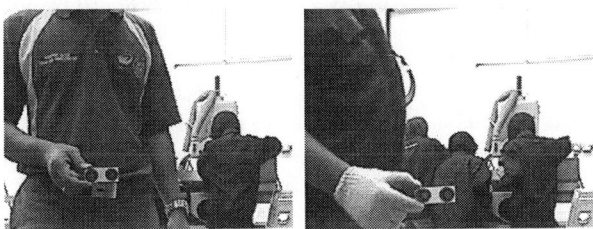

Fig. 11. Detection of marker in a cloud environment.

method. However, it is found that if robot is too close or too far away from the target, the distance computation will not be reliable. The use of optical zoom and obstacle avoidance are not considered in this paper. Our work progresses in these directions.

ACKNOWLEDGMENT

The authors are grateful to Universiti Malaysia Perlis for their supports throughout this research.

REFERENCES

[1] C. Hu, Xudong. Ma and X. Dai, *"A Robust Person Tracking and Approach for Mobile Robot"*, Proceedings of the IEEE International Conference on Mechatronics and Automation, pp. 3571 - 3576, 2007.

[2] J. Zhou and J. Hoang, *"Real time robust human detection and tracking system"*,IEEE Computer Society Conference on Computer Vision and Pattern Recognition, 2005.

[3] M. Kobilarov, G. Sukhatme, J. Hyams, and P. Batavia, *"People Tracking and Following with Mobile Robot Using Omnidirectional Camera and a Laser"*,Proceedings of 2006 IEEE Int. Conf. on Robotics and Automation, pp. 557-562, 2006.

[4] S. M. Yoon and H. Kim, *"Real-time multiple people detection using skin color, motion and appearance information"*,International Workshop on Robot and Human Interactive Communication, pp. 331-334, 2004.

[5] H. Kwon, et al. *"Person tracking with a mobile robot using two uncalibrated independently moving cameras"*,IEEE International Conference on Robotics and Automation, pp. 2877-2883, 2005.

[6] A. Tsalatsanis, K. Valavanis and A. Yalcin, *"Vision Based Target Tracking and Collision Avoidance for Mobile Robots"*,J Intell Robot System, Vol.48, pp. 285-304, 2007.

[7] J. Fritsch, et al. *"Multimodal anchoring for human robot interaction"*,Robotics and Autonomous Systems, Vol. 43, pp.133-147, 2003.

[8] J. Satake and J. Miura, *"Robust Stereo-Based Person Detection and Tracking for a Person Following Robot"*,Proceedings of the IEEE ICRA 2009 Workshop on People Detection and TrackingKobe, Japan, May 2009.

[9] http://www.intel.com.OpenCV: Intel Open Source Computer Vision Library, retrieved on Jan.2010.

[10] L. Luo and X. Li, *"A method to search for color segmentation threshold in traffic sign detection"*,International Conference on Image and Graphics, pp. 774 - 777, 2009.

[11] D. Zhang, et al., *"A New Color-Based Segmentation Method for Forest Fire from Video Image"*, International Seminar on Future BioMedical Information Engineering, pp. 41 - 44, 2008.

[12] R.O. Duda and P.E Hart, *"Use of the Hough transformation to detect lines and curves in picture"*,Commun. ACM, pp: 11-15, 1972

[13] M. Rizon, et al., *"A comparison of circular object detection using Hough transforms and chord intersection"*, Geometric Modeling and Imaging, pp. 115-120, 2007.

[14] N. Bellotto and H. Hu, *"Multimodal People Tracking and Identification for Service Robots"*, Journal of Information Acquisition, Vol. 5, No. 3, pp. 209-221, 2008.

[15] T. Yoshimi, et al., *"Development of a Person Following Robot with Vision Based Target Detection"*, Proceedings of the 2006 IEEE/RSJ International Conference on Intelligent Robots and Systems, pp. 5286 - 5291, 2006.

Performance Improvement of Improved Practical Control Method for Two-Mass PTP Positioning Systems in the Presence of Actuator Saturation

Mohd Fitri Mohd Yakub[1], Rini Akmeliawati[2]
[1] Malaysia-Japan International Institute of Technology (MJIIT)
Universiti Teknologi Malaysia International Campus (UTM IC), Kuala Lumpur, Malaysia
irtif_81@yahoo.com
[2]Department of Mechatronic Engineering
International Islamic University Malaysia (IIUM), Kuala Lumpur, Malaysia
rakmelia@iiu.edu.my

Abstract- **The positioning systems generally need a good controller to achieve a fast response, high accuracy and robustness. In addition, ease and simplicity of controller design structure are very important for practical applications. For satisfying these requirements, NCTF controller has been proposed as a practical control method for two-mass rotary PTP positioning systems. However, the effect of the actuator saturation cannot be completely compensated due to integrator windup when the object parameter varies. This paper presents a method to further improve NCTF controller to overcome the problem of integrator windup. The improved NCTF controller is evaluated experimentally using rotary two-mass positioning system. The effect of the design parameters on the robustness of the improved NCTF with anti-windup integrator controller is evaluated and compared with NCTF without anti-windup integrator and the equivalent PID controller. The results show that the improved NCTF controller is effective to compensate the effect of integrator windup.**

Keywords— **Point-to-point, improved NCTF, two-mass system, integrator windup, Equivalent PID**

I. INTRODUCTION

Positioning systems play an important role in industrial engineering applications such as advanced manufacturing systems, semiconductor manufacturing systems and robot systems. Many effort to improve mechanisnm features of the systemsfor high positioning performance have prove to be costly. Basically, point-to-point (PTP) positioning systems are required to have fast response speed, high accuracy and robustness. However nonlineaarities like friction and saturation which exist in the positioning system may cause slow motion, steady state error and limit cycles.

Friction and saturation are nonlinear phenomena for which the linear control theory like PID controllers is not suitable to handle effectively the frictional system. Friction is characterized by uncertainties due to variations of the lubrication condition or inertia while saturation effect comes from the source of the actuator or electronics power amplifier [1]. Therefore, the positioning systems are also required to have robustness to parameter variation.

Until now many types of controllers have been proposed and evaluated for positioning systems, for example the model following type controller such as controllers with disturbance observer [2], time optimal controllers [3] and sliding mode controllers [4]. These controllers will give good positioning performance in case of the designer have an exact model and value of its parameters. In general, advanced controllers like fuzzy logic controllers [5] and artificial neural network [6] tend to be complicated and required deep knowledge concerning with controller theory and design.

However in practical application, for engineers who are not expert in control theory design will face troublesome and time comsuming to determine exact model and parameter identification for the systems.

In two-mass system applications such as the rolling mill drive, the mechanical part of the drive has a very low natural resonant frequency because of the large roll inertia and the long shaft including the gear box and the spindle. Due to this, the finite but small elasticity of the shaft gets magnified and has a vibrational effect on the load position which may reduce positioning accuracy [7]. Therefore, the existing NCTF controller that has beed proposed in 2002 for one-mass rotary system can not be used directly in the case there is a flexible connection between elements of the positioning systems [8].

In order to overcome this problem, the improvement of Nominal Characteristic Trajectory Following (NCTF) controller has been proposed by authors as a practical controller for two-mass rotary positioning systems in [9]. It has been shown that, the improved NCTF control system has a good positioning performance and robustness. The improved NCTF controller is also effective to compensate the effect of the friction. However, the effect of actuator saturation cannot be completely compensated due to integrator windup when the object parameter varies [10].

This paper describes a method to improve PI compensator of NCTF controller for overcoming the degradation of the positioning performance due to integrator windup. Firstly, NCTF control concept and its controller design procedure are introduced. Then, an improved PI compensator for overcoming the integrator windup is described. Finally, the

978-1-4577-0007-1/11 $26.00 © 2011 IEEE

effect of saturation on positioning performance is evaluated and compared with NCTF without anti-windup integrator and equivalent PID controllers in term of positioning performance and its robustness.

II. NCTF CONTROL CONCEPT

The structure of the NCTF control system is shown in Fig. 1. The NCTF controller consists of a Nominal Characteristic Trajectory (NCT) and a compensator.

The NCTF controller works under the following two assumptions [11]:
a) A DC or an AC servo motor is used as an actuator of the object.
b) PTP positioning systems are discussed, so θ_r is constant and θ_r' =0.

An exact modeling including friction and conscious identification processes is not required in the NCTF controller design. Since the DC motor is used as the actuator, the simplified object can be presented as the following fourth order system:

$$G_o(s) = \frac{\theta_l(s)}{U(s)} = K \frac{\alpha_2}{s(s+\alpha_2)} \frac{\omega_f^2}{s^2 + 2\zeta_f \omega_f s + \omega_f^2} \qquad (1)$$

where $\theta_l(s)$ represents the displacement of the object, $U(s)$, the input to the actuator and K, ζ, α_2 and ω_f are simplified object parameters. The NCT is determined based on the averaged object response which does not include the vibration.

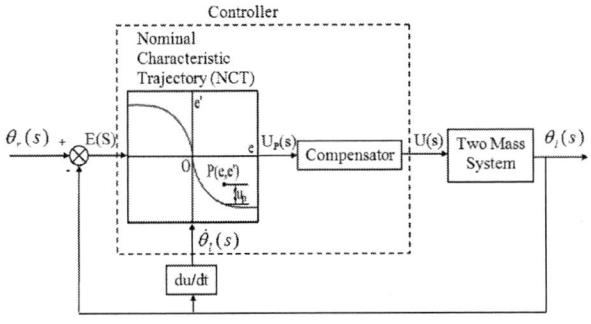

Figure 1: Structure of NCTF control system for PTP positioning

The objective of the NCTF controller is to make the object motion follow the NCT and end at the origin of the phase plane (e, e'). Signal u_p shown in Fig. 1 represents the difference between the actual error rate e' and that of the NCT. The value of u_p is zero if the object motion perfectly follows the NCT. The compensator is used to control the object so that the value of u_p, which is used as an input to the compensator, is zero.

The NCTF controller is designed based on a simple open-loop experiment of the object as follows [11]:
1) Open-loop-drive the object with a stepwise input and measure the displacement and velocity responses of the object.

In order to construct the NCT, a simple open-loop experiment has to be conducted. In the experiment, an actuator of the object is driven with a stepwise input and, displacement and velocity responses of the object are measured. Fig. 2 shows the stepwise input, load velocity and load displacement responses of the object. In this case, the object vibrates due to its mechanical resonance. In order to eliminate the influence of the vibration on the NCT, the object response must be averaged.

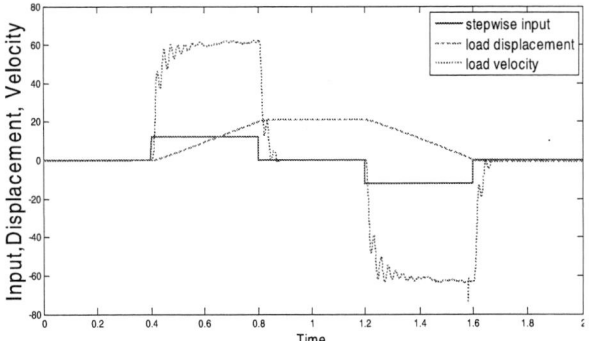

Figure 2: Input and actual object response

2) Construct the NCT by using the object responses. Since the construction is based on the actual responses of the object, the NCT includes effects of nonlinear characteristics such as friction and saturation.

In Fig. 3, moving average filter is used to get the averaged response because of its simplicity [12]. The moving average filter operates by averaging a number of points from the object response to produces each point in the averaged response. The averaged velocity and displacement responses are used to determine the NCT. Since the main problem of the PTP motion control is to stop an object at a certain position, a deceleration process (curve in area A of Fig. 3) is used. Variable h in Fig. 3 is the maximum velocity, which depends on the input step height. From the curve in area A and h in Fig. 3(a), the NCT in Fig. 3(b) is determined. Since the NCT is constructed based on the actual open-loop responses of the object, the NCT includes nonlinearity effects such as friction and saturation.

There are two important parameters in the NCT as shown in Fig. 3(b): the maximum error rate indicated by h, and the inclination of the NCT near the origin indicated by m. As discussed in the following section, these parameters are related to the dynamic parameters of the object. Therefore, the parameters are used to design the compensator.

From the relation between object dynamic in Eq. 1 and NCT in Fig. 3(b), the inclination near the origin, m and the maximum error rate, h relate with parameters of the dynamic object as follows [13]:

$$\begin{aligned} \alpha_2 &= -m \\ K &= -\frac{h}{u_r} \end{aligned} \qquad (2)$$

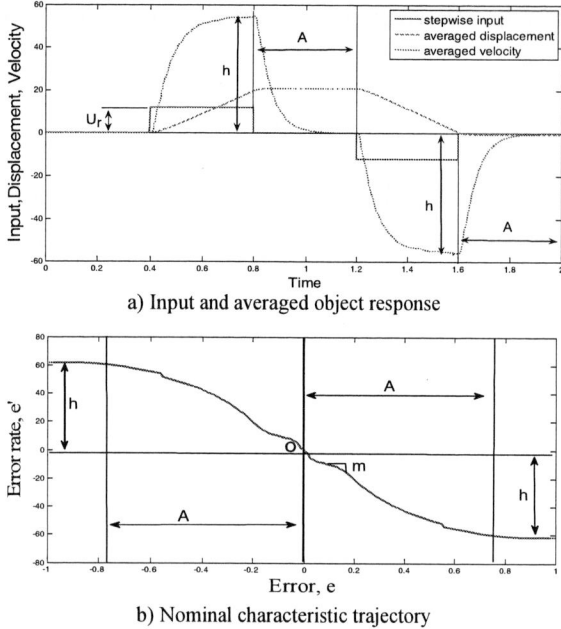

a) Input and averaged object response

b) Nominal characteristic trajectory

Figure 3: Construction of the NCT.

3)   Design the compensator by using the NCT information.

The following PI and notch filter compensator is proposed for two-mass rotary system:

$$G_c(s) = \frac{(K_p s + K_i)}{s} \left(\frac{K_{dc}(s^2 + 2\zeta_f \omega_f s + \omega_f^2)}{(s^2 + 2\zeta_o \omega_o s + \omega_o^2)} \right) \qquad (3)$$

The PI compensator is adopted for its simplicity to forces the object motion to reach the NCT as fast as possible and control the object motion to follow the NCT and end at the origin. The PI compensator parameters K_i and K_p are designed by using ω_n and ζ as the design parameters [13].

Fig. 4 shows the block diagram of the continuous close-loop NCTF control system with the simplified object near the NCT origin where the NCT is linear and has an inclination $\alpha_2=-m$. The signal u_p near the NCT origin in Fig. 4 can be expressed as the following equation:

$$u_p = \dot{e} + \alpha_2 e = \alpha_2 e - \dot{\theta}_l \qquad (4)$$

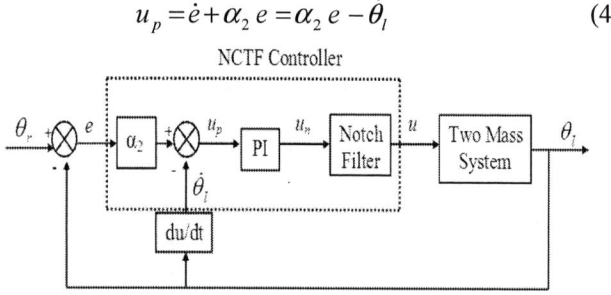

Figure 4: Simplified NCTF control system

A higher ω_n and a larger ζ are preferable in the compensator design. The selection of ω_n and ζ are chosen to have 40% of the values of ζ_{prac}, so that the margin safety of design is 60% [13]. The practical stability limit is determined by driving the mechanism with the NCT and the proportional element only.

The value of the proportional gain is increased until continuous oscillations are generated.

During the design parameter selection, the designer may be tempted to use large values of ω_n and ζ in order to improve the performance. However, excessively large values of ω_n will cause the controller to behave as a pure integral controller, which may lead to instability. Therefore, the choice of ω_n should start with small values and progress to larger values and not from the reverse.

In a two-mass system, the mechanical couplings between the motor, load, and sensor are not perfectly rigid, but instead they act like springs. Here, the motor response may overshoot or even oscillate at the resonance frequency resulting in longer settling time. The most effective way to deal with this torsional resonance is by using an anti resonance notch filter.

According to standard frequency analysis, resonance is characterized by a pair of poles in the complex frequency plane. The imaginary component indicates the resonant frequency, while the real component determines the damping level. The larger the magnitude of the real part, the greater the damping [14].

A notch filter consists of a pair of complex zeros and a pair of complex poles. The purpose of the complex zeros play by ω_f and ζ_f is to cancel the resonance poles of the system. The complex poles which are determined by ω_o and ζ_o, on the other hands, create an additional resonance and to increase a stability of gain margin for the plant. If the magnitude of the real value of the poles is large enough, it will result in a well damped response. The ratio between ζ_f and ζ_o will determine how deep the notch in order to eliminate the resonant frequency of the plant. Parameter K_{dc} will be affected in steady state condition when the transfer function of the notch filter becomes one [13].

Selection of ω_o and ζ_o must made two constraints to make sure the system in stable region:

$$\begin{aligned} \omega_o &> \omega_f \\ \zeta_o &> \zeta_f \end{aligned} \qquad (5)$$

Moreover, large ω_o and ζ_o is preferable but digital implementation of the NCTF controller limit the design parameter to maintain the close loop stability.

Due to the fact that the NCT and the compensator are constructed from a simple open-loop experiment of the object, the exact model including the friction characteristic and the conscious identification task of the object parameters are not required to design the NCTF controller. The controller adjustment is easy and the aims of its control parameters are simple and clear.

III. Compensator Improvement

As the NCTF controller uses PI compensator to force object motion so that it follows the NCT, the integrator windup may occur in connection with large position reference. As discuss in [10], in the case of no parameter variations, there is no significant integrator windup due to the effect of the

978-1-4577-0007-1/11 $26.00 © 2011 IEEE

saturation. The effect of the saturation is successfully compensated by using NCTF controller. However the integrator windup becomes a problem when the parameter varies.

To overcome the problem of integrator windup, the PI compensator is improved by adopting an anti-windup scheme. Hence, an anti-windup PI compensator is proposed to be used instead of pure PI compensator. Here tracking anti-windup, which is a simple and classic anti-windup method, is used [15]. The structure of anti-windup PI compensator is illustrated in Fig. 5, where K_T is called as tracking gain. Based on Fig. 5, once PI compensator output $U(s)$ exceeds the actuator limit, a feedback signal is generated from the difference of the saturated and the unsaturated signals. This signal is used to reduce the integrator input. Mathematically, the output of the anti-windup PI compensator is:

$$U(s) = \left[K_P + \frac{K_I}{s} \right] U_P(s) - \frac{K_T}{s} [U(s) - U_s(s)] \qquad (6)$$

where K_P and K_I are proportional and integral gain obtained from section II. The saturated range from $-u_m$ to $+u_m$ represent the linear range of the actuator. Furthermore, the saturation gain b relates with integral gain K_I and tracking gain K_T as follows:

$$b = \frac{K_T}{K_I} \qquad (7)$$

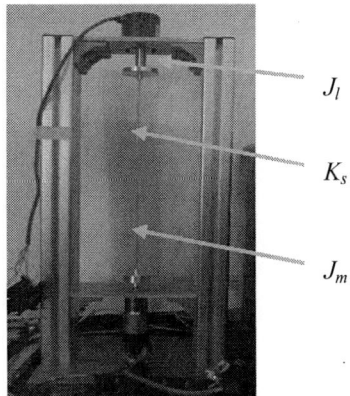

Figure 5: Standard structure of anti windup PI compensator

A rule of thumb for the setting of K_T is often to be $K_T = K_I$ which corresponds to $b=1$, but a higher values may give a further improvement in performances [15].

IV. RESULTS

A. Experimental Setup

To evaluate the effectiveness and robustness of the improved NCTF controller with anti-windup PI compensator to inertia and friction variations, the controllers are applied to an experiment two-mass rotary positioning system shown in Fig. 6. The positioning system consists of a direct current motor, a driver, motor and load mass and low stiffness shaft. The controller is implemented digitally with a 1 ms sampling period. An optical encoder with a 10000 pulse per revolution is used for measuring displacement. The angular load velocity used in the experiment is obtained by applying a backward difference algorithm to the measured angular load

displacement. The control signal is sent to the driver through a digital to analog converter.

Figure 6: The experimental two-mass rotary positioning system

The positioning performance is examined under two conditions namely Normal Object and Increased Inertia Object. Increased Inertia Object has about ten times load inertia than Normal Object.

B. Equivalent PID Controller Design

According to Fig. 3, the inclination, m and maximum error rate, h of the NCT are 81.169 and 61.6, respectively. When designing the PI compensator, design parameters for ζ and ω_n are chosen as 9.5 and 10.5 in order to evaluate the performance of NCTF controller [13].

The performance of the positioning system controlled by NCTF controller is compared with the PID controller. The PID controller is designed based on relation with NCTF at small object motion. The conventional PID controllers have the following transfer function:

$$u = K_P e + K_I \int e \, dt + K_D \dot{e} \qquad (8)$$

By considering Fig. 4, the output of the PI compensator, u of the NCTF controller is:

$$u = K_P u_P + K_I \int u_P \, dt \qquad (9)$$

Moreover based on Eq. (4) and (8), the equation can be rewritten as follows:

$$u = (K_P \alpha_2 + K_I) e + K_I \alpha_2 \int e \, dt + K_P \dot{e} \qquad (10)$$

Hence, by comparing Eq. (8) and (10), its shown at small error the NCTF controller is equivalent to the conventional PID controller has the following parameters:

$$\begin{aligned} K_P &= K_P \alpha_2 + K_I \\ K_I &= K_I \alpha_2 \\ K_D &= K_P \end{aligned} \qquad (11)$$

Table I shows the improved NCTF and equivalent PID controller parameters. In this paper, the PID controller tuned with Ziegler-Nichols and Tyres Luyben method is not discussed again since it gives bad performances in term of robustness to parameter variation [9, 13].

978-1-4577-0007-1/11 $26.00 © 2011 IEEE

TABLE I
CONTROLLER PARAMETERS

Controller	Kp	Ki	Kd
Improved NCTF	4.79e-1	2.65e-1	-
Equivalent PID	6.26e-1	1.64e-1	0.852

C. The effect of the tracking gain, K_T on positioning performances

The crucial problem for anti-windup PI compensator is the value of the tracking gain K_T. Although it is stated that a rule of thumb for the setting of K_T is often $K_T = K_I$ which corresponds to $b=1$, but a higher values may give a further improvement in performances [15]. Therefore, in order to find the appropriate value of K_T, an experimental has been done for different values of b which represent the ratio between tracking gain K_T and integral gain K_I.

Fig. 7 shows the effect of the tracking gain K_T on the positioning performances. It clearly shows that the overshoot and settling time are varies correspond with gain ratio. Therefore, the tracking gain should be decided based on the compromise between the overshoot and settling time. By considering a small enough overshoot and a shortest settling time, the gain ratio $b=4$ is decided for selecting the tracking gain K_T.

D. Comparison with Equivalent PID controller

In this section, the performance of the positioning system controlled by the improved NCTF (NCTF-2) controller with anti-windup PI compensator is compared with that of the improved NCTF (NCTF-1) (i.e. without anti-windup compensator), and equivalent PID (E-PID) controllers. The positioning performance is evaluated based on percentage of overshoot, settling time, and steady-state error of the two-mass rotary positioning system. Fig. 8 shows the step responses to a 0.5 deg step inputs when the controllers are used to control normal object. Their positioning performances are also summarized in Table II. Moreover, for the experimental results, all controllers give positioning accuracy near the sensor resolution, which is 0.036 deg.

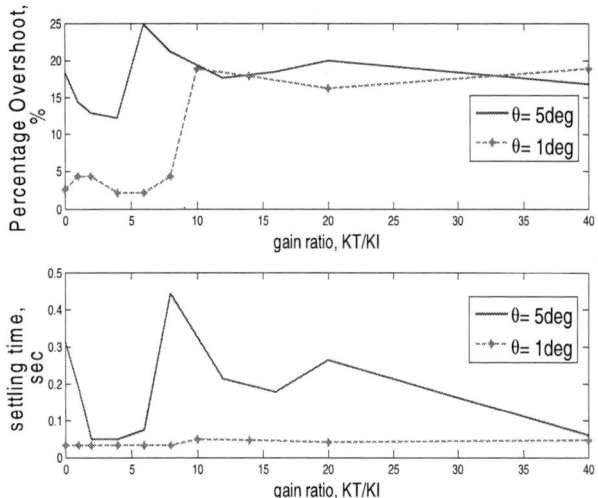

Figure 7: Effect of tracking gain on positioning performance

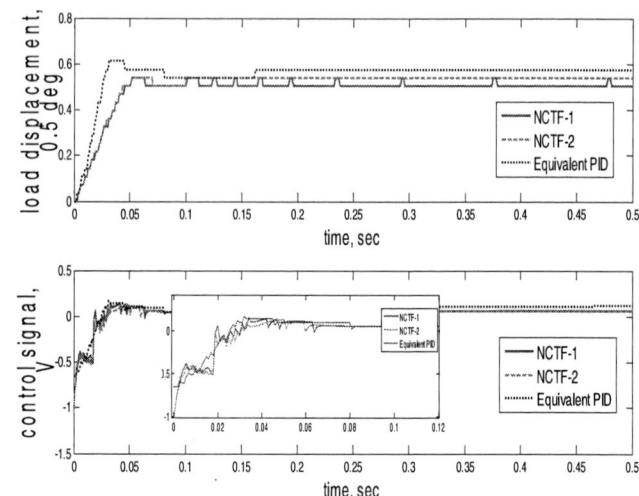

Figure 8: Comparison of response 0.5 deg step input, Normal object

Here, it is clear that all of the controllers produce similar response. Hence, in terms of overshoot and settling time, all of the controllers give similar performance.

In order to evaluate the robustness of the control systems to inertia variation, all of the controllers are implemented on object with inertia is ten time ($10 \times J_l$) of the nominal one. Fig. 9 shows the step responses to 0.5 step inputs when all of the controllers are implemented for controlling increased object inertia. Table II shows the positioning performance resulting from all of the controllers. Fig. 9 and Table II show that both NCTF controllers give a better robustness to inertia variations than the equivalent PID controller and there is no significant saturation of the actuator.

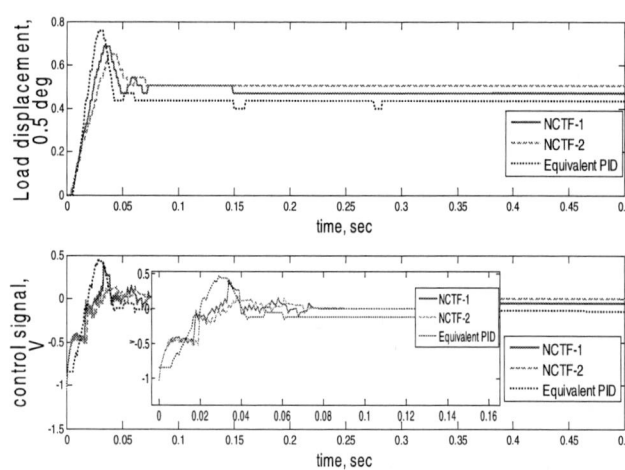

Figure 9: Comparison of response 0.5 deg step input, Increase inertia object
($10 \times J_l$)

Next, experiment was done for a larger step input so that the actuator reached saturation. Fig. 10 shows the step responses to a 10 deg step input when all of the controllers are implemented for controlling increased object inertia. Table II shows the positioning performance resulting from all of the controllers. The saturation of the actuator occurs as shown in Fig. 10. The saturation of the actuator causes an integrator windup when the positioning system is controlled by both NCTF-1 and E-PID controller. Fig. 10 shows that the positioning performance of the positioning system with NCTF-1 is unstable because of the integrator windup. Hence the NCTF-1 becomes less robust to inertia variation when the saturation occurs in comparison with that with NCTF-2 controller. On the other hand, NCTF-2 which uses anti-windup PI compensator can successfully compensate the effect of integrator windup due to actuator saturation. As the results show, the improved NCTF controller (NCTF-2) gives a smaller overshoot and a shorter settling time than the other controllers.

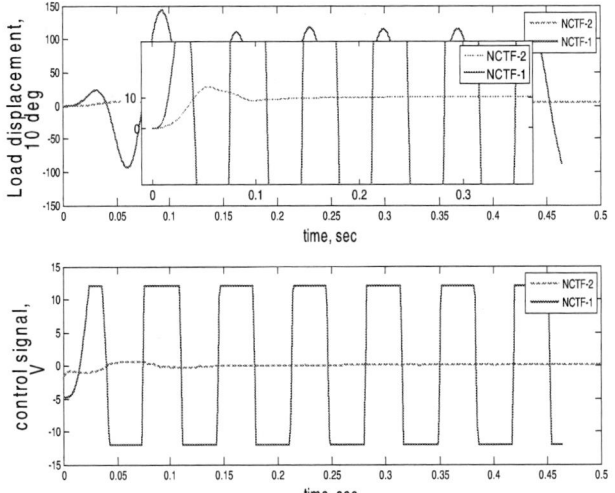

Figure 10: Comparison of 10 deg step response for increased inertia object, (10 x J$_l$)

TABLE II
EXPERIMENTAL POSITIONING PERFORMANCE COMPARISON

Controller			Overshoot (%)	Settling time (sec)	Ess (deg)
Normal object, J$_l$	0.5 deg	NCTF-1	0	0.083	0.036
		NCTF-2	0	0.080	0.036
		E-PID	0	0.050	0.036
	1 deg	NCTF-1	1.7	0.045	0.036
		NCTF-2	1.7	0.042	0.036
		E-PID	7.2	0.03	0.152
	5 deg	NCTF-1	6.92	0.043	0.036
		NCTF-2	5.86	0.042	0.036
		E-PID		unstable	
Increased inertia object, (10 x J$_l$)	0.5 deg	NCTF-1	20.6	0.065	0.036
		NCTF-2	18.6	0.06	0.03
		E-PID	40	0.07	0.072
	5 deg	NCTF-1	36.44	0.06	0.036
		NCTF-2	30.2	0.55	0.036
		E-PID		unstable	
	10 deg	NCTF-1		unstable	
		NCTF-2	40.6	0.75	0.036
		E-PID		unstable	

VI. SUMMARY

This paper has shows the improvement of the PI compensator for NCTF controller to overcome the effect of windup due to actuator saturation. An anti-windup PI compensator is used instead of a conventional PI compensator. Through an experimental using two-mass rotary positioning system, the effectiveness of the NCTF controller with anti-windup PI compensator is evaluated. It proved that the used of anti-windup PI compensator is effective to overcome the problem due to integrator windup. Moreover, the results also show that improved NCTF controller with anti-windup PI compensator is more robust to inertia variation than conventional PI compensator and Equivalent PID controller. Lastly, the application of the NCTF control concept for multi-mass positioning systems may be an interesting topic for further investigation.

ACKNOWLEDGMENT

This research is supported by Ministry of higher education IIUM fundamental research grant scheme (IFRG) no IFRG0702-60.

REFERENCES

[1] Amstrong-Helouvry B, Dupont P. and De Witt C, A Survey of Models, Analysis Tools and Compensation Method for the Control of Machines with Friction, *Automatica*, Vol. 30, pp. 1083-1138, 1994.

[2] Kempf C and Kobayashi S, Disturbance Observer and Feedforward Design for a High speed direct-drive Positioning Table, *IEEE Trans. On Control systems Technology*, Vol. 7, No. 5, 1999, pp. 513-526

[3] Park M.H and Won C.Y, Time Optimal Control for Induction Motor servo System, *IEEE Trans. On Power Electronics*, Vol. 6, No. 3, 1991, pp. 514-524

[4] Li Y.F, Erikson B. and Wilkander J, Sliding Mode Control of Two-mass Positioning systems, *Procedings of the 14th Triennaial World Congress IFAC*, Beijing, China, 1999, pp. 151-156

[5] Shieh M.Y and Li T.H, Design and Implementation of Integrated Fuzzy Logic Controller for a servo motor system, *Mechatronic* Vol. 8, 1998, pp. 217-240

[6] Horgh J.H, Neural Adaptive tracking control of a DC motor, *Information sciences*, Vol. 118, 1999, pp. 1-13

[7] How to work with mechanical resonance in motion control systems. *Control Engineering*, Vol. 47, No. 4, 2000, p. 5.

[8] Sato K, Wahyudi, and Shimokohbe A, Design and Characteristics of Practical Control system for PTP Positioning, *Trans of the Japan Society of Mechanical Engineers*, Vol. 67, No. 664, 2001, pp. 222-228

[9] Fitri M.Y, Wahyudi and R.Akmeliawati, Improved NCTF Control Method for a Two Mass Point to Point Positioning System, *Proceedings of the 2010 IEEE 3rd International Conference on Intelligent and Advanced systems* (icias 2010), Kuala Lumpur, Jun 2010.

[10] Wahyudi and Albagul A (2004). Performance improvement of practical control method for positioning system in the presence of actuator saturation, *Proceedings of 2004 IEEE International Conference on Control Applications*. Taipei, 2-4 September, pp. 296-302.

[11] Wahyudi, Sato K. and Shimokohbe A, Robustness Evaluation of New Practical Control Method for PTP Postioning Systems, *Proceeding of the 2001 IEEE/ASME International Conference on Advanced Intelligent Mechatronics*, pp 843-848, July 2001

[12] Oppenheim A.V. & Schafer R.W, Discrete Time Signal Processing. *Englewood Cliffs*, Prentice Hall, 1999.

[13] Fitri M.Y, Wahyudi and R.Akmeliawati, Performance Evaluation of Improved Practical Control Method of Two-Mass PTP Positioning System, *Proceedings of the 2010 IEEE Symposium on Industrial Electronics & Applications (ISIEA 2010)* Penang, October 2010, pp 550-555.

[14] William East & Brian Lantz, Notch Filter Design, August 29, 2005

[15] S. Crawshaw and G. Vinnicombe. Anti-windup synthesis for guaranteed L2 performance.*Proc. IEEE Conference on Decision and Control*, 2000.

Automated Test Set-up for Reverse Recovery Characterization of Ultrafast Diodes

Juergen Stahl, Daniel Kuebrich, Thomas Duerbaum

Chair of Electromagnetic Fields, Friedrich-Alexander-University of Erlangen Nuremberg

Cauerstr. 7, 91058 Erlangen, Germany

Abstract- **Often data sheets provide only poor information about the recovery behavior of ultra-fast diodes. On the other hand, existing diode models do not predict the real characteristic for all diodes. Nevertheless, due to its importance, this behavior needs to be known and therefore measured. For this purpose, a fully automated measurement set-up for determining the reverse recovery characteristic of ultra-fast diodes in an accurate manner was designed and is described here. All the data obtained is immediately transferred into MATLAB and therefore available for further calculation, model building and model validation. Since the whole set-up is automated, a complete field of variations in the reverse voltage, the forward current, the temperature, and the di/dt can be easily applied to the tested diode. Hence, a lot of information can be obtained effortlessly. This uncomplicated methodology makes it readily available for circuit designers, allowing them to predict the contribution of the reverse recovery of rectifiers to the total losses more accurately. In addition, a real comparison of different diodes at many operation points is made possible.**

I. INTRODUCTION

In power electronics, the loss contributions of rectifiers become increasingly more important [1], particularly in the case of circuits that operate at high frequencies. Losses occur during the conduction time as well as the switching instance. In many practical applications, significant power losses occur during the switching instance as a result of reverse recovery of the rectifier diodes [2], [3]. Hence, for choosing the right diode for each application, it is inevitable for a circuit designer to know the reverse recovery behavior. Since the manufacturers often do not provide such reasonable models for all diodes including the reverse recovery behavior, and data sheets provide only limited information regarding very high di/dt's, different temperatures, reverse voltage or forward current, it is inevitably required to measure the reverse recovery performance of rectifiers. In addition, the only possibility to verify certain models lies in the measurement of diode characteristics [4].

In the following, a few different reverse recovery measurement set-ups are presented. They are compared in terms of their automation possibility. In addition, this paper presents a fully automated measurement set-up for testing the reverse recovery behavior of ultra-fast rectifiers. A few key features of the presented automated set-up are the easily adjustable di/dt, forward current, temperature and reverse voltage. Moreover, the data obtained is immediately transferred to MATLAB, where it is easily available for further calculations. Finally, some measurement results with the presented test set-up are shown.

II. TEST CIRCUITS

Some measurement set-ups for determining the reverse recovery behavior of diodes have been proposed in the literature [2], [3], [4], [5]. Some of them are shortly presented in the following.

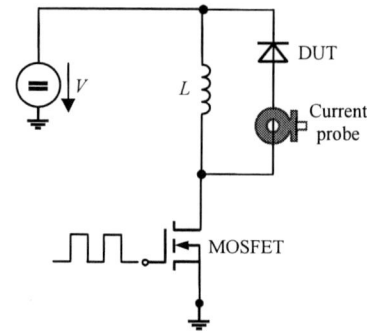

Fig. 1:   Circuit schematic of the clamped inductive switching circuit [3]

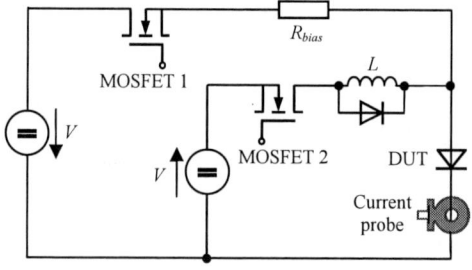

Fig. 2:   Circuit schematic proposed in [3]

978-1-4577-0007-1/11 $26.00 © 2011 IEEE

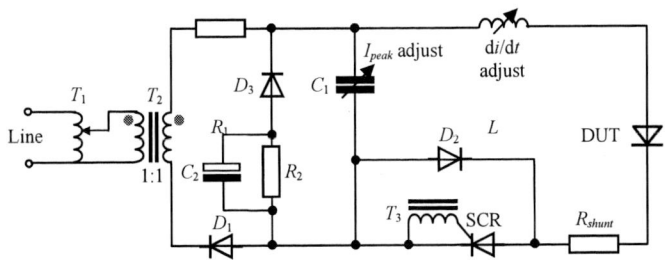

Fig. 3: Circuit schematic of the JEDEC approved rectifier test circuit [3]

Fig. 1 shows the well known clamped inductive switching circuit, and Fig. 2 a second circuit also proposed in [3]. The JEDEC-approved test circuit for determining the reverse recovery characteristics of a diode is depicted in Fig. 3.

Fig. 4 is proposed in [5] and similar to Fig. 2 in the sense that both proposals adjust their di/dt with an inductor L. However, the adjustment of the forward current is different. In Fig. 2 it is done via a resistor R_{bias} and in the circuit of Fig. 4 via an inductor L_1. The JEDEC approved reverse recovery rectifier test circuit is not appropriate for testing diodes at high di/dt levels, as the supplied forward current pulses become too narrow to ensure a full conduction. Hence, this circuit cannot be used for ultra-fast recovery testing. The other three circuits are suitable, including the clamped inductive switching circuit since the drawbacks of this circuit mentioned in [3] and [5] are negligible for high di/dt levels and a proper designed circuit. In addition, this is exactly the way in which diodes go into the reverse biased state in many applications. Therefore, one of these three circuits can be chosen as a set-up for ultra-fast recovery diode testing. However, since a fully automated set-up is desired, there are other constraints that must be taken into account for proper diode testing. These are namely the possibility of a proper automatic adjustment of the reverse voltage, the forward current, the di/dt, and the temperature by a control unit. The first two points are easily done with all three circuits, but the automated control of the current steepness turns out to be rather difficult with the circuits of Fig. 2 and Fig. 4 since they use an inductor for this adjustment. Therefore, we decided to use the clamped

inductive circuit of Fig. 1 in a slightly modified way for our fully automated measurement set-up to characterize ultra-fast recovery diodes. Fig. 5 shows the main circuit of the used set-up. Shunt resistors were utilized instead of a current probe. That was done in order to avoid any loop for insertion of the current probe to minimize the parasitic inductance in the diode path. Hence, the measured voltage across the diode path contains only a minimum value that is caused by the di/dt and the parasitic inductance in this path. Moreover, the arrangement is changed, and the MOSFET is placed in a way that the shunt resistor R_{shunt} is connected to ground. This results in a more difficult MOSFET driving, but on the other hand, it also makes the measurement of the diode current much easier since the two pins of the shunt resistor R_{shunt} can be directly connected to the oscilloscope. Therefore, no differential probe is required which could cause problems or introduce additional constraints. If needed, an external attenuator can easily be added between the oscilloscope and circuit. Hence, the changing of the set-up in Fig. 1 is just to simplify the measurement of the diode current and the diode voltage, and to get rid of problems that occur otherwise.

III. REALIZATION OF THE MEASUREMENT SET-UP

The automated measurement set-up consists of the main circuit (Fig. 5), a controllable current source for driving the switch (MOSFET), an on board control, measurement devices, different sources, filters and the PC with MATLAB as the interface for the user. Fig. 6 shows an overview of the measurement set-up without the main voltage source, PC, and oscilloscope. The main circuit according to Fig. 5 together with the current source, which drives the MOSFET, plus some storage capacitors and voltage probes for measurement, are shown in Fig. 7 in greater detail. Fig. 8 depicts the printed circuit board for control and communication with the PC. All components are explained in greater detail in the following.

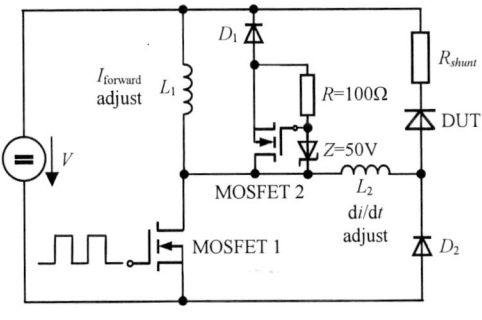

Fig. 4: Circuit schematic proposed in [5]

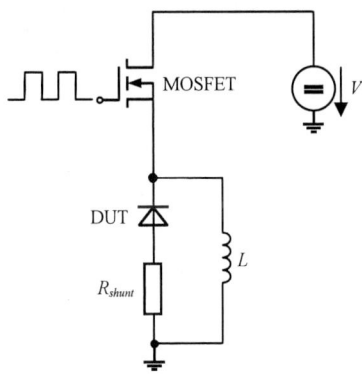

Fig. 5: Proposed diode recovery test circuit for automated measurement set-up

978-1-4577-0007-1/11 $26.00 © 2011 IEEE

Fig. 6: Measurement set-up without external devices

Fig. 7: Main circuit including current source for driving the MOSFET

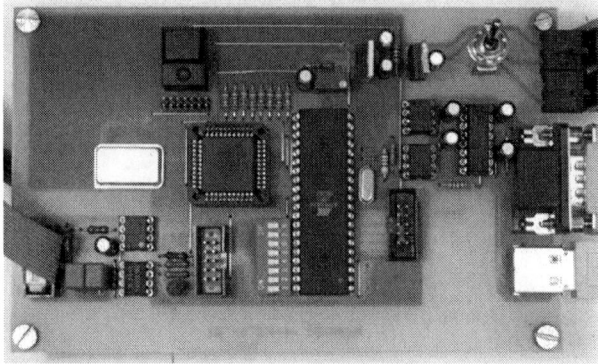

Fig. 8: Board for communication and control

A. Main Circuit

The main circuit (Fig. 5) is a typical clamped inductive switching circuit. However, the arrangement of devices is

Fig. 9: Double pulse sequence for the switching MOSFET

changed in comparison with Fig. 1, in order to suit the measurement task better as already mentioned.

Furthermore, the driving of the MOSFET, and hence the di/dt adjustment for the DUT is done by a current source. One thing of vital importance is that the stray inductance of the reverse recovery path needs to be minimized in order to really regulate the di/dt with the MOSFET and the current source, and also to keep the effects of the stray inductance upon the DUT during the recovery process as low as possible.

The gate of the MOSFET is supplied with a double pulse, which is shown in Fig. 9, and the inductor $L = 1$ mH provides the forward current for the DUT. This forward current builds up during the normally longer first pulse. Therefore, the length of this pulse needs to be controlled since the forward current through the tested diode is adjusted by this on-time of the MOSFET during the first pulse. After the first pulse, the MOSFET is turned off and the inductor current that is built up during the first pulse freewheels through the diode (DUT). Hence, this interval between the two pulses must be long enough to guarantee a full commutation of the inductor current to the DUT. Nevertheless, it should be as short as possible in order to prevent a reduction of the forward current due to present diode losses and to prevent self-heating. The actual reverse recovery incident of the DUT starts with the second pulse. During this second pulse the current in the inductor L rises again from its actual value at this time. Therefore, the length of the second pulse should only be long enough to assure a fully completed reverse recovery action plus some uncertainty margin. If the duration of the second pulse is too long, the set-up or the DUT can be damaged or even be destroyed by a too high current.

The current through the tested diode is measured with a shunt resistor in series to the DUT. Furthermore, the shunt resistor consists of four 4.7 Ω Mini Melf resistors in parallel. Thus, the power dissipation in these resistors does not exceed their limit and the parasitic inductance is kept very small.

B. Current Source

For a proper measurement result it is important to have a reasonably constant di/dt through the DUT during the reverse recovery action. This is achieved by a current source driving the MOSFET with a constant current during the switching instance and therefore during the reverse recovery action. Furthermore, the adjustment of the di/dt is done by setting the current source to a suitable current level. The current source itself mainly consists of a voltage to current

978-1-4577-0007-1/11 $26.00 © 2011 IEEE

converter and a current mirror. In order to avoid transient oscillations in the drain of the switching MOSFET due to oscillations in the actual MOSFET driving current source, the current source is activated before the actual measurement and freewheels in a short circuit, which is realized by means of another switch with very low R_{DSon}, during the startup. This short circuit is released by the control circuit of Fig. 8 in such a way that the double pulse occurs at the gate of the switching MOSFET and a reverse recovery measurement is performed. Since a current source is used for driving the gate of the switching MOSFET, a protection circuit is needed. Otherwise, the maximum gate voltage may be exceeded and a permanent damage would occur to the switching MOSFET. This protection circuit is realized by means of a Zener diode and a transistor.

C. Controlled Parameters and their Proper Adjustment

So far, there have been two things mentioned that need to be controlled in such a measurement set-up that is shown here. The first thing is the current value of the current source that influences the switching speed of the MOSFET in the main circuit and therefore the di/dt value. The second thing that needs to be controlled is the length of the first pulse of the double pulse in order to adjust the forward current to the desired value. These two adjustments are done by the microcontroller ATmega16 from Atmel as an on board control. In addition, this device provides a communication interface with the PC for the main control via MATLAB.

Furthermore, there are two other variables that need to be controlled as well, the reverse voltage and the temperature of the DUT. The reverse voltage on the one hand is adjusted very easily via MATLAB. A GPIB controllable source is used and connected to the PC and the main circuit. Hence, the desired voltage value is transmitted to the voltage source and it is turned on. Therefore, it is no problem at all to automate the adjustment of the reverse voltage. To adjust the temperature of the DUT on the other hand is not that easy, but also possible. We use a small copper block with a resistor inside it as heating device. One side of the copper block has the same area as the outer dimensions of a TO220 casing. The case of the tested diode is connected to this side of the copper block with a good thermal connection and the resistor inside the heating block is connected to an analog controllable voltage source. In order to adjust the temperature accurately, a control loop of the thermal system is realized with a PI controller implemented in MATLAB. We used the digital thermometer GMH 3710 from Greisinger electronic together with a Pt100 1/3 DIN for measuring the temperature of the testing diode and the copper block and for getting the feedback signal in the whole thermal control loop. For measurements above room temperature, the heating block together with the DUT is properly covered with glass wool. This action decouples the thermal system from the environment to some extent.

Hence, slight disturbances like a light airflow caused from the opening of a door for example do not cause an immediate rapid impact onto the heated system. Due to the thermal time constant of the system it takes some time till the temperature reaches the desired equilibrium. However, it is very important to really wait with the measurement till the thermal equilibrium is reached. Otherwise, the measurement result is worthless.

D. Pulse Generation

The pulse generation that is very essential for the whole measurement is not done by the microcontroller. Instead, a complex programmable logic device (CPLD) from Xilinx is used for the pulse generation. A high clock frequency for this device is possible. In such a manner, very narrow and very accurate pulses are possible. The duration of the first pulse that adjusts the forward current through the diode is adjusted by means of a bit pattern at the input of the CPLD. The generation of this bit pattern takes place in the microcontroller. The microcontroller also transmits this bit pattern to the CPLD.

E. Measurement Devices

In order to get any results, measurement devices need to be integrated into the set-up. In total, there are two measurement devices for electrical quantities integrated here, an analog amperemeter from AEG and an oscilloscope TDS5104B 1 GHz from Tektronix.

The amperemeter is only used for a visual indication that the current source is in action and a current is running through the circuit. It is to be noted that this device is not for an accurate measurement. Nevertheless, it displays the topical current value approximately.

The oscilloscope measures the current through the DUT indirectly via the shunt resistor. In addition, the voltage across the tested diode and the shunt resistor is recorded as well with a voltage probe directly at the DUT. Thus, the voltage across the DUT is easily obtained by a subtraction of this measured voltage and the voltage across the shunt resistor. The channel that measures the current is terminated with 50 Ω and the channel that measures the voltage across both, the DUT and the shunt resistor, is terminated with 1 MΩ. Moreover, the oscilloscope gets a trigger impulse from the microcontroller in order to start recording at the right moment.

F. Main Control Unit

All gathered data from the oscilloscope is immediately transferred to the PC and into MATLAB. There it is possible to determine all important information like the peak reverse current I_{RM}, the reverse recovery time t_{rr}, the reverse recovery charge Q_{rr}, and the softness factor S from the data obtained. Additionally, the whole measurement set-up is controlled via a MATLAB user interface.

978-1-4577-0007-1/11 $26.00 © 2011 IEEE

IV. OPERATING SEQUENCE

The fully automated set-up can be controlled via a developed user interface from MATLAB that communicates with the microcontroller on the control board. As such, the adjustments are done in this user interface and the measurement is started after inserting the DUT in the measurement set-up. Finally, the set-up follows the coarse operating sequence of Fig. 10. Afterwards the reverse recovery characteristic of the DUT that is measured with a fast oscilloscope is transferred to the computer. Hence all data is available in MATLAB or any other software and can be used for further calculations, model building, model validation or the extraction of all important and already mentioned diode parameters.

V. MEASUREMENT RESULTS

A. General

A few measurement results are exemplarily shown in this paper. The first two ones are done for a forward current I_F of 3.5 A, a temperature of 25°C, and a reverse voltage V_r of 200 V.

Fig. 11 reveals the reverse recovery behavior of the ultra-fast rectifier MUR160 for the values mentioned above and a di/dt of 1000 A/µs. Fig. 12 presents the result of the ultrafast diode STAA806D for an applied di/dt of 1500 A/µs. This measurement results in a peak reverse current I_{RM} of -16.5 A, a reverse recovery time t_{rr} of 25 ns (here time after the zero crossing of I_r till I_r settles back to 10% of I_{RM}), and a total reverse recovery charge Q_{rr} of 230 nC for the MUR160 ultra-fast rectifier diode. The results for the measurement of the ultrafast diode STAA806D are according to Fig. 12: $I_{RM} = -14.5$ A, $t_{rr} = 20$ ns (here time after the zero crossing of I_r till I_r settles back to 10% of I_{RM}, which is the first minimum after I_{RM}) and $Q_{rr} = 150$ nC. The determined voltage across the diode is also displayed in Fig. 11 and Fig. 12. The voltage overshoot and how it is affected by the abruptness of the turn off of the diode is clearly visible.

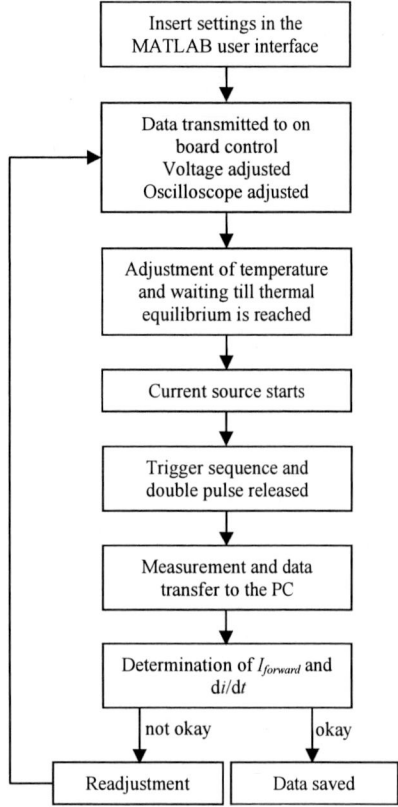

Fig. 10:   Coarse operating scheme

G. Supply Sources

The energy supply for the whole measurement set-up also plays a major role. Three different voltage sources are used in the proposed measurement set-up, the already mentioned GPIB controllable one for the main circuit that determines the reverse voltage of the DUT, a battery of 12 V for the control circuit, and a battery pack of 18 V as supply for the current source. Since the current source and the whole control circuit are floating with respect to ground, sources with no ground connection are necessary. Batteries automatically fulfill such a requirement, and therefore we decided to use batteries for the supply of the current source and the control circuit. The voltage source for the main circuit is set to the desired reverse voltage.

H. Filter Elements

A short look at Fig. 6 reveals that there are quite a number of filter elements involved in the whole measurement set-up that have not been mentioned thus far. These filter elements limit the common mode noise of this complex system and therefore guarantee a proper measurement result with the ground connected oscilloscope. Hence, it is very important to include these filter elements in this measurement set-up.

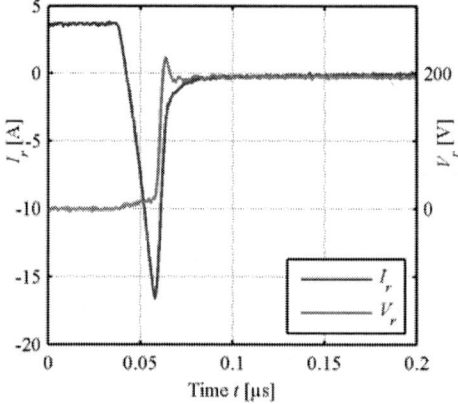

Fig. 11:   Reverse recovery of the diode MUR160

978-1-4577-0007-1/11 $26.00 © 2011 IEEE   102

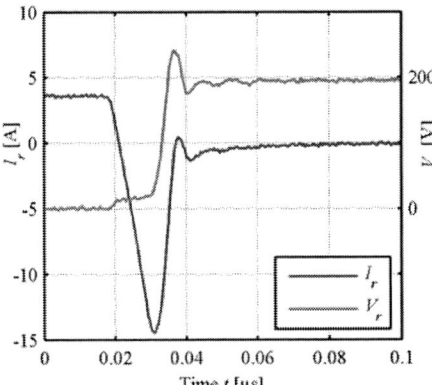

Fig. 12: Reverse recovery of the diode STTA806D

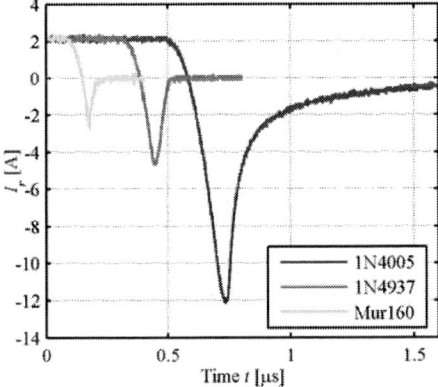

Fig. 13: Comparison of reverse recovery behaviors of different diodes for low di/dt values

A comparison of the measured results between different diodes is possible as well since the measurement is always done in exactly the same set-up. However, it is important to note that the visible and mentioned voltage overshoot is not an absolute value since it also depends on the parasitic components of the circuit. Hence, the voltage overshoot might be higher in the real application.

B. Comparison of Diodes

Also a real comparison with slower diodes at much slower di/dt values is easily possible with this automated measurement set-up. Fig. 13 shows the measured results for the diode current of such a comparison between three diodes for a forward current I_F of 2 A, a temperature of 25°C, and a reverse voltage V_r of 200 V. In addition, the di/dt value for all three different diodes is approximately the same value of 70 A/µs. As Fig. 13 reveals, the comparison is done between the ultra-fast diode MUR160, the fast diode 1N4937, and the very slow mains diode 1N4005. The result is as expected. The slower diodes result in a higher peak reverse current, a higher reverse recovery time, and a higher reverse recovery charge under the same testing conditions such as the same forward current, the same temperature of the device, the same reverse voltage, and roughly the same

di/dt value. Another thing that can be easily compared after having the measurement results is the turn-off behavior of each diode and how abrupt it is. Finally, circuit designers may choose a suitable diode for each application based on the obtained measurement results.

Before finalizing, another measurement result with a different temperature is shown exemplarily. Fig. 14 shows the result of the diode STTA806D for a reverse voltage V_r of 200 V, a forward current I_F of 3.5 A, a di/dt value of around 1000 A/µs, and a temperature of 100°C.

VI. CONCLUSION

A fully automated measurement set-up for determining the reverse recovery characteristic of ultra-fast diodes very accurately was designed and described. With the help of such a set-up, a real comparison of the recovery behavior for a lot of different operating conditions is easily possible. In addition, model building and validation can be done in a straightforward manner since the complete information based on the measured data is immediately available in MATLAB and can be used for further calculations.

REFERENCES

[1] N. Mohan, T. M. Undeland, W.P. Robbins, Power Electronics: Converters, Applications, and Design, 2nd ed. New York: Wiley, 1995.

[2] R. S. Chokhawala, E. I. Carroll, "A snubber design tool for P-N junction reverse recovery using a more accurate simulation of the reverse recovery waveform," IEEE Transactions on Industry Applications, vol. 27, no. 1, January 1991, pp. 74-84.

[3] C. Winterhalter, S. Pendharkar, Krishna Shenai, "A novel circuit for accurate characterization and modeling of the reverse recovery of high-power high-speed rectifiers," IEEE Transactions on Power Electronics, vol. 13, no. 5, September 1998, pp. 924-931.

[4] J. L. Duliere, H. A. Mantooth, R. G. Perry, "A systematic approach to power diode characterization and model validation," IEEE Industry Applications Conference, vol. 2, October 1995, pp. 1069-1075.

[5] Jamie Catt, "An improved method for ultra fast recovery diode testing," IEEE Applied Power Electronics Conference and Exposition, vol. 1, February 1994, pp. 473-479.

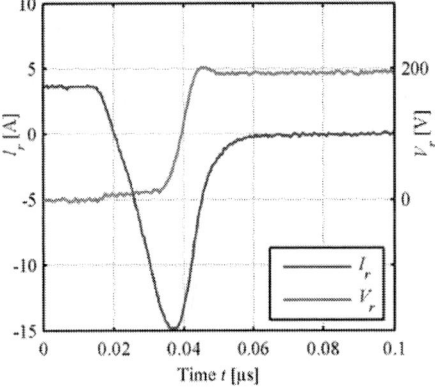

Fig. 14: Reverse recovery of the diode STTA806D with a diode temperature of 100°C and a forward current of 3.5A

Modeling of Lithium Ion Battery with Nonlinear Transfer Resistance

Low Wen Yao and Aziz, J.A.
Faculty of Electrical Engineering,
Universiti Teknologi Malaysia

Abstract—This paper discusses the Lithium Ion battery modeling by using equivalent circuit model with nonlinear transfer resistance. The modeling is done by using MATLAB/Simulink. A constant current pulse test is used to extract the model's parameters and the procedures are presented in details. Then the voltage responses of the conventional model, proposed model and MATLAB Li-ion cell model are compared. From the comparison, it reveals that the proposed model producing a better accuracy compared to conventional equivalent circuit model.

I. INTRODUCTION

Lithium-ion (Li-ion) battery is broadly used as the energy sources in the systems such as portable electronic devices, power systems of aircraft and space and electric vehicle [1, 2]. The widely usage of Li-ion battery is due to its high energy ratio as well as high power ratio compared to lead-acid battery and Nickel-Metal Hydride (NiMH) battery [3]. However, the behavior of Li-ion battery should be predicted in order to optimize the energy usage and prolong the battery's life of usage [4]. Therefore the model of battery is important to let the circuit designer a guide to forecast the behavior of the battery and thus increase the power efficiency of a battery-based system.

State-of-charge (SOC) is the capacity that remaining in a battery and it is considered as a key parameter of a battery [2]. For example in electric vehicle, a battery managing system (BMS) with the function of SOC estimation is required in order to let the user to know how long the EV can be used before it stops working [5, 6, 7, 8, 9]. Moreover, since the Li-ion battery can't be overcharged or over-discharged, an accurate SOC estimation is very important to avoid the system from inadvertent battery abuse and thus ensuring safety [5].

SOC can be estimated with the help of accurate battery modeling [2, 6, 7, 8, 9, 10]. Kalman Filter [5, 6, 7, 8, 9] is a good SOC estimation technique which performed based on battery model. The researches on application of Extended Kalman Filter (EKF) in SOC estimation had been done by Gregory L. Plett [6, 7, 8] and other researchers [5, 9] and the effectiveness of EKF method in SOC estimation had been proven with high accuracy of the result. However, for the EKF SOC estimation method, an accurate battery model is required. The performance of EKF SOC estimation might degrade and become unpractical if a low accuracy battery

model been applied [2]. Therefore, the battery modeling is a crucial issue in order to carry out an accurate SOC estimation.

Researchers have used several techniques on battery modeling. Reference [4], [7], and [11] provide a good review on several battery modeling techniques. Min Chen has categorized modeling techniques into mathematical model, electrochemical (chemical based) model and electrical (circuit based) model [4]. Il-Song Kim has pointed out that circuit based models are more suitable been applied for designing electrical system since the system designer can easily design on battery controller by adopting the mathematical equations which expressed from electrical circuit [2]. On the other hand, Min Chen also has pointed out that electric model is more easy to handle than others since the electrochemical (chemical based) requires microscopic (internal dynamic) data whereas mathematical model only work for certain conditions and may cause inaccurate results [4].

Since electrical model or circuit based model is more suitable been applied in electrical system, therefore it is widely been used in battery modeling. Equivalent electrical circuit model had been used for battery modeling by researchers in [1], [2], [9], [12], [13], [14]. A conventional equivalent electrical circuit model consists of an voltage source which represents open circuit voltage (OCV), a series resistor R_i which represents internal resistance, and a R_dC_d pairs which represent the charge transfer and double layer capacity as shown in Fig. 1 [1, 2, 9, 12, 13]. The value of $OCV_{(SOC)}$ is dependent on SOC. Another version of equivalent electrical circuit models also have been proposed in reference [4], [10], [14], [15], [16] and [17].

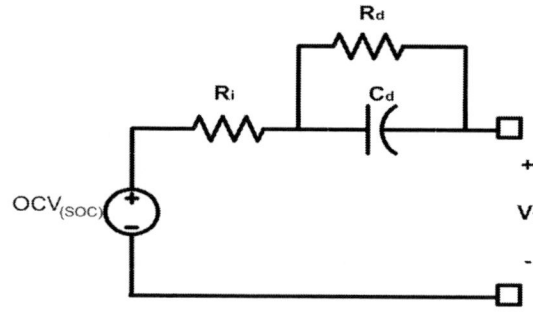

Fig.1. Conventional Equivalent Circuit Model for Battery Modeling

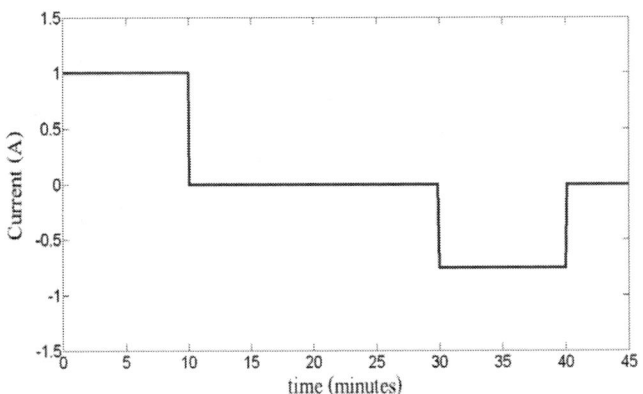

Fig.2. Current Profile for HPPC Test

Current tests such as HPPC (Hybrid Pulse Power Characterization) test [13] and pulse current discharge test [1, 2, 4, 9, 10, 12, 16] have been used in order to extract the parameters of the equivalent electrical circuit model. In HPPC test, battery is charged and discharged with a pulse current profile as shows in Fig.2. On the other hand, pulse current discharge test or sometime called current pulse technique, the battery is discharged by a current pulse (for example 1C, 3 minutes). For both current tests, the battery is required to rest for around one hour before the next cycle of current pulse so that to let the battery approaches equilibrium state. The parameters value and $OCV_{(SOC)}$-SOC relationship can be therefore easily obtained from voltage response of battery when the pulse current been applied [9, 12, 13, 16].

In section II, a new battery modeling method is proposed to improve the conventional modeling methods. After that, the procedures taken to develop proposed model are discussed in section III. The comparison between the voltage response of current pulse test for MATLAB Li-ion cell model, the conventional model and the proposed model are presented and discussed in section IV.

II. PROPOSED MODEL

In this paper, the equivalent circuit model is used on battery modeling. Fig.3 shows the proposed battery model for Lithium Ion battery modeling. In the proposed model, $OCV_{(SOC)}$ is a dependent voltage source which affected by SOC of the battery. R_i is the internal resistance of the battery. $R_{d(SOC)}$ and C_d are the RC parallel circuit used to describe the transient response of the voltage response where $R_{d(SOC)}$ is a transfer resistance and C_d is the double layer capacity.

The proposed model has slightly different from conventional equivalent circuit model since $R_{d(SOC)}$ is modeled as a nonlinear transfer resistance which depends on SOC. Nonlinear transfer resistance is a concept which is used to improve the performance of conventional model which using the constant RC parallel network. From reference [4], constant RC parallel network only works for certain SOC and temperature. The results from reference [4] show that a nonlinear resistance exists in battery model. Moreover, the

concept of nonlinear RC parallel network also been used in the PNGV model in reference [15] and integrated battery model in reference [15]. Therefore, the concept of nonlinear $R_{d(SOC)}$ is adopted in this proposed model.

For the Li-ion battery, the self-discharge rate is extremely lower than lead-acid battery and NiMH battery [10], therefore the self-discharge issue is ignored in this proposed model. The effect of temperature also been ignored by assuming that the battery been applied in the environment with narrow range of temperature. The battery relaxation effect is not been considered in the proposed model by assuming the battery can achieve equilibrium state in a very short time. However, in order to improve the battery model, the relaxation effect still can be modeled by adding another RC parallel network in series with $R_{d(SOC)}$ and C_d pairs.

III. MODEL DEVELOPMENT

The parameters of the model can be extracted from the current pulse test [9, 12, 13, 16]. In the constant current pulse test, the battery is discharge with a constant current for a certain period so that the SOC of battery drops to the predicted level. For example, if the battery discharges with 1C for 180s, the SOC of the battery will drops 5%. After that, the battery will keep in rest for 3420s to let the battery approaches equilibrium state. The equilibrium state is used to obtain the value of $OCV_{(SOC)}$.

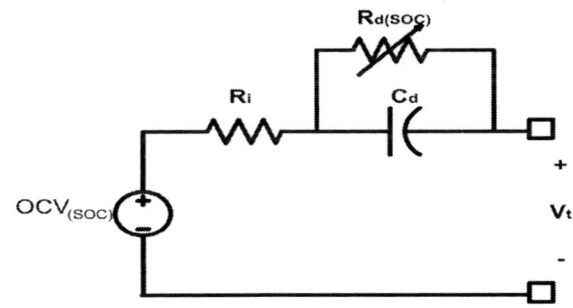

Fig.3. Lithium Ion Battery Model

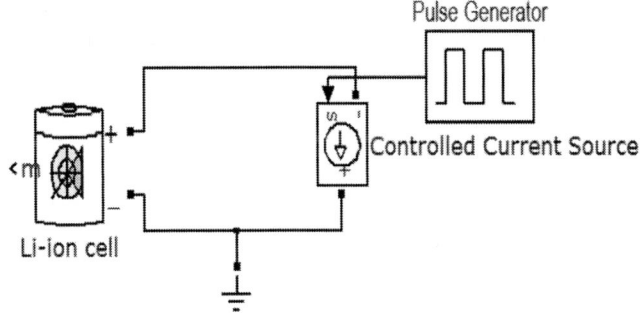

Fig.4. Simulation Circuit for Constant Current Pulse Test

978-1-4577-0007-1/11 $26.00 © 2011 IEEE

TABLE I
SETTING FOR LI-ION CELL

Battery type	Lithium-Ion
Nominal Voltage (V)	3
Rated Capacity (Ah)	18
Maximum Capacity (Ah)	18
Fully Charged Voltage (V)	3.5
Nominal Discharge Current (A)	9
Internal Resistance (Ohms)	0.001
Capacity(Ah) @ Nominal Voltage	15.6
Exponential zone [Voltage(V), Capacity(Ah)]	[3.26 0.25]

TABLE II
SETTING FOR PULSE GENERATOR

Amplitude	18
Period (secs)	3600
Pulse Width (%)	5%
Pulse Type	Time-Based
Time(s)	Use Simulation time

A. Simulation Circuit for Battery Modeling

Fig.4 shows the simulation circuit used for extracts the parameters of battery model. Controlled Current Source is controlled by a Pulse Generator so that the battery will discharge with 1C for 180s and then rest for 3420s. Table I shows the setting of Li-ion cell whereas Table II shows the setting of Pulse Generator. For this simulation, 18Ah Li-ion battery is used, therefore constant current of 18A (1C) is discharged. For this simulation, the initial state-of-charge of the Li-ion cell is set to 100% and the simulation time is set to 64800s so that the voltage response for the Li-ion cell in the range of 100% state-of-charge to 10% state-of-charge can be studied.

B. Extraction of Parameters' value

The value of model's parameters can be extracted from the voltage response of the Lithium Ion battery when constant current pulse test been performed. The current pulse and the voltage response of the battery are show in Fig.5 and Fig.6 respectively.

From the voltage response of the pulse test as shown in Fig.6, it can noticed that there is an instantaneous voltage drop (ΔV) when the constant current started to discharge. Besides, there is an instantaneous voltage raise once constant current stopped to discharge. The voltage response shows the existence of the internal resistance, R_i. The value of R_i can be calculated by using equation (1), i.e.

$$R_i = \frac{\Delta V}{I} = \frac{0.018}{18} = 0.001\Omega \qquad (1)$$

The relationship curve between $OCV_{(SOC)}$ and SOC can be obtained from the voltage response. $OCV_{(SOC)}$ is the terminal voltage of battery when the equilibrium state achieved (no chemical reaction in the battery). The $OCV_{(SOC)}$-SOC relationship is profiled and plotted by using Microsoft Excel. The function of $OCV_{(SOC)}$ also can be obtained by using a *polynomial trend line* which fit the

$OCV_{(SOC)}$-SOC curve. In this example, the $OCV_{(SOC)}$-SOC curve can be fit by a *5th order polynomial trend line* as shown in Fig.7. The function of $OCV_{(SOC)}$ is written as:

$$OCV_{(SOC)} = (3.82 \times 10^{-10})SOC^5 - (1.21 \times 10^{-7})SOC^4$$
$$+ (1.51 \times 10^{-5})SOC^3 - (9.3 \times 10^{-4})SOC^2$$
$$+ 0.0295SOC + 2.85 \qquad (2)$$

C. Nonlinear Transfer Resistance

From the voltage response as shown in Fig.6, when the current pulse been applied, the battery experiences the instantaneous voltage drop that caused by internal resistance. After that, the voltage response will go through a transient response and after that achieve a "steady state". The transient response (100s to 150s in Fig.5) can be modeled by using a $R_{d(SOC)}$ and C_d pair.

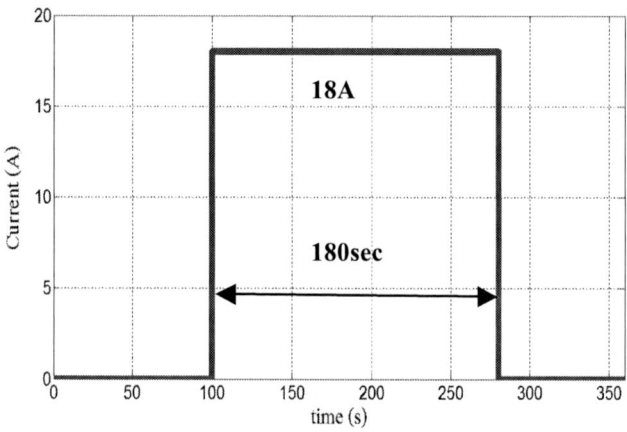

Fig.5. Lithium ion battery discharge with constant current of 1C

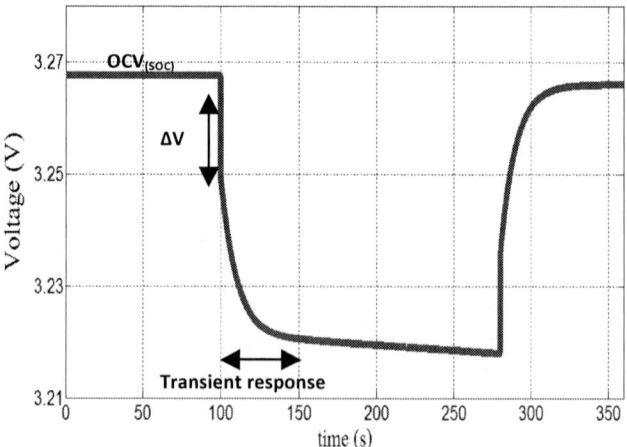

Fig.6. Voltage response of Li-ion battery when discharge with constant current of 1C

OCV(SOC)- SOC Relationship

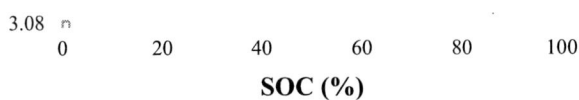

Fig.7. OCV(SOC)-SOC Relationship

However, the voltage drop of the transient response is not constant. These phenomena can be identified by a constant current discharge test. Constant current discharge test is performed by discharging the Li-ion cell with a constant current. The constant current discharge test is repeated for constant current of 18A (1C), 13.5A (0.75C), 9A (0.5C) and 4.5A (0.25C). The relationship between V_t and SOC for 0.25C, 0.5C, 0.75C and 1C for constant current discharge test are profiled and plotted by using Microsoft Excel as shown in Fig.8. From Fig.8, the spaces between one V_t line and another are the same. The results shows that the voltage differences between $OCV_{(SOC)}$ and V_t are linearly increase when the discharge current increases linearly. Since the voltage difference is contributed by both R_i and $R_{d(SOC)}$, thus the value for both resistances are independent on the value of current.

Besides, the V_t-SOC relationship in Fig.8 also shows that the voltage difference between $OCV_{(SOC)}$ and V_t is larger at the lower SOC compare to the higher SOC. Since the value of R_i is a constant, hence we can conclude that the nonlinear voltage difference is caused by a nonlinear transfer resistance. Therefore, the concept of nonlinear transfer resistance, $R_{d(SOC)}$ is applied in this proposed model.

The $R_{total(SOC)}$ for a certain state-of-charge (SOC_k) can be calculated by using equation (3).

$$R_{total(SOC_k)} = R_{d(SOC_k)} + R_i$$
$$= \frac{OCV_{(SOC_k)} - V_t(SOC_k)}{I} \qquad (3)$$

The relationship between the $R_{total(SOC)}$ and SOC is obtained by profiling the values of $R_{total(SOC)}$ for several SOC. The curve in Fig.9 shows the relationship between the $R_{total(SOC)}$ and SOC. The curve can be formulated with a *power trend line* using Microsoft Excel.

In this example, the *power trend line* shows that the $R_{total(SOC)}$-SOC relationship is:

$$R_{total(SOC)} = 0.092SOC^{-0.8} \qquad (4)$$

Therefore,

$$R_{d(SOC)} = 0.092SOC^{-0.8} - R_i$$
$$= 0.092SOC^{-0.8} - 0.001 \qquad (5)$$

Vt-SOC Relationship

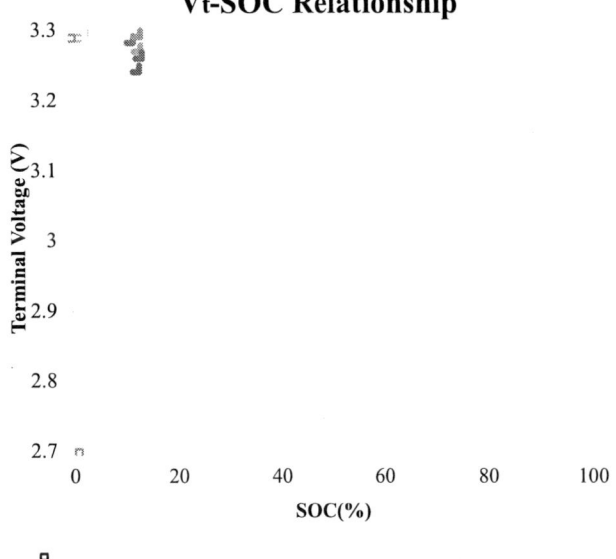

Fig.8. V_t-SOC Relationship

R(total(SOC)) - SOC Relationship

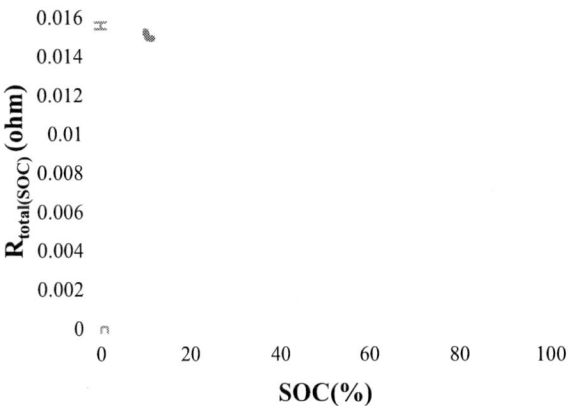

Fig.9. $R_{total(SOC)}$-SOC Relationship

D. Double Layer Capacity

After the transfer resistance, $R_{d(SOC)}$ had been obtained, the double layer capacity can be calculated by equation (6).

$$C_d = \frac{\tau}{R_{d(SOC)}} \qquad (6)$$

For this example, the value of time constant is 10s. The modeling of double layer capacity can be made by forming a nonlinear capacitor. However, in order to simplify the modeling method and also due to the short transient response, the constant value of capacitor is used in modeling of double layer capacity. In this aspect, the mean value of transient resistance is chosen from the SOC range of 30% to 80% (see Fig.9). In this example, the $R_{d(SOC)}$ with 0.003Ω is chosen. Hence, C_d is calculated as 3333F.

IV. RESULTS AND DISCUSSION

After all the parameters' value been selected, a modeling circuit can be set up. By referring to Fig.2, the $OCV_{(SOC)}$ voltage source can be modeled by a controlled voltage source which controlled by the SOC or a lookup table. The nonlinear $R_{d(SOC)}$ resistor also can be modeled by using a controlled current source since nonlinear resistor block is not available in MATLAB/Simulink. The supplied current for the controlled current source is controlled by using the equation (7), where V_{Cd} is voltage across C_d.

$$I = \frac{V_{Cd}}{0.092SOC^{-0.8} - 0.001} \qquad (7)$$

Fig.10, Fig.11 and Fig.12 show the comparison of voltage response for MATLAB Li-ion cell model, proposed model and conventional model for constant current pulse test with of 1C, 0.5C and 2C respectively.

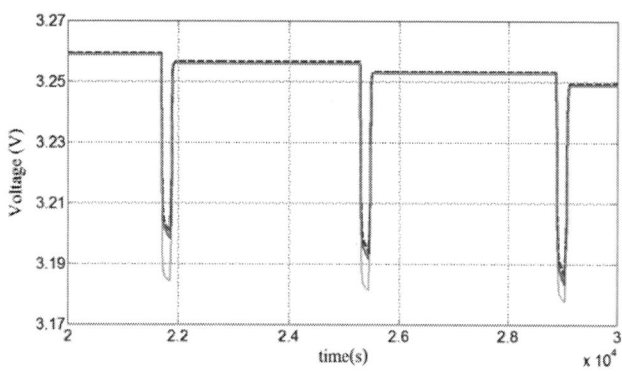

Blue Colour: Voltage Response of MATLAB Li-ion cell model
Red Colour: Voltage Response from proposed model
Green Colour: Voltage Response from conventional model

Fig.10. Comparison between Voltage Response from MATLAB Li-ion cell Model, Proposed Model and Conventional Model (1C, 180s pulse)

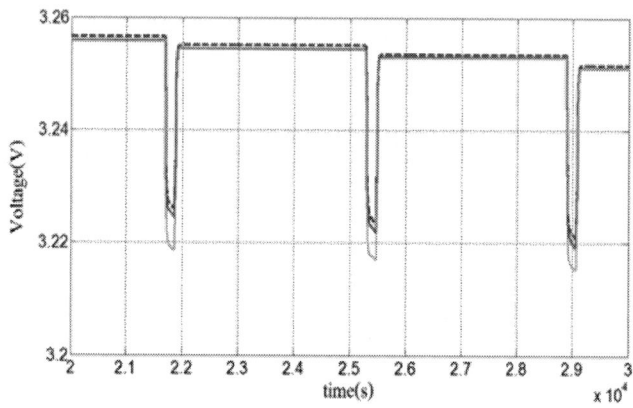

Blue Colour: Voltage Response of MATLAB Li-ion cell model
Red Colour: Voltage Response from proposed model
Green Colour: Voltage Response from conventional model

Fig.11. Comparison between Voltage Response from MATLAB Li-ion cell Model, Proposed Model and Conventional Model (0.5C, 180s pulse)

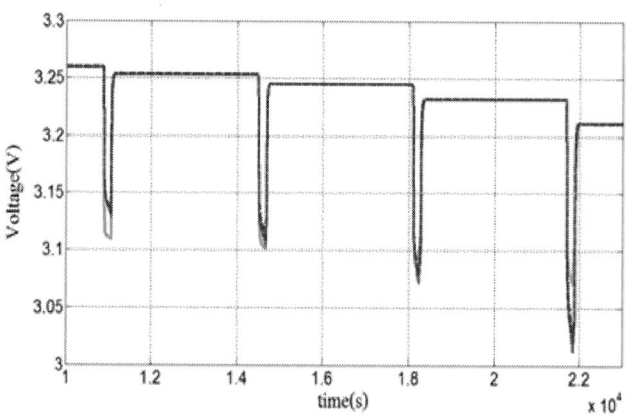

Blue Colour: Voltage Response of MATLAB Li-ion cell model
Red Colour: Voltage Response from proposed model
Green Colour: Voltage Response from conventional model

Fig.12. Comparison between Voltage Response from MATLAB Li-ion cell Model, Proposed Model and Conventional Model (2C, 180s pulse)

From Fig.10, Fig.11 and Fig.12, it can be compared that the proposed model achieving greater accuracy than the conventional model with fixed R_d. From Fig.12, the result shows that although conventional model with fixed R_d has voltage response that satisfied for a certain SOC (for pulse at 18,000sec) but the voltage is deviate for the others. However, for the proposed model, the voltage responses are nearly equal to the voltage response from the MATLAB Li-ion cell model. Therefore, this proposed modeling is potentially resulting better battery modeling since the proposed model did not results in the significant voltage error.

V. CONCLUSION

In conclusion, the nonlinear transfer resistance concept which been applied in battery modeling has resulting better accuracy than the conventional model. It is expected that the proposed modeling method can be implemented practically. However, in order to make a better modeling on battery, the battery relaxation effect should be considered in practical approaches. Another RC parallel network is needed to connect in series with $R_{d(SOC)}C_d$ parallel network for modeling on the battery relaxation effect.

VI. ACKNOWLEDGMENT

This work was financially supported by Fundamental Research Grant Scheme (FRGS) of Malaysia and Universiti Teknologi Malaysia under the project "Fundamental Study of Electric Vehicle Charging System".

REFERENCES

[1] L. Gao, S. Liu and R.A. Dougal, "Dynamic lithium-ion battery model for system simulation," *IEEE Transactions on Components and Packaging Technologies, vol. 25, No.3*, 495-505, September 2002.

[2] I-S. Kim, "The Novel State of Charge Estimation Method for Lithium Battery Using Sliding Mode Observer,"*Journal of Power Source 163 (2006)*, 584-590. 2006.

[3] Minxin Zheng, Bojin Qi, and Hongjie Wu, "A Li-ion Battery Management System Based on CAN-bus for Electric Vehicle," *Beijing University of Aeronautics and Astronautics,* China, 2008.

[4] M. Chen and G. Rincon-Mora, "Accurate electrical battery modelcapable of predicting runtime and i-v performance," *Energy Conversion,IEEE Transactions on*, vol. 21, no. 2, pp. 504–511, June 2006.

[5] B.P. Divarkar et.al, "Battery Management System and Control Strategy for Hybrid and Electric Vehicle,"*2009 3rd International Conference on Power Electronics System and Applications.* 2009.

[6] G. Plett, "Extended Kalman Filtering for Battery Management Systems of LiPB-based HEV Battery Packs. Part 1. Background."*Journal of Power Sources* 134 (2004), 252-261. 2004.

[7] G. Plett, "Extended Kalman Filtering for Battery Management Systems of LiPB-based HEV Battery Packs. Part 2. Modeling and Identification."*Journal of Power Sources* 134(2004), 262-276. 2004.

[8] G. Plett, "Extended Kalman Filtering for Battery Management Systems of LiPB-based HEV Battery Packs. Part 3. State and Parameter Estimation."*Journal of Power Sources* 134(2004), 277-292. 2004.

[9] S.J Lee et.al, "State-of-Charge and Capacity estimation of Lithium-ion Battery Using New Open-Circuit Voltage versus State-of-Charge,"*Journal of Power Sources* 185 (2008) 1367-1373. 2008.

[10] I-S Kim, "Nonlinear State of Charge estimator for Hybrid Electric Vehicle Battery," *IEEE Transactions on Power Electronics*, vol.23, No.4, July 2008.

[11] P. Singh and A. Nallanchakravarthula,"Fuzzy Logic Modeling of Unmanned Surface Vehicle (USV) Hybrid Power System,"*Proceedings of the 2005 Intelligent Systems Application to Power Systems*, Arlington, VA, November 2005.

[12] Ralf Benger et.al, "Electrochemical and thermal modeling of lithium-ion cells for use in HEV and EV application*,"World Electric Vehicle Journal* Vol.3-ISSN2032-6653, 2009.\

[13] S. Bernhard et.al."Modeling of High Power Automotive Batteries by the Use of an Automated Test System,"*IEEE Transactions on Instrumentation and Measurement*, Vol 52, No. 4, August 2003.

[14] M. C.Glass, "Battery Electrochemical Nonlinear/Dynamic Spice Model," Proc. Energy Convers. Eng. Conf. vol.1, 1996, pp292-297.

[15] S.Li and C.Zhang, "Study on Battery Management System and Lithium-ion Battery," *IEEE International Conference on Computer and Automation Engineering*, 2009.

[16] S. Abu-Sharkh and D.Doerffel, "Rapid Test and Non-linear Model Characterization of Solid-state Lithium-ion Batteries," *J. Power Sources*, vol.130.pp.266-274, 2004.

[17] M. Knauff, J. McLaughlin, C. Dafis, D. Niebur, H. Kwatny, C. Nwankpa,and J. Metzer, "SimulinkModel of a Lithium-ion Battery for the HybridPower System Testbed," *Proceedings of the ASNE Itelligent Ships Symposium,*May 2007

Modeling and Simulation of Adaptive Neuro-Fuzzy Controller for Chopper-Fed DC Motor Drive

Yousif I. Al-Mashhadany, MIEEE.

Electrical Engineering Dept. College Engineering.
University of Al-Anbar
Al-Anbar, Iraq
Email: yousif_phd@hotmail.com

Abstract— **The classical controllers algorithm is both simple and reliable, and has been applied to thousands of control loops in various industrial applications over the past 60 years (89%-90% of applications). This paper presents the neuro-fuzzy controller incorporates fuzzy logic algorithm with a five-layer artificial neural network (ANN) structure. The conventional controller is replaced by Adaptive Neuro-Fuzzy Inference System (ANFIS) before that made the identification process of ANFIS controller by the data base of classical controller to be consider as initial condition for controlling process with this system.**

The simulation of the design is achieved by using Matlab Ver. 2010a. Chopper-Fed DC Motor Drive (Continuous / Discrete) are consider as case study. Satisfactory results are obtained explaining the ability of ANFIS controller to control with the dynamic high nonlinear system and can be get very good results by tunes the fuzzy controller.

Index Terms—**Adaptive Neuro-Fuzzy Inference System (ANFIS) ,Chopper fed-DC motor drive.**

I. INTRODUCTION.

Most industrial processes exhibit highly nonlinear dynamics but are never the less controlled with the classical PID linear technique, a widespread methodology with the advantage of few parameters to tune according to established guidelines. However, the use of the PID linear technique in the control of a nonlinear process makes the tuning configuration strongly dependent on the particular steady-state working condition. Thus, a fixed tuning of the PID algorithm cannot guarantee good system performances for any operating point. A methodology overcoming this limitation is a hierarchical control strategy, consisting of a fuzzy supervisor and of the PID controller itself: the fuzzy supervisor modifies the PID tuning on-line, according to rules with one or more input variables, usually the set point, the error and/or the actual control action. Together with adaptively, the foremost advantages of this control strategy are the ability to cope with incomplete knowledge of the process and to provide smooth transitions from an operating region to another. We remark that the last two characteristics do not apply to conventional gain scheduling or CGS (fuzzy supervision of PIDs is often referred to in literature as fuzzy gain scheduling or FGS), for the simple reason that CGS, unlike FGS, is not rule-based. For further insights on

fuzzy supervision of a PID controller and its theoretical aspects see [1,2].

The speed of DC motor can be adjusted to a great extent as to provide controllability easy and high performance. The controllers of the speed that are conceived for goal to control the speed of DC motor to execute one variety of tasks, is of several conventional and numeric controller types, the controllers can be: PID Controller, Fuzzy Logic Controller; or the combination between them Fuzzy-Genetic Algorithm, Fuzzy-Neural Networks, Fuzzy-Ants Colony, Fuzzy-Swarm [3,4,5].

The source is a constant-voltage DC, such as a battery or diode-bridge rectified AC supply, a different type of converter is required to convert the fixed voltage into a variable-voltage/variable-current source for the speed control of the DC motor drive. The variable DC voltage is controlled by chopping the input voltage by varying the on- and off-times of a converter, and the type of converter capable of such a function is known as a chopper.

A schematic diagram of the chopper is shown in Fig.1. The control voltage to it's gate is v_e. The chopper is on for a time t_{on} and its off time is t_{off} . Its frequency of operation is:

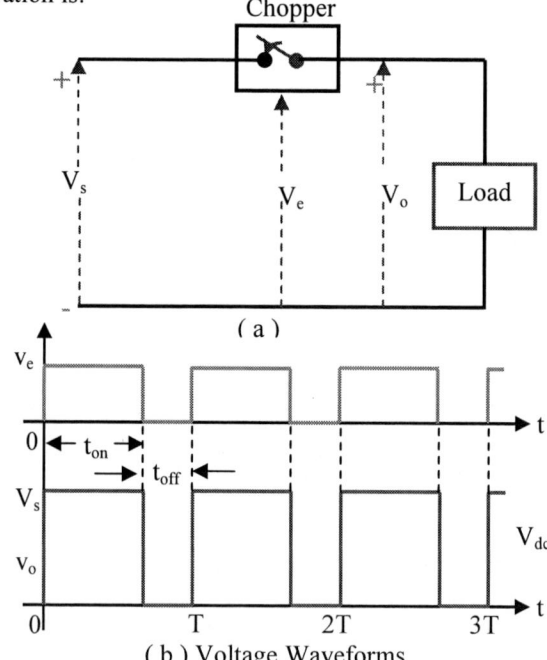

Fig.1. Chopper schematic and its waveforms.

$$f_c = \frac{1}{(t_{on} + t_{off})} = \frac{1}{T} \qquad \ldots\ldots 1$$

And its duty cycle is defined as

$$d = \frac{t_{on}}{T} \qquad \ldots\ldots 2$$

The output voltage across the load during the on-time of the switch is equal to the difference between the source voltage Vs and the voltage drop across the power Switch. Assuming that the switch is ideal, with zero voltage drop. the average output voltage V_{dc} is given as

$$V_{dc} = \frac{t_{on}}{T} V_s = d V_s \qquad \ldots\ldots 3$$

where Vs is the source voltage.

II. MODELING OF CHOPPER MOTOR:

The closed-loop speed-controlled separately-excited dc motor drive is shown in Fig.2. for analysis, With inner current loop alone, the motor drive system is a torque amplifier. The commanded value of current is compared to the actual armature current and its error is processed through a current controller. The output of the current controller in conjunction with other constraints, determines the base drive signals of the chopper switches. The current controller can be either of the following types [7]:

A. Pulse-Width-Modulation (PWM) controller.

B. Hysteresis controller.

In the Pulse-Width-Modulation (PWM) controller The current error is fed into a controller, which could be proportional (P), proportional plus integral (PI), or proportional, integral, and differential (PID). The most commonly used controller among them is the PI controller. The current error is amplified through this controller and emerges as a control voltage, v_c. It is required to generate a proportional armature voltage from the fixed source through a chopper operation. Therefore, the control voltage is equivalent to the duty cycle of the chopper. Its realization is as follows. The control voltage is compared with a ramp signal to generate the On- and Off-times. On signal is produced if the control voltage is greater than the ramp (carrier) Signal off signal is generated when the control signal is less than the ramp signal [8].

This logic amounts to the fact that the duration for which the control signal exceeds the ramp signal determines the duty cycle of the chopper. The on- and off-time signals are combined with other control features, such as interlock, minimum on- and off-times, and quadrant selection. The interlock feature prevents the turning on of the transistor (top/bottom) in the same leg before the other transistor (bottom/top) is turned off completely. This is ensured by giving a time delay between the turn-off instant of one device and the turn-on instant of the other device in the same phase leg. - Simultaneous

conduction of the top and bottom devices in the same leg results in a short circuit of the de source it is known as shoot-through failure in the literature. The de motor equations, including its field, are [9]:

$$\left. \begin{array}{l} V_a = R_a i_a + L_a \dfrac{di_a}{dt} + K\phi_f w_m \\[2mm] V_f = R_f i_f + L_f \dfrac{di_f}{dt} + \\[2mm] T_e - T_l = J \dfrac{dw_m}{dt} + B_1 w_m \\[2mm] T_e = K\phi_f i_a \end{array} \right\} \qquad (4)$$

Where : $K\phi_f = Mi_f$, and M is the mutual inductance between the armature and field windings. The tacho generator and the filter are combined in the transfer function as:

$$\frac{w_{mr}(s)}{w_m(s)} = \frac{H_w}{1 + sT_w} \qquad (5)$$

Generation of base-derive signals from current error for forward motoring when one is using the chopper. The input to the speed controller is the speed error. and the controller is of a PI type given by [10]

$$\frac{T_e^*(s)}{(w_r^* - w_{mr})} = \frac{K_s(1 + sT_s)}{sT_s} \qquad G_c(s) = K_c. \qquad (6)$$

The current-error amplifier is modeled as a again and is given by:

The chopper is modeled as a first-order lag, with a gain given by:

$$G_c(s) = \frac{K_r}{(1 + \dfrac{sT}{2})}. \qquad (7)$$

The PWM current controller has a delay of half the time period of the carrier waveform, and its gain is that of the chopper. Hence, its transfer function, including that of the chopper, is

$$G_c(s) = \frac{K_c K_r}{(1 + \dfrac{sT}{2})}. \qquad (8)$$

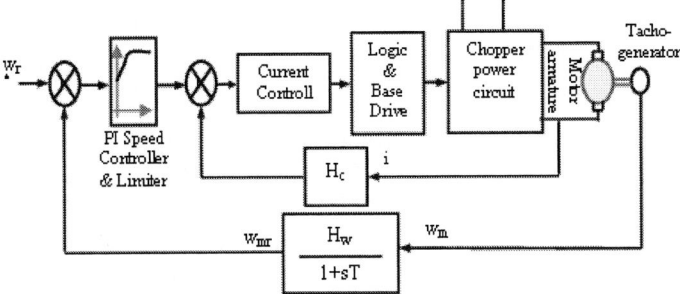

Fig.2. The closed-loop speed-controlled separately-excited dc motor [6].

978-1-4577-0007-1/11 $26.00 © 2011 IEEE

Where Ke is the gain of the PWM current controller, Kr is the gain of the chopper, and the time constant T is given by,

$$G_c(s) = \frac{1}{carrier\ \ frequency} = \frac{1}{f_c}. \tag{9}$$

The gain of the PWM current controller is dependent on the gain of the current error amplifier. For all practical purposes, the PWM current control loop can be modeled as a unity-gain block if the delay due to the carrier frequency is negligible. The hysteresis controller has instantaneous response: hence, the current loop is approximated as a simple gain of unity.

III. A DAPTIVE NEURO FUZZY CONTROLLER

The neuro-fuzzy controller incorporates fuzzy logic algorithm with a five layer artificial neural network (ANN) structure as shown in Fig.3 [4].

A tuning block is utilized to adjust fourth layer's parameters in order to correct any deviation of control effort. The speed error and the rate of change of actual speed error are the inputs of the neuro-fuzzy controller, which are given by

$$\left. \begin{array}{l} Input1 = \varepsilon_w = w^* - w \\[2mm] input2 = \Delta\varepsilon_w = \dfrac{\varepsilon_w(n) - \varepsilon_w(n-1)}{T} \times 100\% \end{array} \right\} \tag{10}$$

Where: w^* is the command speed and T is the sampling time.

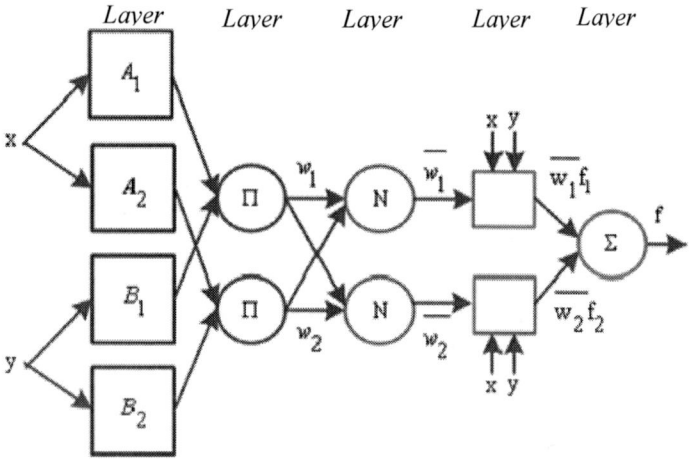

Fig.3. ANFIS architecture of 2-input Sugeno fuzzy model with 2 rules [4].

Sugeno fuzzy model with five- layer ANN structure is used in proposed controller. In this five-layer ANN structure the first layer represents for inputs, the second layer represents for fuzzification, the third and fourth layers represents for fuzzy rule evaluation and the fifth layer represents for defuzzification. A two input first order Sugeno fuzzy model with two rules is depicted in fig. 2. In

layer 1, every node i is an adaptive node with a node function

$$\left. \begin{array}{ll} O_{1,i} = \mu_{Ai}(x) & for \quad i = 1,2 \quad or \\[2mm] O_{1,i} = \mu_{Bi-2}(y) & for \quad i = 3,4 \end{array} \right\} \tag{11}$$

(here we denote the output of the *ith* node in layer l as $O_{1,I}$) where x (or y) is the input to node i and *Ai (or Bi-2)* is a linguistic label such as 'small' or 'large' associated with this node. The membership function for A can be any appropriate parameterized membership function. In proposed scheme generalized bell function is used as a membership function given by equation (12) [11,12,13].

$$\mu_A(x) = \left(1/(1 + \left|\frac{x - c_i}{a_i}\right|^{2bi})\right) \tag{12}$$

where $\{ai, bi, ci\}$ is the parameter set. As the values of these parameters changes, various forms of bell shaped membership functions can be get for fuzzy set A. Parameters in this layer are referred to as premise parameters. In layer 2, every node is a fixed node labeled Π, whose output is the product of all the incoming signals.

$$O_{2,i} = w_i = \mu_{Ai}(x)\mu_{Bi}(y), \qquad i = 1,2 \tag{13}$$

Each node output represents the firing strength of a rule. In layer 3, every node is a fixed node labeled N. The outputs of this layer are normalized firing strengths given by equation (14).

$$O_{3,i} = \overline{w}_i = \frac{w_i}{w_1 + w_2}, \qquad i = 1,2 \tag{14}$$

In layer 4, every node i, is an adaptive node with a node function given by equation (15).

$$O_{4,i} = \overline{w}_i f_i = \overline{w}_i(p_i x + q_i y + r_i) \tag{15}$$

where w_i is a normalized firing strength from layer 3 and $\{p_i, q_i, r_i\}$ is the parameter set of this node. Parameters in this layer are referred to as consequent parameters. Layer 5 is the single node layer with a fixed node labeled Σ, which computes the overall output as the summation of all incoming signals.

$$O_{5,i} = \sum \overline{w}_i f_i = \frac{\sum i w_i f_i}{\sum i w_i} \tag{16}$$

The hybrid approach converges much faster since it reduces the search space dimension of the original pure back propagation. In hybrid learning, for back propagation, objective function to be minimized is defined by (17) [14,15].

$$E_p = \sum_{m=1}^{l} (T_{m,p} - O_{m,p})^2 \tag{17}$$

where $T_{m,p}$ is the m_{th} component of p_{th} target output vector and O_{mp} is the m_{th} component of actual output vector produced by the presentation of the p_{th} input vector. Hence the over all error measure is given by (18).

$$E = \sum_{p}^{p} = 1^{E_p} \qquad (18)$$

Learning rules can be derived as follows:

$$\left. \begin{aligned} a_i(n+1) &= a_i(n) - \eta_{ai}(\partial E/\partial a_i) \\ b_i(n+1) &= b_i(n) - \eta_{bi}(\partial E/\partial b_i) \\ c_i(n+1) &= c_i(n) - \eta_{ci}(\partial E/\partial c_i) \end{aligned} \right\} \qquad (19)$$

Where η_{ai}, η_{bi}, and η_{ci} are the learning rates of the NN see[16,17].

IV. CASE STUDY:

The simulation of case study is achieved by using MATLAB Ver. 2010a. Two forms of simulation for the Chopper-Fed DC Motor Drive was used:
- ➢ **Chopper-Fed DC Motor Drive in Continuous.**
- ➢ **Chopper-Fed DC Motor Drive in Discrete.**

The description of the circuit The DC motor is fed by the DC source through a chopper which consists of GTO thyristor and free-wheeling diode D1. The motor drives a mechanical load characterized by inertia J, friction coeficient B, and load torque TL. The hysteres is current controller compares the sensed current with the reference and generates the trigger signal for the GTO thyristor to force the motor current to follow the reference. The speed control loop uses a proportional-integral controller which produces the reference for the current loop. Current and Voltage Measurement blocks provide signals for visualization purpose. The two circuits are shown in Fig.s(4 & 5).

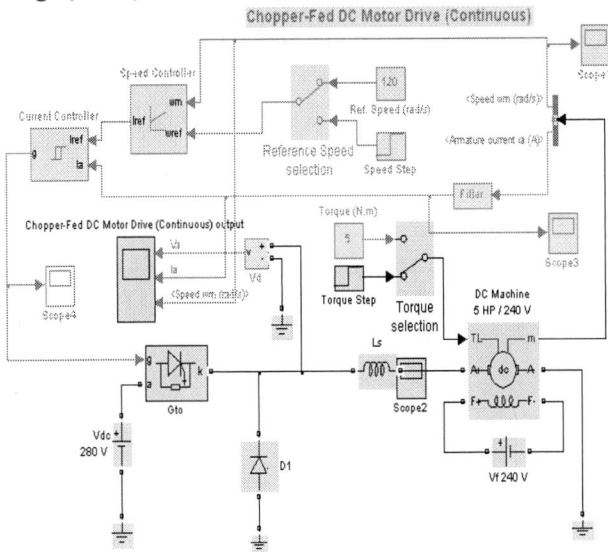

Fig.4. Chopper-Fed DC Motor Drive in Continuous.

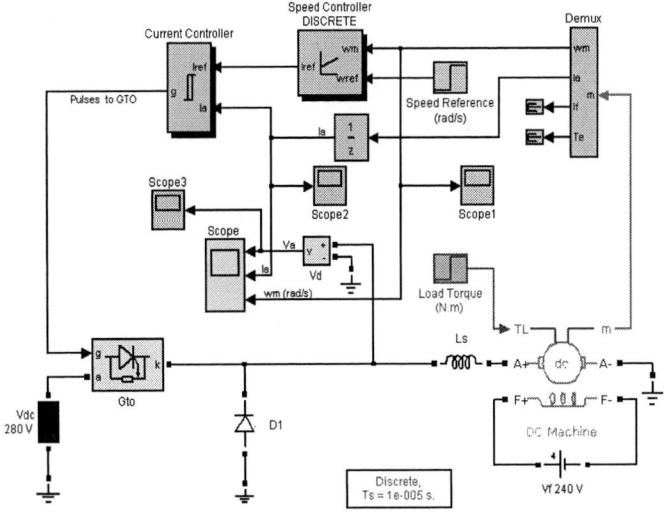

Fig.5. Chopper-Fed DC Motor Drive in Discrete.

Start the simulation. Observe the motor current, voltage, and speed during the starting on the scope. At the end of the simulation time (1.5 s), the system has reached its steady-state. *Response to a change in reference speed and load torque.*

The initial conditions state vector 'xInitial' to start with wm = 120 rad/s and TL = 5 N.m has been saved in the 'power_dc drive_init.mat' file. This file is automatically loaded in your workspace when you start the simulation (see Model Properties). In order to use these initial conditions you have to enable them. Check the Simulation/Configuration Parameters menu , then select "Data Import/Export" and check "Initial state".

Now, double click the two Manual Switch blocks to switch from the constant "Ref. Speed (rad/s) " and "Torque (N.m)" blocks to the Step blocks. (Reference speed wref changed from 120 to 160 rad/s at t = 0.4 s and load torque changed from 5 to 25 N.m at t= 1.2s). Restart the simulation and observe the drive response to successive changes in speed reference and load torque.

V. Simulation Results:

The simulation of the design is achieved by using ANFIS (Fuzzy Logic Controller (FIS) with rule reviewer) controller Matla Ver. 2010a toolbox. This simulation can be summarized as follow:
- ➢ *Firstly,* setting the configuration of ANFIS controller is shown in Fig.6.a. and rearrange the set of input's data and output data and the type FIS function that be used in simulation.
- ➢ After that by using ANFIS editor sets all parameters of identification of Neuro controller and load the input/output data. The training of Neoru-Fuzzy controllers of Chopper-Fed DC Motor Drive (continuous / discrete) and error signal for this training are shown in Fig's(6.a,7 ,8). The training of network is achieved by using back propagation algorithm with epoch length 30.

978-1-4577-0007-1/11 $26.00 © 2011 IEEE

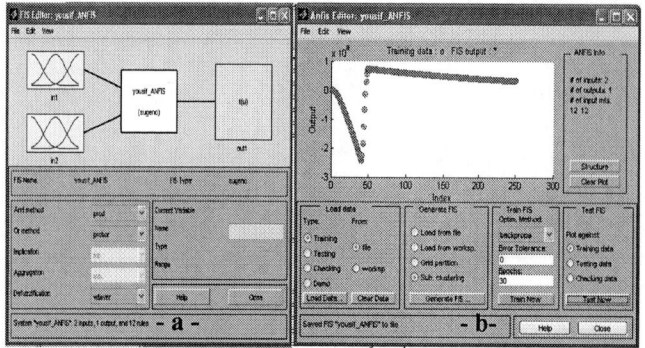

Fig.6. Training of ANFIS controller for Chopper-Fed DC Motor Drive Continuous.

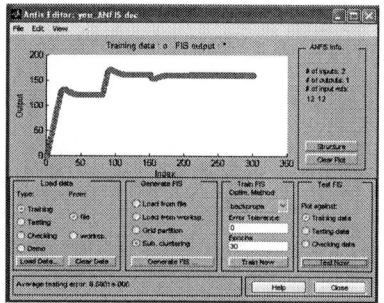

Fig.7. Training of ANFIS controller for Chopper-Fed DC Motor Drive Discrete.

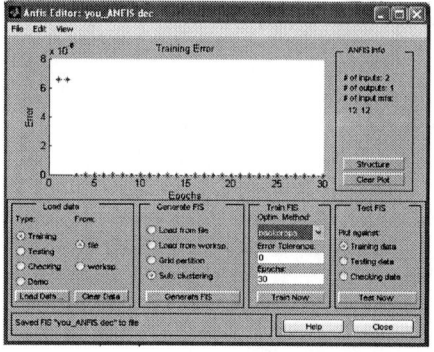

Fig. 8. Error signal of training network.

➢ The structure of ANFIS controller in continuous/discrete are used 12 rules, the structure is shown in Fig.9.a. and the form of rule is shown in Fig.9.b, while the membership functions for the first input is shown in Fig10.a.. and for second input is shown in Fig.10.b.

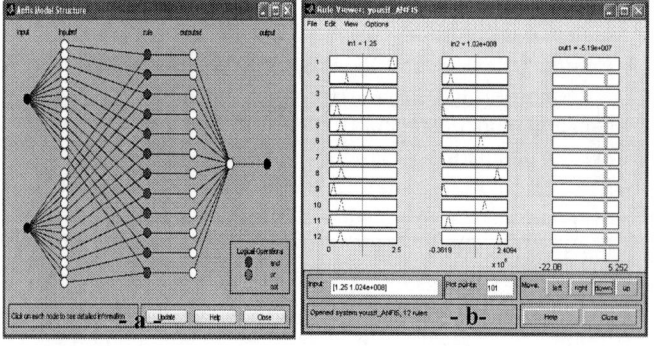

Fig.10. ANFIS Controller Structure and Rules for ANFIS controller.

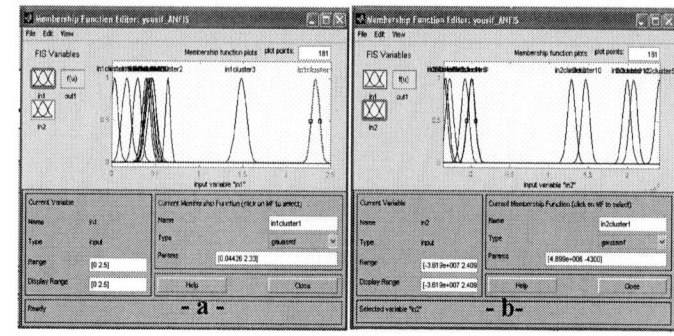

Fig.11.a,b. The membership functions for first and second inputs.

➢ The output of this design after complete and connect to the system is shown in Fig.12.a. for Chopper-Fed DC Motor Drive Discrete and Fig.12.b. for the Chopper-Fed DC Motor Drive Continuous. This results is describing the speed of motor (w rad/s), the armature current (I_a) and voltage (V_a). Are showing in the Fig. 12.c. and Fig.12.d. for two simulated circuit (continuous / Discrete).

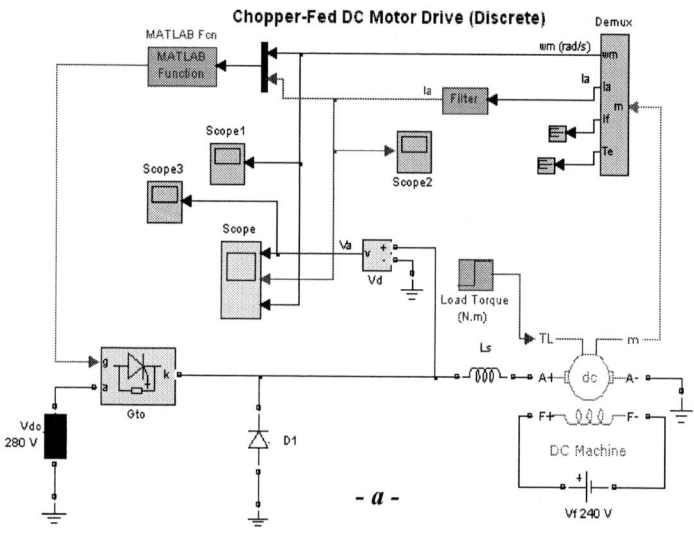

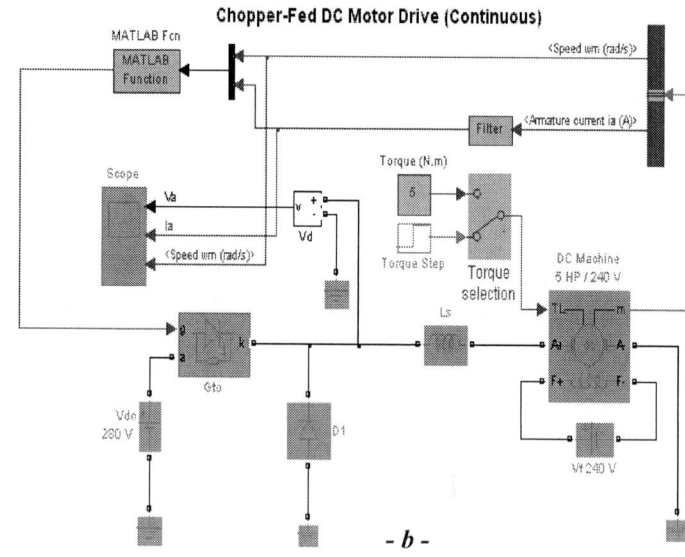

978-1-4577-0007-1/11 $26.00 © 2011 IEEE   114

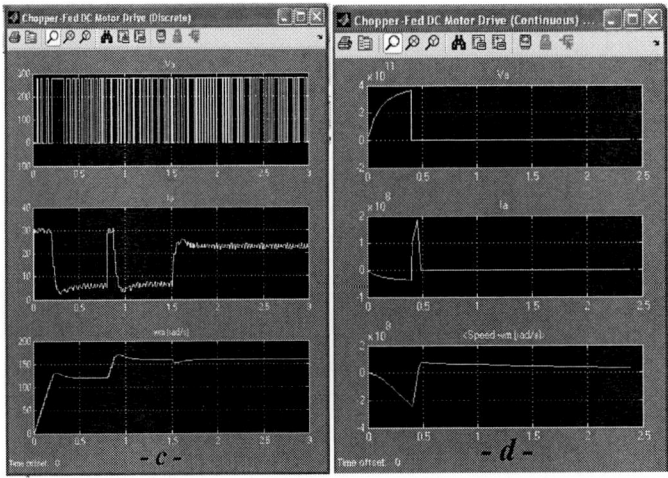

Fig.12. Simulation results

CONCLUSION

From the results were showing in above can be concluded that the ANFIS has the ability to compressed the power system (Chopper-Fed DC Motor drive continuous / discrete) to followed the desired output. This ability is depending on the accuracy of identification, where the knowledge of ANFIS controller with the performance of system depends on identification process. FIS controller Matlab toolbox is sufficient for design of ANFIS with any system by building the FIS function for this plant.

The parameters of ANFIS controller must be choose and then retuning by using trail and error to get optimal value of parameters. The error signal in identification process can be depends as the cost function of optimally parameters value. The mathematical model and knowledge with the performance of system are must be available to get the good results by busing ANFIS controller.

REFERENCES

[1]. M. Dotoli, B. Maione, B. Turchiano, " . Fuzzy-Supervised PID Control: Experimental Results", Via Re David, 200, I-70125 Bari, Italy, 2008.

[2]. P. Haiguo, W. Zhixin, " Simulation Research of Fuzzy-PID Synthesis Yaw Vector Control System of Wind Turbine", Wseas Ransactions on Systems and Control Manuscript received June 27, 2007; revised Sep. 22, 2007

[3] S. Chen, B.L. Lukb, C.J. Harrisa, L. Hanzoa, " Fuzzy-logic tuned constant modulus algorithm and soft decision-directed scheme for blind equalization", Digital Signal Processing 20 (2010) 846–859, www.elsevier.com/locate/dsp.

[4] R. Kumar, R. Gupta, R.S. Surjuse, " Adaptive Neuro-Fuzzy Speed Controller for Vector Controlled Induction Motor Drive ", this paper first received 17 Sept. 2008 and in revised form 10 July 2009, Digital Ref:A17050211, will published Kumar Rajesh et.al: Asaptive Neuro-Fuzzy…2011.

[5] S. Joseph Jawhar, N.S. Marimuthu, S.K Pillai and N. Albert Singh, " Neuro-Fuzzy Controller for a Non Linear Power Electronic Buck & Boost Converters", Asian Power Electronics Journal, Vol. 1, No. 1, Aug 2007.

[6] R. Krishnan," Electric Motor Drives Modeling, Analysis, and Control ", Prentiot HaJl, USA, 2001.

[7] D. Konaté," Mathematical Modeling, Simulation,Visualization and e-Learning", Proceedings of an InternationalWorkshop held at Rockefeller Foundation's Bellagio Conference Center, Milan, Italy, 2006.

[8] M. A. El-Sharkawi," Fundamentals of Electric Drives ", Brooks/Cole publishing Company, USA, 2002.

[9] A. Hughes ,"Electric Motors and Drives Fundamentals, Types and Applications", Third edition, Austin Hughes. Published by Elsevier Ltd, 2006.

[10] M. George, "Speed Control of Separately Excited DC Motor",American Journal of Applied Sciences 5 (3): 227-233, ISSN 1546-9239, 2008.

[11] Z. Liu, "Modeling and Simulation of Self-Tuning PI Control for Electrical Machines", Ingersoll-Rand Co. Ltd, USA, The paper first received on 19 March 2006 and in revised form 29 March 2007.

[12] B. Butkiewicz, " About Robustness of Fuzzy Logic PD and PID Controller under Changes of Reasoning Methods", Aachen, Germany, ESIT 2000, 14-15 September 2000.

[13] M.F.Maher, A. Yhya, R.M.Kuraz, " Design Fuzzy Self Tuning of PID Controller for Chopper-Fed DC Motor Drive", Al-Rafidain Engineering, PP, 54-66, Vol.16 No.2 2008.

[14] K. Tanaka, H. O. Wang, "Fuzzy Control Systems Design and Analysis: A Linear Matrix Inequality Approach",Copyright _ 2001 John Wiley & Sons, Inc.,ISBNs: 0-471-32324-1 ŽHardback.; 0-471-22459-6 ŽElectronic.

[15] J. Jantzen, " Design Of Fuzzy Controllers ", Technical University of Denmark, Department of Automation, Bldg 326, DK-2800 Lyngby, DENMARK.Tech. report no 98-E 864 (design), 19 Aug 1998.

[16] G. Chen,T.T. Pham, " Introduction to Fuzzy Sets, Fuzzy Logic, and Fuzzy Control Systems ", 2001 by CRC Press LLC.

[17] V. Kumar, K.P.S. Rana, V. Gupta, "Real-Time Performance Evaluation of a Fuzzy PI + Fuzzy PD Controller for Liquid-Level Process", International Journal of Intelligent Control and Systems, Vol. 13, NO. 2, JUNE 2008, 89-96

Direct Torque Control of Induction Machines Utilizing 3-level Cascaded H-Bridge Multilevel Inverter and Fuzzy Logic

A. Mortezaei[1], N. A. Azli[2], N. R. N. Idris, S. Mahmoodi and N. M. Nordin
Power Electronics and Drive Research Group (PEDG), Energy Research Alliance
Universiti Teknologi Malaysia, 81310 UTM Johor Bahru, Malaysia
mortezaei_ali@yahoo.com[1]
naziha@ieee.org[2]

Abstract—This paper proposes the use of a 3-level Cascaded H-Bridge Multilevel Inverter (CHMI) topology which results in further torque ripple minimization compared to the 2-level inverter-based Direct Torque Control (DTC). This is due to the increase in the inverter switching voltage vectors that allows minimization of the torque error. This in turn can reduce the Total Harmonic Distortion (THD) of the output voltage and current as well. This paper also presents two different control methods in selecting the appropriate output voltage vector for reducing the torque and flux error to zero. The first is based on the conventional DTC scheme using a pair of hysteresis comparators and look-up table to select the output voltage vector for controlling the torque and flux. The second is based on a new fuzzy logic controller (FLC) with Sugeno as its inference method to select the output voltage vector by replacing the hysteresis comparators and look-up table in the conventional DTC scheme. The latter has solved the problem of variable switching frequency which is the main characteristic of the former. By using FLC DTC not only the flux ripples reduce significantly but also the THD of the phase current decreases since a more sinusoidal current waveform is achieved. The simulation results have proven that by using the 3-level CHMI, torque ripple reduction is obtained compared to the 2-level inverter-based DTC while fuzzy DTC shows reduction in the stator flux ripples and the THD of the phase current.

I. INTRODUCTION

Direct Torque Control (DTC) has first been proposed by Takahashi in 1986. The basic of this high performance induction motor drive is limit cycle control and both fast torque response and efficiency operation are provided [1]. In DTC strategy the control of torque and speed are directly based on the electromagnetic state of the motor [2]. The main features of DTC are decoupled control of torque and flux, absence of mechanical transducers, very simple control scheme with low computational time, reduced parameter sensitivity and current regulator, PWM pulse generation as well as PI control of flux and torque and co-ordinate transformation are not required [3][4]. It only needs to know the stator resistance and terminal quantities (*v* and *i*) in order to perform the torque and stator flux estimations. The first industrial, speed-sensorless DTC induction motor drive has been introduced by ABB in 1996. This simple control scheme has gained popularity over the years and it is expected that they will soon replace the vector control drives commonly found in industrial applications [5].

It is well known that by using the multilevel inverter topology, the number of inverter states increases which results in having more inverter switching voltage space vectors to regulate instantaneously the electromagnetic torque and stator flux magnitudes in DTC strategy. Multilevel inverter also generates low dv/dt leading to low electromagnetic interference (EMI) and winding insulation stress which is desirable for high power and voltage applications [6]. By applying the 3-level cascaded H-bridge inverter, the number of effective switching voltage vectors is increased to nineteen including four different groups of voltage vectors. These different voltage vectors are available to be used to minimize the torque ripples of the induction machine.

Two different control methods are introduced to control the torque and flux in a 3-level CHMI-based DTC. The first is based on the conventional DTC scheme using a pair of hysteresis comparator and look-up table while the second is based on an FL DTC scheme using Sugeno as its inference method to select the output voltage vector for minimizing the torque and flux error to zero.

II. 3-LEVEL CASCADED H-BRIDGE MULTILEVEL INVERTER TOPOLOGY

Fig. 1 shows the schematic diagram of a 3-level CHMI. The CHMI is composed of three phases, of which in each phase the H-bridge inverter is fed by an independent DC source. Fig. 2 shows the voltage space vector diagram of the 3-level CHMI. It has 19 effective voltage vectors which is divided into 4 different groups according to their magnitudes as follows:

(V_0), $(V_1, V_4, V_7, V_{10}, V_{13}, V_{16})$, $(V_3, V_6, V_9, V_{12}, V_{15}, V_{18})$, $(V_2, V_5, V_8, V_{11}, V_{14}, V_{17})$.

Note that a 2-level inverter is only capable of producing 7 effective voltage vectors.

978-1-4577-0007-1/11 $26.00 © 2011 IEEE

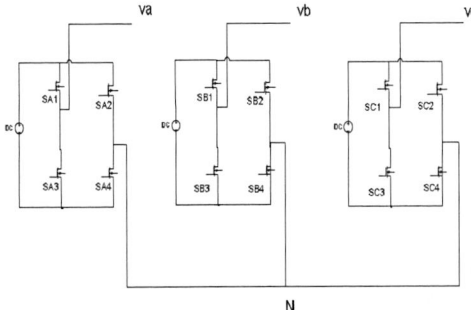

Fig. 1. 3-level CHMI

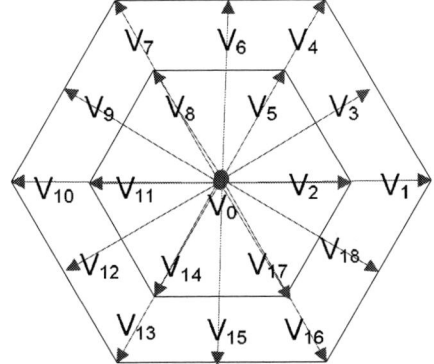

Fig. 2. Voltage space vector diagram of the 3-level CHMI

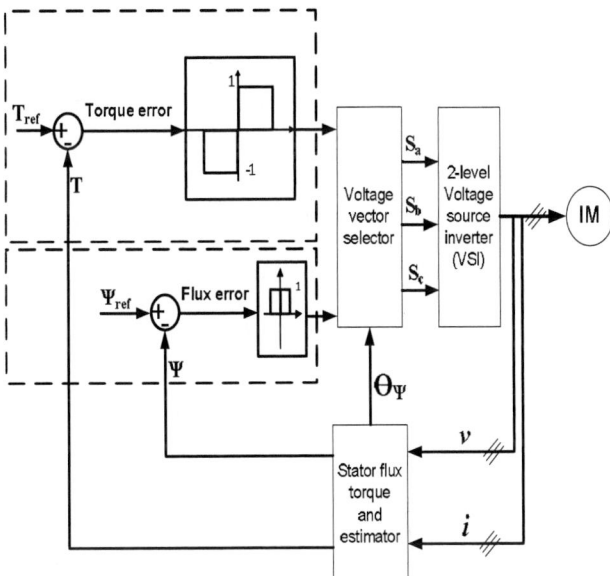

Fig. 3. 2-level inverter-based conventional DTC scheme

III.   3-LEVEL CHMI-BASED CONVENTIONAL DTC

The basic configuration of the conventional DTC drive presented by Takahashi is as shown in Fig. 3. It composed of a 2-level Voltage Source Inverter (VSI), torque and flux estimators, voltage vector selector and two hysteresis comparators [1].

The principle of the conventional DTC is to minimize the torque and flux errors to zero by using two hysteresis comparators. The hysteresis comparators are fundamental to the DTC scheme because they are responsible of both determining the appropriate voltage vector selection and the period of the voltage vector selected.

It has been mentioned that a 3-level CHMI has 19 effective voltage vectors that can be chosen to minimize the error of torque and flux. To take the most advantages of using the 3-level CHMI due to the increased number of space vectors, the following strategies are applied:

1. Dividing the stator flux plan into twelve sectors of 30° degrees, starting with the first sector situated between -30° and 0°. By increasing the number of sectors to twelve, a more accurate selection of the inverter switching voltage vectors to minimize the error of the torque and flux to zero can be obtained, resulting in improvement of the responses of the flux and torque [6].

2. Applying a 7-level hysteresis comparator to control the torque. Increasing the levels of torque hysteresis controller means defining more levels of error. This allows the controller to differentiate between small and large torque errors. This means that the voltage vectors chosen for large errors that happen during start up or due to a step change in torque or flux responses are different from those that are chosen during smaller errors or at steady state.

Fig. 4 shows the torque and flux hysteresis comparator of a 3-level CHMI based-conventional DTC. It can be seen that a 2-level hysteresis comparator is applied to control the flux and a 7-level hysteresis comparator is applied to control the torque. The output of the flux hysteresis comparator which is the flux error status has 2 values of 0 and 1 and the output of the torque hysteresis comparator which is the torque error status has 7 integer values starting from -3 to 3.

The optimum selection of the switching voltage vectors in all sectors of the stator flux plane in the 3-level CHMI - based conventional DTC can be tabulated in the so-called optimum switching voltage vector selection table given by Table I. The table is used to select the voltage vectors depending on flux error, torque error and the stator flux orientation. The terms (TES), (FES), and (SEC) are equivalent to stator flux error status, torque error status and the sectors respectively. When the flux error status is 0 the flux decreases and when it is 1 the flux increases. When the torque error status has the value of between 3 to 1, the torque increases; 3 has the highest value of increase while 1 has the lowest value of increase. When the torque error status is 0 the torque maintains and when it has the value of -3 to -1, the torque decreases; -3 has the highest value of decrease and -1 has the lowest value of decrease.

978-1-4577-0007-1/11 $26.00 © 2011 IEEE   117

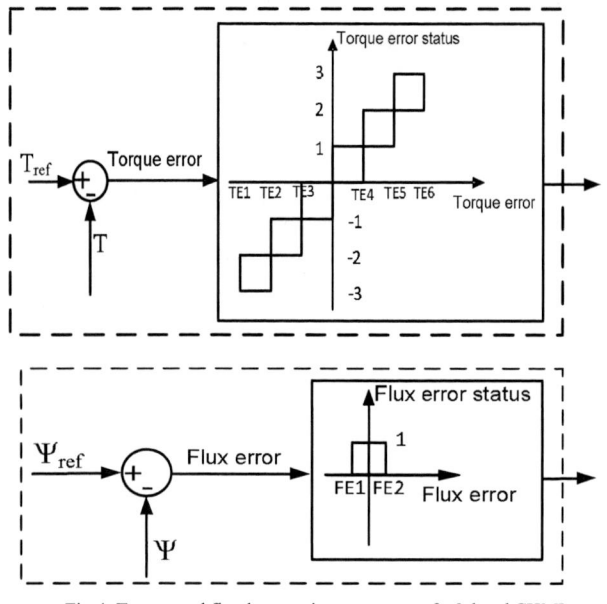

Fig.4. Torque and flux hysteresis comparator of a 3-level CHMI

TABLE I
VOLTAGE VECTOR SELECTION TABLE

FES	SEC / TES	S1	S2	S3	S4	S5	S6	S7	S8	S9	S10	S11	S12
0	3	7	7	10	10	13	13	16	16	1	1	4	4
	2	6	9	9	12	12	15	15	18	18	3	3	6
	1	8	8	11	11	14	14	17	17	2	2	5	5
	0	0	0	0	0	0	0	0	0	0	0	0	0
	-1	14	14	17	17	2	2	5	5	8	8	11	11
	-2	12	15	15	18	18	3	3	6	6	9	9	12
	-3	13	13	16	16	1	1	4	4	7	7	10	10
1	3	4	4	7	7	10	10	13	13	16	16	1	1
	2	3	6	6	9	9	12	12	15	15	18	18	3
	1	5	5	8	8	11	11	14	14	17	17	2	2
	0	0	0	0	0	0	0	0	0	0	0	0	0
	-1	17	17	2	2	5	5	8	8	11	11	14	14
	-2	15	18	18	3	3	6	6	9	9	12	12	15
	-3	16	16	1	1	4	4	7	7	10	10	13	13

IV. 3-LEVEL CHMI-BASED FL DTC

The application of FLC using Mamdani as the inference method to select the output voltage vector has already been investigated in DTC [7][6]. However an FLC using Sugeno as the inference method for selecting the output voltage vector is introduced in this paper which has resulted in the solving of the problem of variable switching frequency in conventional DTC due to the replacement of the hysteresis comparators and look-up table. By using FL DTC not only the flux ripples

reduce significantly but the Total Harmonic Distortion (THD) of the phase current also decreases since a more sinusoidal current waveform is achieved. The control scheme using the FLC is shown in Fig. 5.

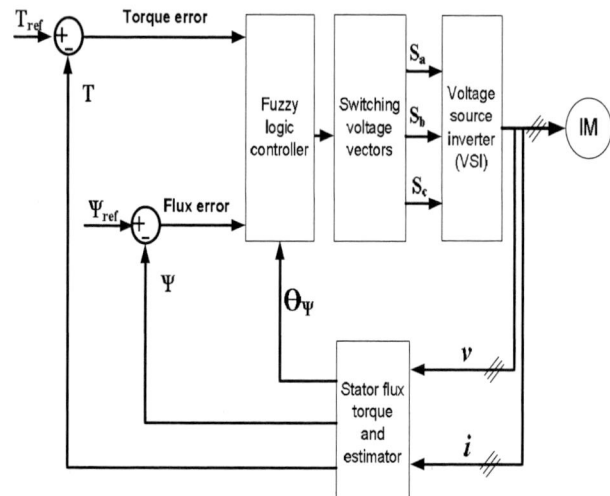

Fig. 5. DTC scheme based on FLC

There are 3 inputs and 1 output variables for the FLC. The inputs variables are the error of the stator flux, the error of the torque and the angle of the stator flux. The membership functions for the torque error, stator flux error and stator flux position are as shown in Fig. 6. The linguistic terms used for the error of stator flux and torque membership functions correspond to the stator flux and torque error status of Fig. 4 and Table I.

The universe of discourse of the stator flux position has been divided into 12 Fuzzy sets (S_1 to S_{12}) corresponding to the sectors in Table I. The output variable is discrete and includes 19 membership functions having the linguistic terms of (0, 1, 2 … 18) corresponding to 19 effective inverter switching voltage vectors of the 3-level CHMI (V_0 - V_{18}) of Fig. 2. It means that, for example, when the output of the FLC is 10, the inverter switching voltage vector V_{10} is selected for application to the inverter.

The total number of rules is 168. The inference method used is Sugeno's procedure based on prod-probor decision and the Wtaver method is used for defuzzification. With this method, each value of the stator flux and toque errors and the angle of stator flux are located in a specific membership function. The control output obtained is one of the membership functions of the output variable fed to the inverter. The controller rules are based on Table I. In the description of the membership functions the correspondence between the linguistic terms and the labels utilized in Table I has been explained. The control rules for this controller are presented utilizing the 3 input and 2 output variables. The i_{th} rule R_i is written as:
R_i: if the flux error is A_i, the torque error is B_i and the flux position is C_i while the inverter switching vector is N_i.

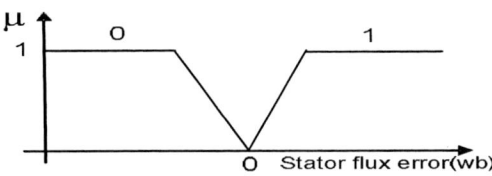

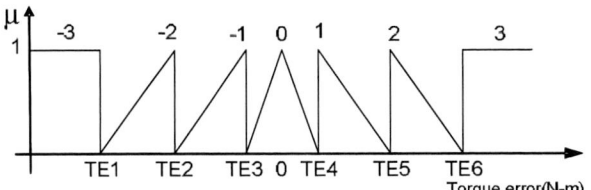

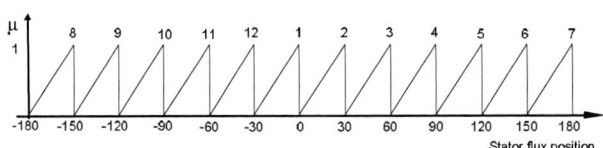

Fig. 6. Membership functions for the input variables of the FLC

TABLE II
MOTOR PARAMETERS AND REFERENCE VALUES

Nominal power,	4 KW
Nominal voltage (line-line)	400 V 50 Hz
Nominal frequency	50 Hz
Nominal speed	1430 RPM
Number of poles	4
Stator resistance and inductance	1.405Ω 0.005839H
Rotor resistance and inductance	1.395Ω 0.005839H
Mutual inductance	0.1722 H
Reference torque	10 Nm
Reference flux	1 Wb
Step time	0.01s

V. SIMULATION RESULTS

A total of 3 systems; 2-level inverter-based conventional DTC (2LC-DTC), 3-level CHMI-based conventional DTC (3LC-DTC), and 3-level CHMI-based FL DTC (3LF-DTC), have been simulated using MATLAB/Simulink to evaluate their performances. A 10 µs sample time has been set for all simulations conducted. Table II shows the induction motor parameters and reference values of torque and stator flux. The simulation carried out is the system response to a torque and stator flux step from 0 to the reference values for both magnitudes.   Fig. 7 shows the simulation results of conventional DTC of an induction motor using a 2-level inverter. Fig. 8 shows the simulation results of conventional DTC of an induction motor using a 3-level CHMI while Fig. 9 illustrates the simulation results of FL DTC of an induction motor using a 3-level CHMI.

By comparing the simulation results of (2LC-DTC) with (3LC-DTC), it is observed that the torque ripples reduce significantly in the latter. By comparing the simulation results of (3LC-DTC) with (3LF-DTC), it is seen that in the latter the

ripples of the flux as well as the percent THD of the phase current have reduced and a sinusoidal phase current has been achieved due to solving of the variable switching frequency problem created by the hysteresis controller.

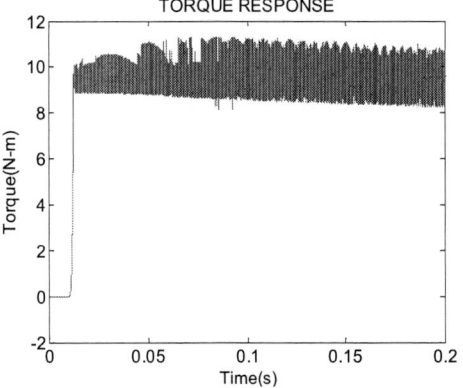

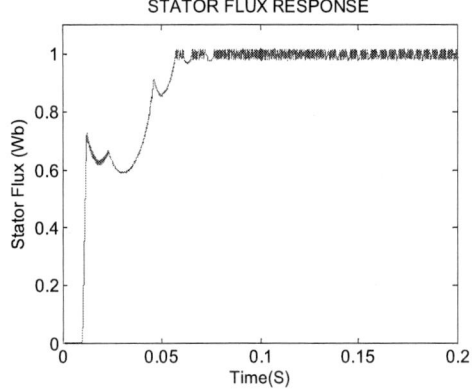

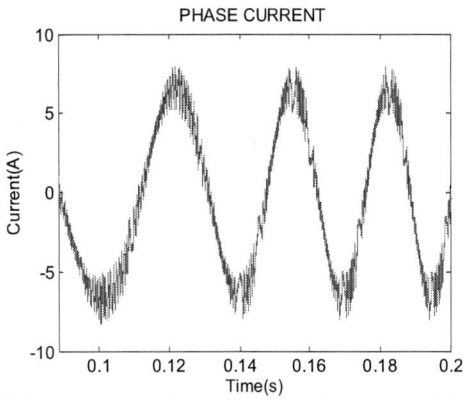

Fig. 7. Torque and stator flux responses and phase current of the 2-level inverter-based conventional DTC

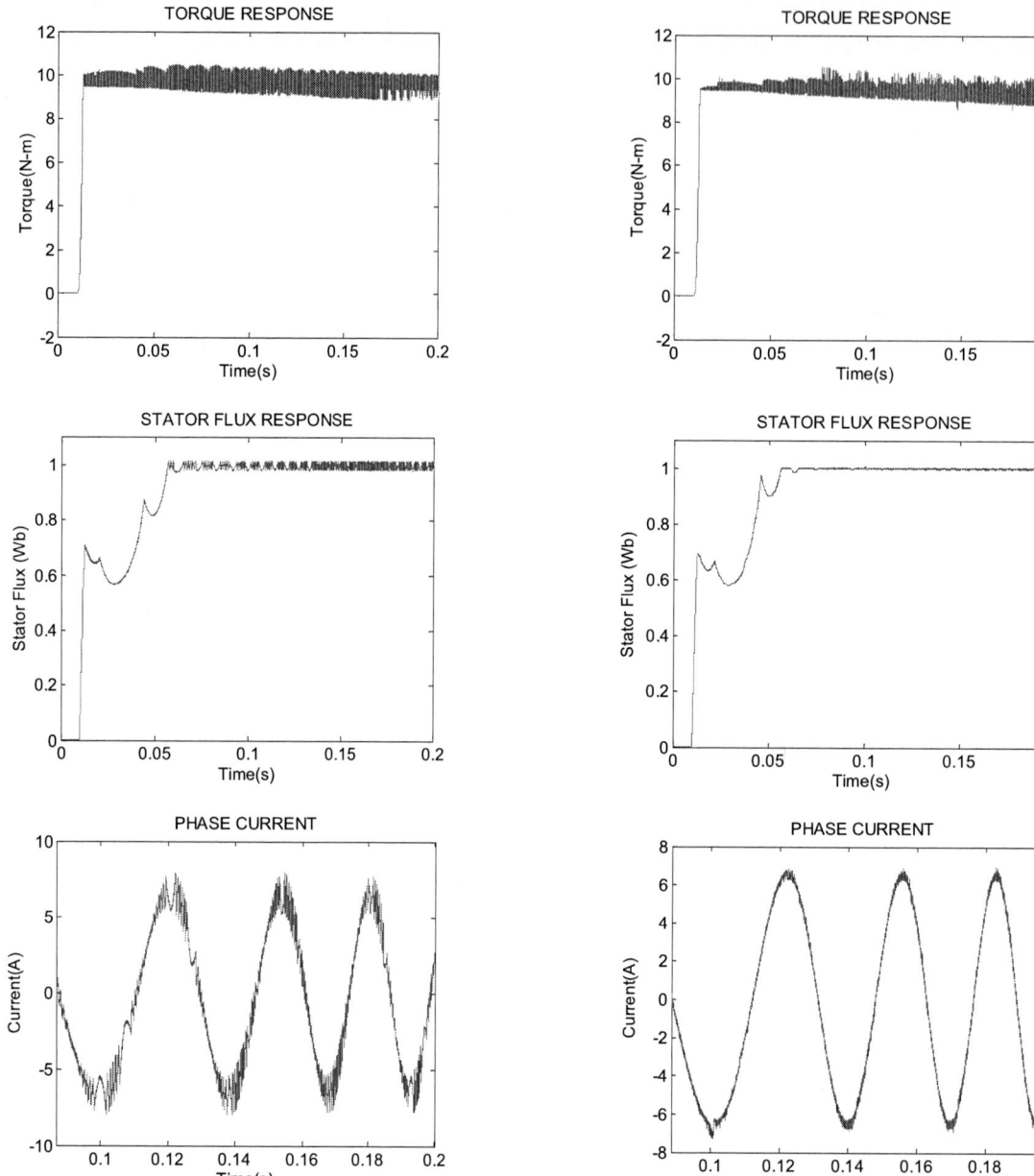

Fig. 8. Torque and stator flux responses and phase current of the 3-level CHMI-based conventional DTC

Fig. 9. Torque and stator flux responses and phase current of the 3-level CHMI-based FL DTC

VI. Conclusion

DTC enables both quick torque response and efficient operation and has many advantages compared to Field Oriented Control (FOC), such as less machine parameter dependence, simpler implementation and quicker dynamic torque response It only needs to know the stator resistance and terminal quantities (v and i) in order to perform the stator flux and torque estimations. Although DTC is gaining popularity, there are a few drawbacks that need to be addressed. Variable switching frequency and high torque and flux ripples are two major problems, which draws full attention of most researchers.

The simulation results have clearly shown that the 3-level CHMI-based DTC is capable of significantly reducing the torque ripples compared to the 2-level inverter-based DTC. The simulation results have also verified that utilizing the FLC that employs Sugeno as the inference method for selecting the output voltage vector instead of the hysteresis controller has resulted in the reduction of the flux ripples significantly as well as reduces the THD of the phase current by making it sinusoidal.

.

VII. References

[1] I. Takahashi, T. Noguchi, (1986) "A new quick-response and high efficiency control strategy of an induction motor", IEEE Trans. Ind. Appl., Vol. IA-22, No 5, pp. 820-827.

[2] John R G Schofield, (1995) "Direct Torque Control - DTC", IEE, Savoy Place, London WC2R 0BL, UK.

[3] L. Tang, L. Zhong, M. F. Rahman, Y. Hu, (2002) "An Investigation of a modified Direct Torque Control Strategy for flux and torque ripple reduction for Induction Machine drive system with fixed switching frequency", 37th IAS Annual Meeting Ind. Appl. Conf. Rec., Vol. 1, pp. 104-111.

[4] N. R. N. Idris, A. H. M. Yatim, (2000) "Reduced torque ripple and constant torque switching frequency strategy for Direct Torque Control of induction machine", 15th IEEE-Applied Power Electronics Conference and Exhibition 2000 (APEC 2000), Vol. 1, pp. 154-161.

[5] P. Tiitinen and M. Surandra, (1996) "The next generation motor control method, DTC direct torque control", Proceeding of the 1996 International Conference on Power Electronics Drives and Energy System for Industrial Growth, N. Delhi, India, Vol.1, pp. 37-43. 111

[6] del Toro, X., Calls, S., Jayne, M.G., Witting, P.A., Arias, A., Romeral, J.L., (2004) "Direct torque control of an induction motor using a three-level inverter and fuzzy logic", IEEE Conf., Vol.2, pp. 923-927

[7] Mir, S. A.: Zinger. D. S.: Elbuluk. M. E. "Fuzzy Controller for Inverter Fed Induction Machines." IEEE Transactions ont Itndustrial Applications. vol. IA-21, no. 4. Jani/Feb 1993. pages 1009-1015.

FPGA Based High Precision Torque and Flux Estimator of Direct Torque Control Drives

Tole Sutikno, Nik Rumzi Nik Idris, Aiman Zakwan Jidin, Mohd Zaki Daud
Department of Energy Conversion, Faculty of Electrical Engineering, Universiti Teknologi Malaysia (UTM)
email: thsutikno@gmail.com, nikrumzi@ieee.org

Abstract- **This paper presents an improved FPGA-based torque and stator flux estimators for direct torque control (DTC) induction motor drives, which permit very fast calculations. The improvements are performed by 1) using two's complement fixed-point format approach to minimize calculation errors and the hardware resources usage in all operations, 2) calculating the discrete integration operation of stator flux using backward Euler approach, 3) modifying the non-restoring method to calculate complicated square root operation of stator flux, 4) introducing a new sector judgment method, and 5) reducing the sampling frequency down to 5μs. To avoid saturation due to DC offset present in the sensed currents, the LP Filter is applied. The simulation results of DTC model in MATLAB/Simulink, which performed double-precision calculations, are used as references to digital computations executed in FPGA implementation. The Hardware-in-the-loop (HIL) method is used to verify the minimal error between MATLAB/Simulink simulation and the experimental results, and thus the well functionality of the implemented estimators.**

I. INTRODUCTION

Direct Torque control (DTC) was first introduced by Takahashi (1986) [1] and Depenbrock (1988) [2] as an alternative for controlling induction machines. It has simple structure with fast torque response. Furthermore, it does not require PWM pulse generator, coordinate transformation, position encoder as well as current regulators [3-5].

The DTC algorithm is frequently implemented by serial calculations based on a Microcontroller or Digital Signal Processor (DSP) [3-4, 6]. These hardware are truly software-based platform and is not suitable for an implementation that require high-speed computation. As a solution, FPGA is proposed to perform very fast executions [5, 7]. Moreover, the high sampling frequency in FPGA allows the minimization of torque ripple [8-10].

Unfortunately, it is not easy to implement DTC in FPGA device, especially in performing the torque and flux estimation Complex digital computations are involved, such as binary multiplications, integral operation and also a square root calculation targeted for FPGA implementation; the difficulties of which have been addressed in several researches [11-15].

According to [1-2], sampling time is the crucial part of torque and flux estimation in DTC. Toh [12] has implemented all parts of the DTC in FPGA hardware except for torque and flux estimator whereby it was estimated using the DSP; as a result, the sampling time was reduced to 55μs. On the other hand, Monmasson [5] has developed the implementation of

DTC in FPGA hardware. However since it was implemented using a Xilinx System Generator fixed-point toolbox, the sampling time is limited to 50μs. Ferreira [13] also has difficulty in increasing the sampling frequency, and he only succeeded in reducing the sampling time to 25μs. Other work include [16] with a sampling time of 25μs, and [14, 17] of 150 μs and 100μs respectively.

The main contribution of the research presented in this paper is the new design of the torque and flux estimator for DTC implementation in FPGA, with a sampling time reduced to 5μs. In the design, new implementation architecture, improved digital properties, new square root algorithm and new sector identification method is introduced. They are implemented by using two's complement fixed-point representation with variable words' sizes. The design is prepared for fast computation, and therefore there is no need of using CORDIC algorithm [5, 15], soft-core CPU [16], as well as transformation from Cartesian to polar coordinates [18]. By using the proposed method, a simple control structure of DTC as introduced in [1] can be retained. The implementation technique is verified experimentally.

II. DIRECT TORQUE CONTROL DRIVES

Fig. 1 represents the scheme of a DTC drive. The estimated flux magnitude and torque are compared with their references values. Torque and flux comparators consisted of three and two-level hysteresis comparators respectively. The sector judgment is used to evaluate the position of the stator flux vector in DQ coordinates.

The switching table produces the switching status according to the outputs of torque and flux comparators and the sector judgment. These switching status are connected to the inverter, which is connected to the motor. They are also used as the inputs to torque and flux estimator.

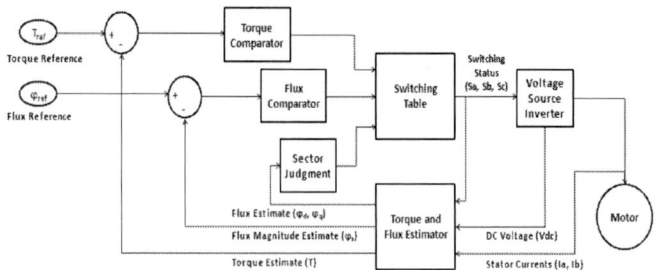

Fig. 1. DTC scheme

In order to estimate the stator flux and the electromagnetic torque, several variables need to be determined. Firstly, the stator currents from the motor I_a and I_b, are transformed into DQ coordinates, which are used in DTC algorithm, as follows:

$$I_D = I_a \tag{1}$$

$$I_Q = \frac{\sqrt{3}}{3}(I_a + 2I_b) \tag{2}$$

At the same time, by using the switching status (Sa, Sb and Sc) produced by the switching table, the stator voltages in DQ components are determined:

$$V_D = \frac{V_{dc}}{3}(2S_a - S_b - S_c) \tag{3}$$

$$V_Q = \frac{\sqrt{3}}{3}V_{dc}(S_b - S_c) \tag{4}$$

Then, using the calculated I_d, I_q, V_d and V_q, the estimation of the stator flux in DQ coordinates are performed as follows:

$$\varphi_D = \varphi_{D_{old}} + (V_D - R_S I_D)T_s \tag{5}$$

$$\varphi_Q = \varphi_{Q_{old}} + (V_Q - R_S I_Q)T_s \tag{6}$$

Finally, equation (7) calculates flux magnitude by using a square root calculation, whereas the electromagnetic torque is estimated in equation (8).

$$\varphi_s = \sqrt{\varphi_D^2 + \varphi_Q^2} \tag{7}$$

$$T = \frac{3}{4}P(I_Q\varphi_D - I_D\varphi_Q) \tag{8}$$

III. PROPOSED METHOD TO IMPROVE TORQUE AND FLUX ESTIMATOR

The algorithm of torque and flux estimation is implemented in an architecture consisted of six main blocks, as shown in Fig. 2. This architecture has six inputs: two 21-bit currents (I_a and I_b), 12-bit high voltage DC-supply (V_{dc}) and three switching status Sa, Sb and Sc. At the end, it produces three outputs: the estimation values of torque (T), flux (φ_s) and sector φ_s. The sampling time is set to 5 µs, which is limited by the ADC used.

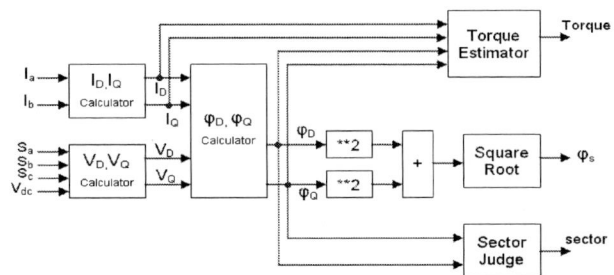

Fig. 2 Block Diagram of torque and flux estimators

A. Torque and Flux Estimator Architecture

All the equations which modelled the motor behavior are implemented in a two-stage-pipelined architecture, as presented in Fig. 3. Several mathematical operations are performed in parallel. At the first stage, stator currents and voltages in DQ-coordinates are calculated in parallel so that those results can be used to estimate the stator flux in the same stage. The resulted currents and flux are used to determine the flux magnitude and the torque estimation in the second stage. A 62-bit non-restoring square root is implemented in order to compute the flux magnitude.

As a matter of fact, [13] proposed that three-stage-pipelined architecture should be implemented in this module, by separating the computation of stator currents and voltages from the estimation of the stator flux. However, the former can be considered as an immediate calculation and thus, those

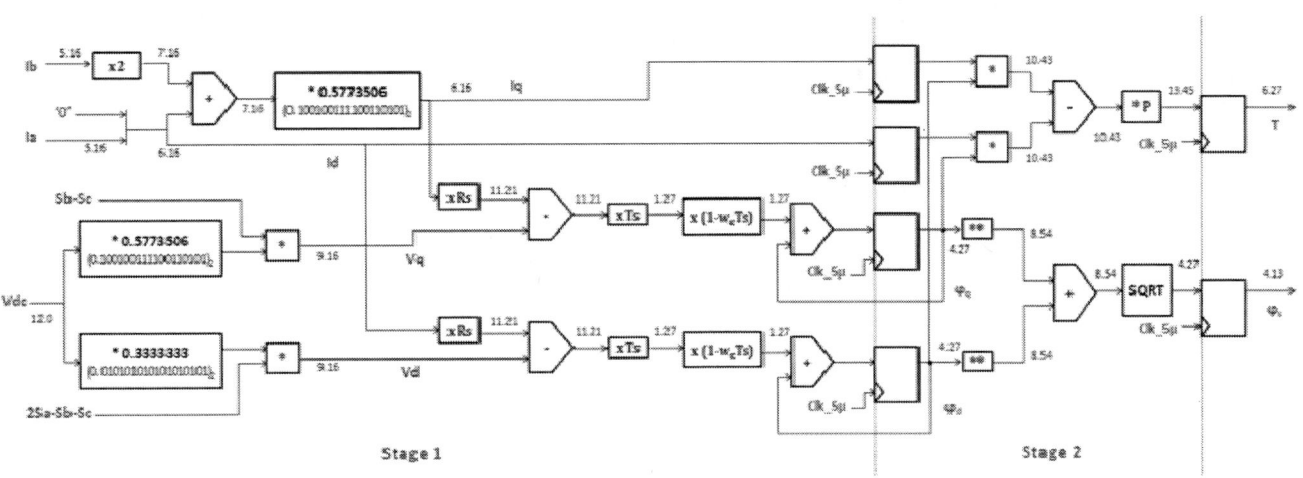

Fig. 3. The architecture of torque and flux estimators

978-1-4577-0007-1/11 $26.00 © 2011 IEEE

calculations can be merged into one single stage. As a consequence, the latency of the estimator is reduced from 15 μs to 10 μs.

B. The Digital Properties

To achieve a good implementation, several digital properties need to be considered when designing this estimator. Adopted binary format, quantization and sampling time are among those key factors.

1). Binary format representation

In this implementation, two's complement fixed-point representation is used in all operations, except for the square root calculation. In this particular case, unsigned fixed-point representation is applied, since its operand and its results are always positive. The major advantage of using two's complement fixed-point representation for real numbers is that it adheres to the same basic arithmetic principles as integers. The representation makes the system both simpler to implement and capable of easily handling higher precision arithmetic that does not require examining the signs of the operands to determine whether to add or subtract [14, 19].

Recent DTC implementation as in [5, 16] generally used 32-bit format where some bits might be left unused, while 16-bit format is not appropriate to achieve good DTC implementation [13, 16]. Therefore, variable word-size approach [13] is adopted for this implementation and so, all the redundant bits can be eliminated by truncating process to minimize the hardware resources usage. In additional, in this implementation, extended word-size more than 32-bit for the magnitude of flux and square root operation is used.

2). Quantization

The determination of word size is one of the critical parts in FPGA implementation. On one hand, insufficient number of bits used may reduce the precision or cause the calculation error, which can unstabilize the whole system. On the other hand, larger words used may increase the hardware area used for the implementation.

Since two's complement fixed-point format is used for the implementation, at least 2 things that need to be verified. Firstly, the size of the integer must be properly chosen to avoid the problem of overflow. Secondly, the number of the fractional bits used must be sufficient in order to minimize the calculation error. For example, due to the fact that the input currents I_a and I_b are varied from -10A to 10A, at least 5 bits are necessary for the integer bits. While 16 fractional bits used can result in a very good precision, since the step change per bit is very small ($\simeq 15$ μA).

One of the critical parts in this architecture is the stator flux estimation, where the integration is performed. This operation can easily produce errors if the sampling time Ts is not properly scaled; in this case, Ts = 5 μs = 0.000005 s. In fact, 21 bits are necessary at the minimum to represent Ts. In this case, Ts = 0.00000476837 s $(0.000000000000000001010)_2$. However, 27-bit representation is chose to have a better precision and thus, Ts = 0.00000499934 s $(0.000000000000000001010011111)_2$.

Fig. 4 shows the estimated torque taken during the steady state. From this figure, it can be seen that the torque estimation for 21-bit Ts is imprecise, when compared to MATLAB/Simulink double precision estimation, which is the ideal case. In fact, the number of bits is increasing after each operation in order to avoid calculation errors or imprecision. This will result in the rising of the hardware area used. Therefore, truncation process must be performed avoid excessive increase of the number of bits used

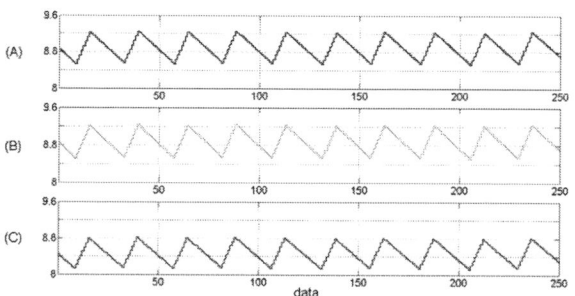

Fig. 4. The torque estimation during steady state. (A) Estimated torque for Matlab double precision; (B) Estimated torque for Ts in 27 bits; (C) Estimated torque for Ts in 21 bits.

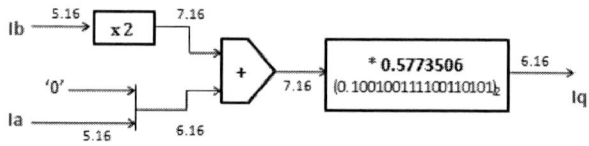

Fig. 5. The example of Iq calculation

In the example of I_q calculation, as shown in Fig. 5, when I_b is multiplied by 2, the result should be in 6.16 bits (6 integer bits plus 16 fractional bits). Nevertheless, it is stored in 7.16 bits to avoid overflow which may happened during the addition operation. Next, when the addition result is multiplied by $\frac{1}{3}\sqrt{3}$, which is coded in 1.18 bits, I_q should be represented in 8.34 bits, but it is truncated to 6.16 bits.

3). Sampling time

The sampling time Ts is limited to 5 μs by the ADC used. Therefore all the operations involved in this model were performed within this sampling time.
It should be noted that the use of high sampling frequency is important in DTC implementation, for the purpose of minimizing the torque ripple. The sampling time used for DSP implementation is normally much larger than Ts, which is not

less than 50 µs. Therefore, it is reduced by a factor of 10 for this FPGA implementation and thus, lower torque ripple is produced, as shown in Fig. 6.

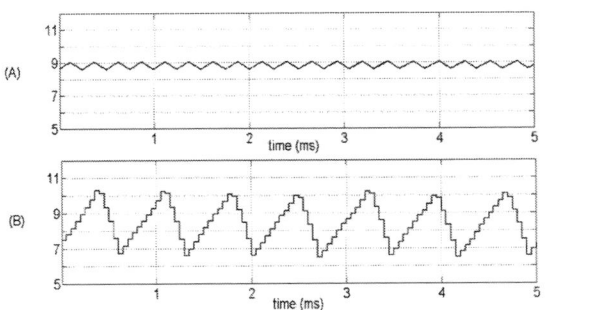

Fig. 6. The effect sampling time to torque ripple
(A) Estimated torque for Ts=5µs; (B) Estimated torque for Ts=50µs.

C. Backward Euler Approach

The discrete backward Euler formula is $y(n) = y(n-1) + k.T.u(n)$. It is simpler for implementation in FPGA hardware than the forward Euler and Trapezoidal method that required the register to store the previous value of $u(n-1)$ function. The backward Euler integration method is also capable of maintaining the system stability in the large step size. Therefore, the discrete backward Euler integration method is chosen to calculate the quadrature flux (φ_D and φ_Q).

D. LP Filter

In the stator flux estimator ((5) and (6)), R_s is the estimated stator resistance, while Ts is the implementation sampling time. These equations correspond to the integration using Backward Euler Method. In [20-21], it was shown that a filter should be added to the integrator in the practical implementation to avoid integration drift problem due to DC offset present in the sensed currents. Thus, equation (5) and (6) become:

$$\varphi_D = (\varphi_{D_{old}} + (V_D - R_S I_D)T_s)(1 - \omega_c - T_s) \quad (9)$$

$$\varphi_Q = (\varphi_{Q_{old}} + (V_Q - R_S I_Q)T_s)(1 - \omega_c - T_s) \quad (10)$$

Choosing an appropriate cut-off frequency for the LP Filter is very important to optimize steady state operation, which depends on the operating frequency. By setting a cut-off frequency closer to the operating frequency, the dc offset in the estimated stator flux can be reduced. However, it will introduce phase and magnitude errors. The introduction of these errors will not be discussed in this paper.

E. Non-restoring Square Root Algorithm

The stator flux (φ_s) in DTC drive is calculated from the square root of quadrature flux magnitude. To calculate the stator flux (φ_s), the non-restoring square root algorithm as proposed in [22] is modified as below (D=radicand, q=quotient and r=remainder):

$$r_0 = -1 \quad (^n/_2 + 2bits)$$
$$q_0 = 0 \quad (^n/_2 + 1bits)$$
For i=0 to n-1 do:
If $r_i \geq 0$ then
$$r_{i+1} = (4r_i + D_{(n-2i)-1}D_{(n-2i)-2}) - (4qi + 1)$$
else
$$r_{i+1} = (4r_i + D_{(n-2i)-1}D_{(n-2i)-2}) + (4qi + 3)$$
If $r_{i+1} \geq 0$
$$q_{i+1} = 2q_i + 1$$
else
$$q_{i+1} = 2q_i$$

The final result of the square root is equal to $q_n(^n/_2 -1\ down\ to\ 0)$, coded in $^n/_2$ bits.

F. New Sector Identification

The present work has created a simpler method to judge the sectors of voltage vector based on comparison between $\varphi_Q, \sqrt{3}\varphi_D, -\sqrt{3}\varphi_D$ and 0 refer to Table 1, which is modified from [23]. With the comparison, it is simpler to determine the sector of voltage vector compared to conventional method through arc tan of angle, three stages comparison based on φ_D, φ_Q or determination of angle using CORDIC algorithm [15].

Table 1. The proposed simpler identification of the sectors

Sector	Vector Angle	$\varphi_Q > 0$	$\varphi_Q > \sqrt{3}\varphi_D$	$\varphi_Q > -\sqrt{3}\varphi_D$
I	$(0^0, 60^0)$	1	0	1
II	$(60^0, 120^0)$	1	1	1
III	$(120^0, 180^0)$	1	1	0
IV	$(180^0, 240^0)$	0	1	0
V	$(240^0, 300^0)$	0	0	0
VI	$(300^0, 360^0)$	0	0	1

1: satisfy, 0: not satisfy

The proposed method is a single stage with three comparisons performed in parallel without calculation of angle; so it is a fast computation and hence incorrect voltage vector selection can be reduced.

II. EXPERIMENT SETUP

The validation of designed torque and flux comparators was

978-1-4577-0007-1/11 $26.00 © 2011 IEEE

performed based on Hardware-in-the-Loop (HIL) method. The DTC model in MATLAB/Simulink was simulated and then, the same data I_a, I_b, S_a, S_b, and S_c used for the simulation were copied from MATLAB workspace to VHDL codes, as the inputs for the targeted FPGA. The VHDL codes were simulated in ModelSim-Altera before being synthesized and implemented in FPGA. The test design flow is presented in Fig. 7.

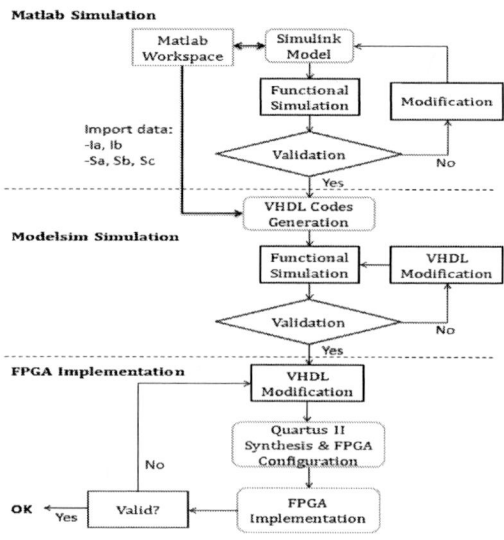

Fig. 7. Top-down test design flow

In the implementation of DTC based on FPGA, there are many digital properties that need to be considered, such as binary format representation (data type), quantization, sampling time and word length (size). In fact, error or inaccuracies due to these digital problems can seriously affect the estimation. For example, the torque ripple can be reduced from 2N-m to 0.2N-m corresponding to a change in sampling time from 50µs to 5µs, as shown Fig. 6. In this paper, these digital properties have been considered in the implementation.

It is shown in that the calculation of torque and flux are can be performed with high precision and small sampling time. The new architecture utilized the backward Euler approach, LP Filter-based estimator, modified non-restoring square root algorithm and new sector identification, which are the essential elements of high performance DTC motor drives. Furthermore, the design is simple and flexible and the VHDL source code can be easily modified. For example, the truncation process can be adjusted to avoid excessive increase in the hardware resource of FPGA, but its accuracy is still sufficient.

III. RESULTS AND DISCUSSION

The experiments were conducted on Altera APEX

EP20K200EFC484-2x, and used 2093 logic elements for the implementation. The tests were performed when the motor was operated in steady state condition, and the results were observed on the oscilloscope. For validation purposes, the results were compared with the MATLAB/Simulink simulation results. Fig. 8 presents the input test, while Fig. 9 and Fig. 10 shows the comparisons between MATLAB/Simulink simulations and the experimental results. The results show that experimental results are in agreement with the simulation in Simulink, where it is conducted in double-precision computation.

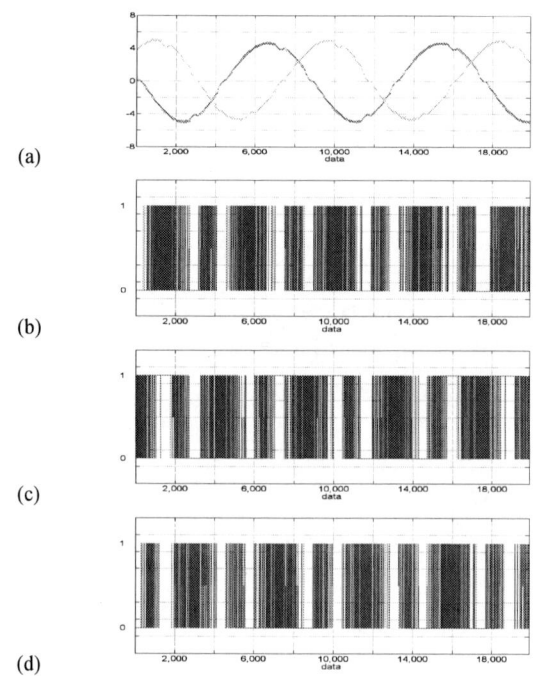

Fig. 8. The inputs test. (a) the stator currents Ia & Ib; (b) Sa; (c) Sb; (d) Sc.

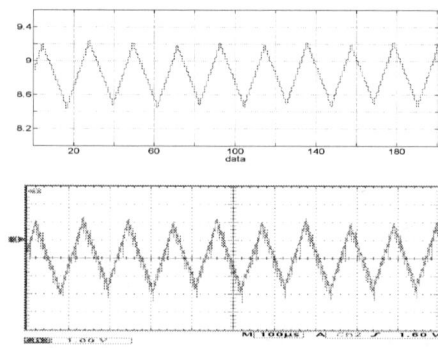

Fig. 9. The comparison between MATLAB/Simulink simulation and experimental result for torque estimation.

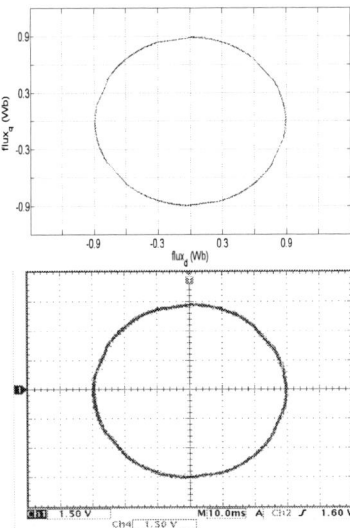

Fig. 10 The comparison between MATLAB/Simulink simulation and the experimental result for flux locus.

IV. CONCLUSION

FPGA is suitable for high-performance DTC implementation owing to its high speed processing. The sampling time for the stator flux and torque estimation is extremely small, which cannot be obtained using DSP or fast microcontroller. The choice of word sizes, the binary format and the sampling time used are very important in order to achieve good implementation of the estimators. To get simpler implementation and fast computation, the backward Euler approach to calculate the discrete integration operation of stator flux, the modified non-restoring method to calculate complicated square root operation of stator flux and a new sector judgment method were introduced. The design, which was coded in synthesizable VHDL for the implementation on Altera APEX20K200EFC484-2x device has produced very precise estimations with minimal error and was verified with MATLAB/Simulink double-precision simulation.

ACKNOWLEDGMENT

The authors would like to thank Universiti Teknologi Malaysia for providing the funding for the research.

REFERENCES

[1] I. Takahashi and T. Noguchi, "A New Quick-Response and High-Efficiency Control Strategy of an Induction Motor," *IEEE Transactions on Industry Applications*, vol. Vol.IA-22, No.5, pp. 820-827, Sept/Oct 1986.

[2] M. Depenbrock, "Direct self control (DSC) of inverter-fed induction machine," *IEEE Trans. on Power Electronics*, vol. 3 (4), pp. 420–429, 1988.

[3] B. K. Bose and P. M. Szczesny, "A microcomputer-based control and simulation of an advanced IPM synchronous machine drive system for electric vehicle propulsion," *Industrial Electronics, IEEE Transactions on*, vol. 35, pp. 547-559, 1988.

[4] S. M. A. Cruz, *et al.*, "DSP implementation of the multiple reference frames theory for the diagnosis of stator faults in a DTC induction

motor drive," *Energy Conversion, IEEE Transactions on*, vol. 20, pp. 329-335, 2005.

[5] E. Monmasson and M. N. Cirstea, "FPGA Design Methodology for Industrial Control Systems: A Review," *Industrial Electronics, IEEE Transactions on*, vol. 54, pp. 1824-1842, 2007.

[6] L. Lianbing, *et al.*, "A high-performance direct torque control based on DSP in permanent magnet synchronous motor drive," in *Intelligent Control and Automation, 2002. Proceedings of the 4th World Congress on*, 2002, pp. 1622-1625 vol.2.

[7] V. D. Colli, *et al.*, "Design of a System-on-Chip PMSM Drive Sensorless Control," in *Industrial Electronics, 2007. ISIE 2007. IEEE International Symposium on*, 2007, pp. 2386-2391.

[8] M. Mizuochi, *et al.*, "Multirate Sampling Method for Acceleration Control System," *Industrial Electronics, IEEE Transactions on*, vol. 54, pp. 1462-1471, 2007.

[9] R. Morales-Caporal and M. Pacas, "A Predictive Torque Control for the Synchronous Reluctance Machine Taking Into Account the Magnetic Cross Saturation," *Industrial Electronics, IEEE Transactions on*, vol. 54, pp. 1161-1167, 2007.

[10] C. L. Toh, *et al.*, "Constant and high switching frequency torque controller for DTC drives," *Power Electronics Letters, IEEE*, vol. 3, pp. 76-80, 2005.

[11] K. Piromsopa, *et al.*, "An FPGA Implementation of a fixed-point square root operation," presented at the Int. Symp. on Communications and Information Technology (ISCIT 2001), ChiangMai, Thailand, 2001.

[12] C. L. Toh, *et al.*, "Implementation of a New Torque and Flux Controllers for Direct Torque Control (DTC) of Induction Machine Utilizing Digital Signal Processor (DSP) and Field Programmable Gate Arrays (FPGA)," in *Power Electronics Specialists Conference, 2005. PESC '05. IEEE 36th*, 2005, pp. 1594-1599.

[13] S. Ferreira, *et al.*, "Design and prototyping of direct torque control of induction motors in FPGAs," in *Integrated Circuits and Systems Design, 2003. SBCCI 2003. Proceedings. 16th Symposium on*, 2003, pp. 105-110.

[14] L. Charaabi, *et al.*, "FPGA-based real-time emulation of induction motor using fixed point representation," in *Industrial Electronics, 2008. IECON 2008. 34th Annual Conference of IEEE*, 2008, pp. 2393-2398.

[15] C. T. Kowalski, *et al.*, "FPGA Implementation of DTC Control Method for the Induction Motor Drive," presented at the EUROCON, 2007. The International Conference on Computer as a Tool, 2007.

[16] Y. Utsumi, *et al.*, "Comparison of FPGA-based Direct Torque controllers for Permanent Magnet Synchronous Motors," *Journal of Power Electronics*, vol. 6, April 2006.

[17] R. Morales-Caporal and M. Pacas, "Digital implementation of a direct mean torque control for AC servo drives based on a hybrid DSP/FPGA controller system," in *Power Electronics Congress, 2008. CIEP 2008. 11th IEEE International*, 2008, pp. 77-83.

[18] C. T. Kowalski and J. D. Lis, "Speed sensorless DTC control of the induction motor using FPGA implementation," *COMPEL: The International Journal for Computation and Mathematics in Electrical and Electronic Engineering*, vol. 29, pp. 109-125, 2010.

[19] B. W. Robinson, *et al.*, "Fixed and Floating-Point Implementations of Linear Adaptive Techniques for Predicting Physiological Hand Tremor in Microsurgery," *Selected Topics in Signal Processing, IEEE Journal of*, vol. 4, pp. 659-667, 2010.

[20] N. R. N. Idris and A. H. M. Yatim, "An improved stator flux estimation in steady state operation for direct torque control of induction machines," in *Industry Applications Conference, 2000. Conference Record of the 2000 IEEE*, 2000, pp. 1353-1359 vol.3.

[21] N. R. N. Idris and A. H. M. Yatim, "An improved stator flux estimation in steady-state operation for direct torque control of induction machines," *Industry Applications, IEEE Transactions on*, vol. 38, pp. 110-116, 2002.

[22] S. Samavi, *et al.*, "Modular array structure for non-restoring square root circuit," *Journal of Systems Architecture*, vol. 54, pp. 957-966, 2008.

[23] T. Sutikno, *et al.*, "New approach FPGA-based implementation of discontinuous SVPWM," *Turk J Elec Eng & Comp Sci*, vol. 18, p. 6, 2010.

Inter-Turn Stator Winding Fault Diagnosis and Determination of Fault Percent in PMSM

M. A. Shamsi nejad and M. Taghipour
Department of Electrical and Computer Engineering
University of Birjand
Birjand, IRAN

Abstract- **Fault occurrence in electrical machines is avoidable but, quick fault detection can increase system reliability. According to studies of electrical power research institute (EPRI), most of the occurred faults in electrical motors are created in bushes (41%) and winding (37%).**

According to high probability of winding fault occurrence in motor, its fault diagnosis is so important. Source of this fault is often winding isolators. Inter-turn winding fault creates short circuit in phase winding and changes phase impedance. As a result, machine becomes unbalance. Rate of impedance reduction depends on intensity of fault. Prominent characteristic of winding fault is second harmonic creation in sequence components, motor currents, torque and speed. Therefore, the most common approaches rely on frequency analysis. In this paper, to diagnosis winding fault two approaches based on frequency analysis studied, MCSA and EPVA and to determine percentage of fault second harmonic of i_d has been used.

Index Terms— **EPVA, Fault detection, MCSA, Permanent Magnet Synchronous Motors (PMSM), Winding fault.**

I. INTRODUCTION

Permanent Magnet Synchronous Motors (PMSM) are one of the most important electric machines. Two important advantages of PMSMs are high power density and easy control of external torque by stator's current control. In the past, due to high manufacturing cost of permanent magnetic material, application of this kind of motors was limited in so important cases. But nowadays, because of progress in new magnetic material, they are widely used in industry such as; in traction, automobiles, robotics and aerospace technology [1] and also, in general, they can be used in electrical vehicles and ship propulsion systems [2]. In this kind of motors, magnetic field of rotor is created by one permanent magnet. It should be mentioned that, since controlling external torque by tuning stator's current is possible; therefore, PMSM Motors are desirable for industrial application. As we know, something which is harmful for any electrical machine is damage or fault in their constitutive components. Therefore, diagnosis of these faults is so important. Fault detection and effective diagnosis

of fault can improve the reliability of the system and can prevent expensive maintenance and reparations. As mentioned before, because of sensitive application of PMSMs in aerospace technology and ships, fault detection in time is so important. Moreover, determining percentage of the fault is important. It gives us an idea to change the control strategy in presence of fault.

In this field some work has been done, such as; in [2], a negative sequence analysis coupled with a fuzzy logic based approach has been used for fault diagnosis of a PMSM. In [3], a fault model has been proposed for inter-turn fault of stator winding in a PMSM. And according to the fault model, a series of algorithms have been proposed for fault detection and diagnosis. In [4], a Permanent Magnet AC drive and machine has been tested and the most critical areas have been analyzed for reliability, also, steady-state techniques e.g. Motor Current Signatures Analysis (MCSA) and the Extended Park's Vector Approach (EPVA) as well as a new transient technique have been applied. That is a combination of the EPVA, the Discrete Wavelet Transformation and statistics, to the detection of inter-turn faults in a Doubly-Fed Induction Generator (DFIG).

In this paper, we focus on inter-turn fault of winding. To detect this fault two methods proposed. These methods are Motor Current Signature Analysis (MCSA) and Extended Park's Vector Analysis (EPVA) which work based on frequency analysis and in order to determine percentage of this fault second harmonic of x_q has been utilized.

II. MOTOR CURRENT SIGNATURE ANALYSIS APPROACH (MCSA)

MCSA method, analyze current waveform of motor using signal processing algorithms (such as; FFT, Wavelet and etc.) and other artificial intelligence techniques. Current and voltage components and power signals can be used in MCSA method.

Frequency analysis of stator current components (x_q, x_d), torque and speed in PMSM shows that sensibility of motor speed, x_q and x_d signals to second harmonic is more than the other signals. Therefore, we take these signals for frequency analysis in MCSA approach. Comparing these values with threshold values in steady state operation can be a sign of fault outbreak. This approach is completely complicate from mathematic point of view and needs a large number of statistic data.

M. A. Shamsi nejad, Assistant professor, Department of Electrical Engineering, University of Birjand, Birjand, Iran (e-mail: mashamsinejad@birjand.ac.ir).

M. Taghipour, Master student, Department of Electrical Engineering, University of Birjand, Birjand, Iran (e-mail: mtaghipour@birjand.ac.ir).

978-1-4577-0007-1/11 $26.00 © 2011 IEEE

Simulation results on x_q, x_d and motor speed shows that second harmonic of x_q is more sensitive to fault occurrence and this component can detect winding fault more powerful than other signals and is so faster.

III. EXTENDED PARK'S VECTOR ANALYSIS (EPVA)

EPVA is a new technique that has been applied to detect permanent fault of rotor, stator winding fault and unbalanced voltage source. This approach is based on equation (1) which is obtained from extension of Park's vector. By extending Park's vector, component of d axis is always sinusoidal and component of q axis is always in cosine form. In healthy state, values of these two vectors are equal and curve of x_q versus x_d is a circle.

$$\begin{bmatrix} x_d \\ x_q \end{bmatrix} = \begin{bmatrix} \sqrt{\dfrac{2}{3}} & -\dfrac{1}{\sqrt{6}} & -\dfrac{1}{\sqrt{6}} \\ 0 & \dfrac{1}{\sqrt{2}} & -\dfrac{1}{\sqrt{2}} \end{bmatrix} \begin{bmatrix} i_a \\ i_b \\ i_c \end{bmatrix} \quad (1)$$

In inter-turn fault mode, second harmonic appear in current components x_d and x_q. Therefore, the currents; i_a, i_b and i_c convert to its components x_q and x_d using (1). Second harmonic of amplitude of x_q and x_d can be a criteria to inter-turn fault detection.

$$M_2 = \sqrt{x_d^2 + x_q^2} \quad (2)$$

In this equation M_2 is second harmonic of amplitude of d-q currents.

IV. SIMULATION RESULTS

Trace of x_q versus x_d in healthy mode are presented in Fig. 1.(a). It should be a circle and figure 1b is variation of M_2. These values are normalized to unit.

Fig. 2 is trace of x_q versus x_d in fault mode. In this figure inter-turn fault is 50% which occurred at t=3sec. Fulfilled tests in different speeds and fault percentages show that after fault occurrence, it is created noticeable amount of second harmonic in component x_d which is sufficient for fault detection. Therefore, in order to detect winding fault calculation of id or magnitude of x_q and x_d (M_2) and evaluation of its second harmonic is sufficient.

Detour of M_2 from threshold values represents fault occurrence. Fig. 3 presents threshold values of M_2 in different speeds. Comparing M_2 and threshold values can be a criterion to detect inter-turn winding fault. Line ($y=0.0002 \times (x-0.103)$) could be a good estimation threshold values which calculated by Matlab simulator. Using this approximate equation, threshold values can be obtained in different speeds and can increase accuracy of the system.

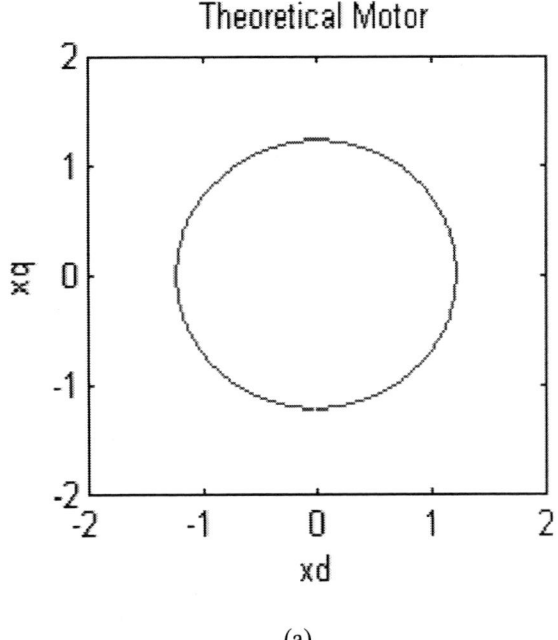

(a)

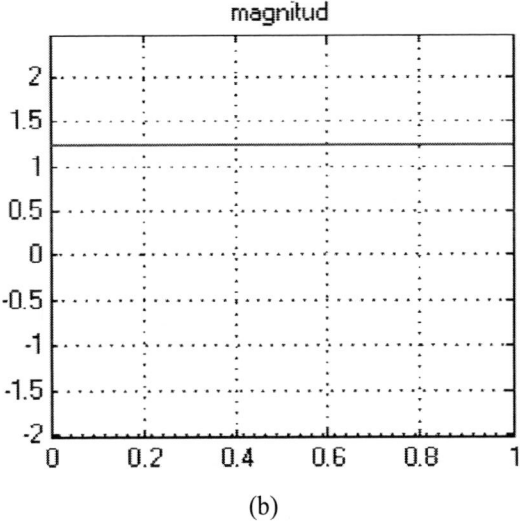

(b)

Fig. 1. (a) Curve of x_q versus x_d and (b) Value of M_2 in healthy state.

V. DETERMINATION OF FAULT PERCENTAGE

After fault detection, percentage of fault detection can be obtained. By increasing of fault intensity, amplitude of second harmonic of motor's components increases non-linearly. In this research, we try to find a relationship between second harmonic of motor's components and percentage of inter-turn winding fault using mathematical methods. By different accomplished simulation on signals x_d, x_q, τ, i_{abc} and speed, the signals; x_q and speed which was more sensitive to second harmonic has been selected in this work.

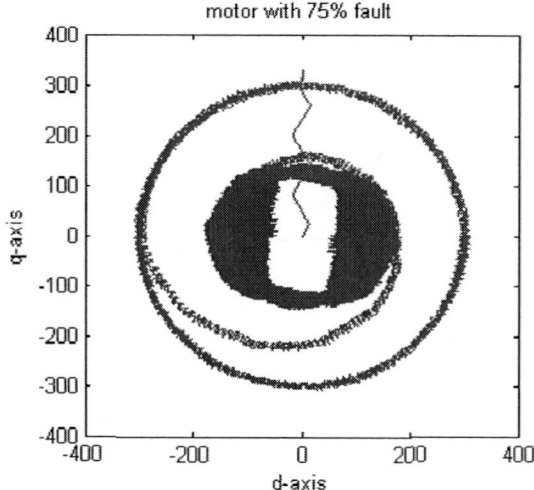

Fig. 2. Curve of x_q versus x_d (in 50% fault).

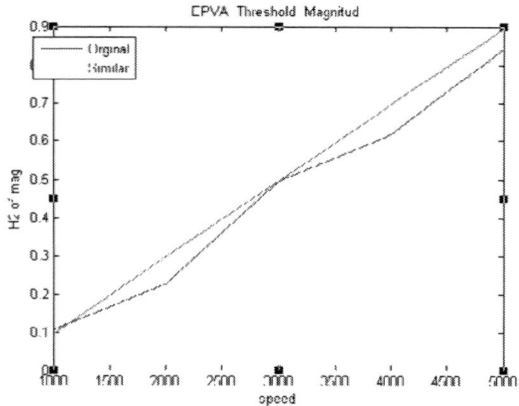

Fig. 3. Threshold values of second harmonic of x_q and x_d
Red curve: real value
Blue curve: approximated curve

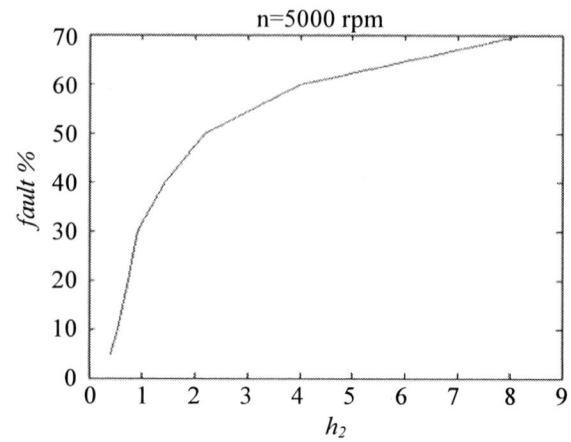

Fig. 4. Variation of second harmonic of xq versus inter-turn stator winding fault

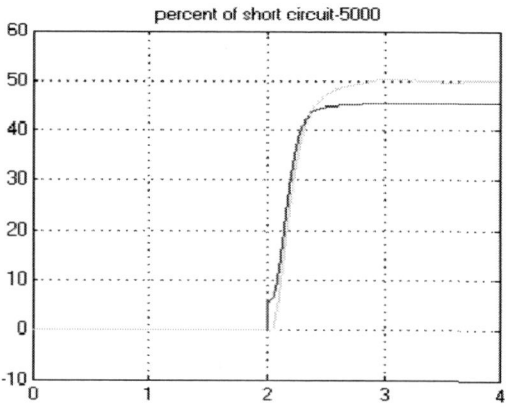

Figure 5. Determination of 50% fault at t=2s
Green curve: Based on second harmonic of i_q
Blue curve: Based on second harmonic of speed

In order to precise determination of percentage of winding fault and in order to reduce effect of noises on measurement, it is better to filter second harmonic of motor's components and utilize filtered data. Curve of variation of filtered second harmonic of x_q versus inter-turn winding fault in 5000 rpm has been shown in Fig. 4. Using Curve fitting in MATLAB simulator we obtain (2). This equation is for 5000 rpm.

$$fault = 0.026h_2^5 - 0.79h_2^4 + 8.36h_2^3 - 38.18h_2^2 + 86.h_2 \qquad (2)$$

Where, fault is percentage of inter-turn stator winding fault and h_2 is second harmonic of x_q. This equation has compatibility with curve of variation of filtered second harmonic of x_q versus inter-turn winding fault in wide range.

Fig. 5 present simulation results to determining the percentage of inter-turn winding fault. In this simulation, fault occurred at t=2s with 50% inter-turn winding fault in 5000 rpm. We can see accuracy of proposed model to approximate the error percentage.

VI. CONCLUSION

In MCSA method, component x_q is sufficient for frequency analysis and its amplitude of second harmonic can be applied for fault detection. Additional to high sensitivity of x_q, this component can be obtained by Park transform and is always available in control memory; therefore, this component can be a good choice for this work. If MCSA system is designed well, it will be capable to prepare very quick and correct diagnosis of faults for machine. Accuracy of this method depends on load of machine, noise and variation of speed. As shown, in approach EPVA only calculation of x_d and evaluation of amplitude of second harmonic are adequate for fault detection. In order to determine percentage of winding fault, two components; x_q and motor's speed has been considered and results of simulation show that x_q has more sensitivity to second harmonic with respect to speed and is more accurate criterion for determining percentage of winding fault.

VII. REFRENCES

[1] J. Rais, M.P. Donsión, "Permanent magnet synchronous motors: Influence of the parameters on the control of a PMSM", The International Journal for Computation and Mathematics in Electrical and Electronic Engineering, Vol. 27 Iss: 4, pp.946 – 957, 2008.

[2] J. Quiroga, Li Liu, D. A. Cartes, "Fuzzy Logic based Fault Detection of PMSM Stator Winding Short under Load Fluctuation using Negative Sequence Analysis", 2008 American Control Conference Westin Seattle Hotel, Seattle, Washington, USA June 11-13, 2008.

[3] LI. LIU, "Robust Fault Detection And Diagnosis For Permanent Magnet Synchronous Motors,". jun 2006, Florida, USA.

[4] J. DAVID NEELY, "Fault Types And Reliability Estimates In Permanent Magnet AC Motors,". 2005, Michigan, USA.

[5] H. DOUGLAS, P. PILLAY, P. BARENDSE, "The Detection Of Interturn Stator Faults In Doubly-Fed Induction Generators", 2005 IEEE.

VIII. BIOGRAPHIES

Mohammad Ali Shamsi Nejad received the B.Sc. degree in electrical engineering in 1990 and the M.Sc. degree in 1996, both from the Sharif University of Technology, Tehran, Iran. He received the Ph.D. degree from the Institut National Polytechnique de Lorraine (INPL), Nancy, France in 2006.

During 1996–2003, he was an Assistant Professor at Birjand University, Iran. Since 2006 he is Assistant Professor at Birjand University, Iran. His research activities in Birjand University deal with control of electrical machines and power electronics

Mehran Taghipour-Gorjikolaie was born in sari, IRAN, in 1986. He received a B.Sc. degree in electrical engineering from University of Mazandaran, Babol, IRAN in 2008 and He is presently pursuing the M.Sc. degree at the Birjand University, Birjand, IRAN.

His research interests consist of modeling and simulating of power-electronic systems, electrical machines and artificial intelligence algorithms.

High Reliability for Electric Machines Driving Critical Loads: A Review

J. Nandan and Gobbi. R, *SMIEEE*
Department of Electrical and Electronic Engineering
The University of Nottingham
Semenyih, Selangor - MALAYSIA
(keyx1nga@nottingham.edu.my, gobbi.ramasamy@nottingham.edu.my)

Abstract- **The term reliability is a criterion that is a must in the procurement and selection process of an electrical machine. As the cost of machines varies proportionally to its reliability status, studies have been conducted throughout the world to achieve one ultimate goal – to produce a machine with higher reliability and minimum cost spent in manufacturing it. Several methods to improve the current reliability of the machines were reviewed here. The reviews were done on the subject of reliability of electrical machines with highlights of its importance and contribution to the real-time applications. A new direction is proposed here for the detection and remedial of incipient faults particularly in critical load systems.**

Keywords- **Electric machines, realibility, condition monitoring, fault-tolerant, critical load.**

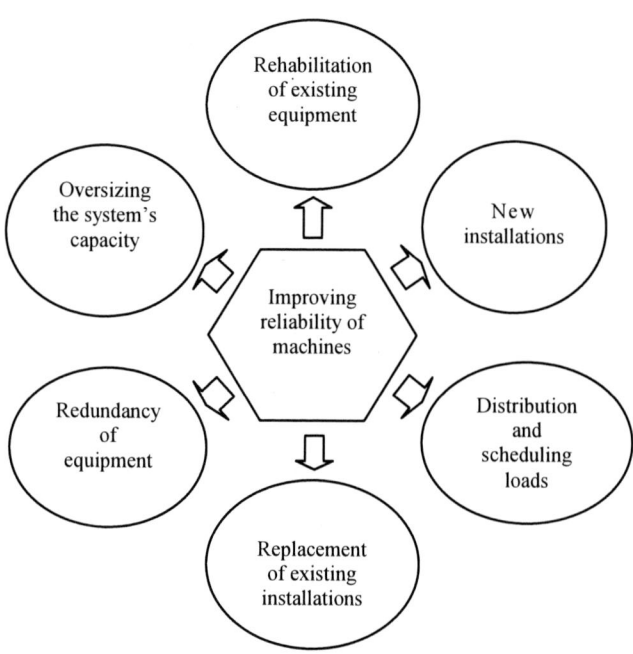

Fig. 1. Methods of improving the reliability of machines

I. INTRODUCTION

Achieving higher operational reliability in any of the electrical machines is one of the greatest challenges faced by and of interest to many researchers in the field of Electrical Engineering for the past few decades [1]. Even the conventional method used in condition monitoring of electrical machines tested at its highest potential could not penetrate the possibility of gaining higher reliability with minimization of cost in installing sensors and upgrading current drives. However, current suggestions have indicated that the addition of condition monitoring functions to exisisting systems is neither expensive nor complicated [2-6].With the current improvement of designs in drives, an incipient failure or future fault can be detected much earlier than what is expected in a conventional condition monitoring system. The advancement in sensors that are integrated with the drives has allowed the inherent characteristics of abnormal behaviours in machines before fault to be utilised in prediction of faults , a technique which was not used most of the time in the past. This allows the precaution measures to be implemented before the occurrence of fault to avoid major failures and reduce the downtime of a machine. Several methods in improving the reliability of machines are shown in Fig. 1 [7].

The major faults that usually occur in electrical machines can be classified into two types, mechanical and electromagnetic faults [8]. The former is usually detected during or after fault wherelse the latter is normally located before hand with the help of current advancement in analysis tools. Stator faults like turn-to-turn, coil-to-coil, line-to-line and line-to-ground fault are common electromagnetic faults. One of these faults is well depicted in [9]. Apart from that, abnormal connnections of the stator winding or supply at the input terminals and demagnetization of the permanent magnet in PMSM are other less common faults [8]. Any form of faults, either major or minor could cause a catasthropic failure especially in the case of critical load systems which require continuous quality power supply. Reliability is often regarded as the key element in a critical load system which has many sections under it demanding continuous power supply especially under any abnormal conditions. Fig. 2 shows some of the general facilities with critical load systems [7].

978-1-4577-0007-1/11 $26.00 © 2011 IEEE

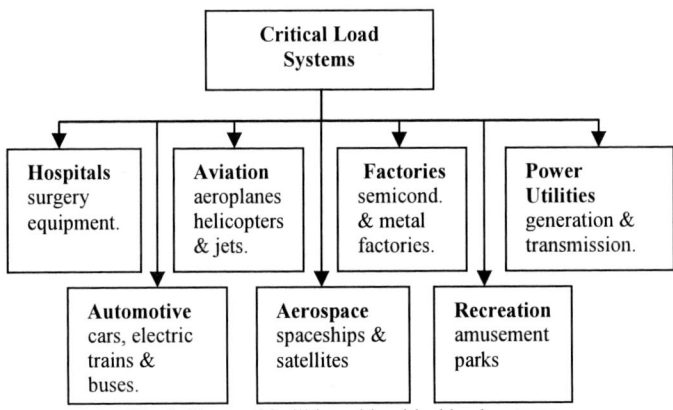

Fig. 2 Types of facilities with critical load systems.

Infact, reliability of machines is actually much more appreciated in sectors like aviation, automotive, and aerospace where the safety of living things is considered the first of priorities. Any failure of machines in these sectors could give instant impact and destruction to the entire system leaving the lives of the associated humans in jeopardy.

II. LITERATURE REVIEW

Artificial Intelligence (AI) tools have been of much importance to the diagnostic procedures of faults in electrical machines for the past few decades as strongly mentioned by the author in [10]. The paper has also classified the types of faults in electrical machines mainly to:

a) Stator winding faults which constitute 30% to 40% of induction motors' faults most of the times.
b) Rotor broken bar and end ring faults ranging from 5 to 10% of the induction motors' fault.
c) Bearing damage faults that are categorised under the mechanical faults, taking the equal share of faults as the stator winding faults.
d) Rotor eccentricity faults which is sectioned into static and dynamic both occuring simultenously due to nonuniform airgap formed by manufacturing defects.

That being said, applications and concepts of AI tools used in diagnosing the faults were well discussed here on the surface side as shown below:

a) Artificial Neural Networks (ANN) which mimic the human brain's simple arithmetic units performed in a complex layer architecture are being used entensively also in condition monitoring.
b) Fuzzy and Adaptive Fuzzy Systems which technically represent the way humans think in prioritising options in decision making.
c) Expert Systems that emulate the human mind's thoughts that are represented in a bound domain of knowledge which will produce inference mechanism without the presence of a human or an operator at the instant of the fault diagnosis.

The mathematical models of vector control of machines, fault tolerant operation and thermal effect of turn faults are explained and shown well here. More case studies and data intepretations using the AI tools for condition monitoring and fault diagnosis of the electrical machines are discussed in [11-17].

One of the most important techniques used in the diagnostic of faults is the motor current signature analysis (MCSA) which consist of a sensor to measure the stator current and a data acquisition system to acquire the signal waveform. Faults such as broken bars, shorted turns, and air gap eccentricity in machines can be identified using this method. The author has provided adequate analysis and information on MCSA system for the broken bars fault in [18-21]. A new digital/analog canceling technique based on recursive Discrete-time Fourier Transform (DFT) is discussed here, replacing the conventional method which uses analog notch filters . The sensitivity of the latter to temperature variations which may shift the filter resonance frequency and the inadequacy of the 12 bits A/D converters to cope with the magnitude differences between the fundamental and sideband components are its disadvantages. With both errors combined, an undesired response is produced. These setbacks have prompted the author to come out with the new technique. The new filter introduced here filters out the fundamental frequency component even before the analog to digital (A/D) conversion and later obtaining only the sidebands in the current signal, a presence that is evident in the case of broken bar faults. The fundamental frequency is first syntesized by estimation of the amplitude and the phase of the stator current fundamental. Proper techinuques in estimating using the DFT is well explained here with mathematical formula and graphics modelling. More estimation methods are also given in [22-24]. Since the fundamental component is already filtered out, 12 or 16-bit A/D converters can be replaced with an 8-bit to cope only with sideband components.

Another method in analysing faults caused by broken bars and stator inter-turn short circuit by means of measuring the vibration acceleration and the airgap force distribution patterns is introduced in [25]. Under the forced faults mentioned, the radial electromagnetic force distribution along the airgap is determined through simple numerical simulations by a system which consists of a data intepretation setup and vibration sensors connected to the machines. The signal is obtained from the sensors which are attached to the top and non-drive end of the machine. The paper also presented that the the the voltage measurement from the Pulse Width Modulation (PWM) was also used in the simulation to determine fault patterns. The data from the vibration sensor and the PWM are collected from the induction motor under three conditions – healthy, broken bars and inter-turn short circuit. The conditions also considered for no-load, half-load and full-load situations. The simulation by the DFT method produces both higher and lower frequency spectras for the vibration acceleration of the motor which are used to record the

978-1-4577-0007-1/11 $26.00 © 2011 IEEE

difference. It is noted here that the higher frequency spectra shows slight difference in ampitude which is quite impossible to study the pattern for during fault condition as shown in Fig. 3. On the other hand, the lower frequency shows a visible difference in the vibration pattern for all the fault and load conditions as noticed in Fig. 4. Hence, one of the signatures of electrical faults is determined and noticed here. Similar studies related to torsional vibration can be seen in [24,26-30]. Apart from the method above, another mathematical tool, the Wavelet transform is also discussed in [31- 35] .The wavelet transform particularly the multi-resolution analysis (MRA) has proven to be more accurate and easy to compute than the Fourier transform. The availability of data from this transform is significant and evident for classification of faults in electrical machines. This, when integrated with the AI tools could provide a hybrid version of fault detection method with superior manipulation and interpretation of data to show the signatures of any incipient fault that is expected to take place in the future. This scope is widely investigated and focused to achieve higher efficiency and reliability of any electrical system.

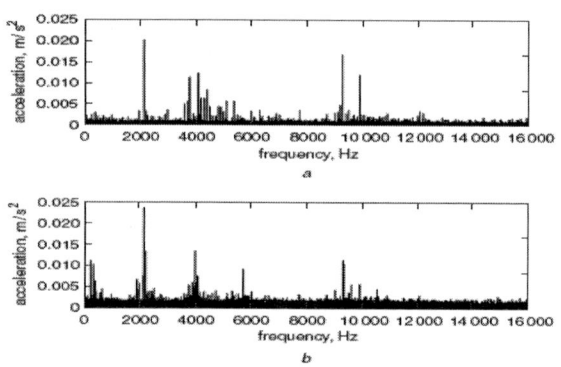

Fig. 3 High frequency spectra of the vibration acceleration under half load operation (1.9% slip) for (a) Healty motor and (b) Motor with three broken bars [25].

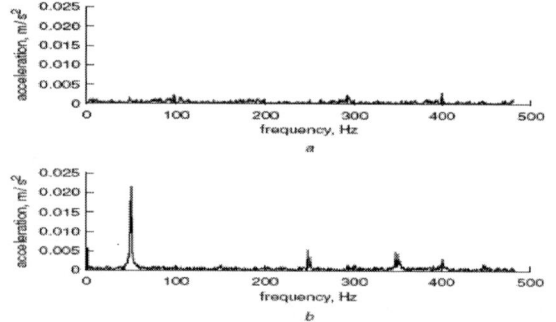

Fig. 4 Low frequency spectra of the vibration acceleration under half load operation (1.9% slip) for (a) Healty motor and (b) Motors with three broken bars [25].

Alternately, a method for monitoring the aging of stator winding insulation in machines is introduced in [36]. A small variation of the turn-to-turn capacitances due to the dielectric stress produces high frequency resonance which is used to determine the aging of the winding insulation. A kind of twisted pair of magnet wires is used as a specimen to perform thermal and electrical accelerated aging. The specimen selection is based on IEC standards as mentioned in the paper. A nearly 200 specimens were used in accelerated thermal tests. The breakdown voltage of each cycle is recorded for ten randomly chosen specimens. The data is presented as a mean value of breakdown voltage of ten specimens versus aging time under stress temperature of 270°C. At the same time, the the capacitance values for the same cycles are measured depicted in a graph. It can be clearly noted the objective of the test that as the breakdown voltage reduces, the capacitance value increase indicating aging near insulation failure of the twisted wires. This variation can indicate a shift in a winding's High Frequency (HF) resonance than be simply mesured by an impedance analyser during off-line. A similar method was caried out for electrical accelerated aging but at a temperature of 120°C. The testing on the 'motorette' were also performed to show that the external magnetic field magnitude and phase can be easily measured to detect the shift in winding resonance frequency.

III. DISCUSSION

The calssification of the Artificial Intelligence (AI) tools were explained accordingly by the author in [10] with briefing of the types of faults in electrical machines. Although the experimental application of the proposed technique is performed in real time in [18-19], the defects of the machine were artificially created. This method is proven only in the worst case situation when the bars are already broken. If this occurs in practical, it could be at the expense of the reliability of the machine where the fault is only located at its extreme level. The proposed technique, if focused on the detecting incipient faults or possibly even before that, a great breakthrough in achieving high reliabiltity of machines is reckoned to be achieved in the near future. The real time test in the case studies shown in [18-19,27,36] is only specificly for the three-phase induction motor. If the same test conducted on a permanent magnet synchronous motor (PMSM), data may vary proportionally to its rotor core structure which is made of magnet and has different electromagnetic properties than the usual coil wires. The emphasis on this is not shown there. More tests should be conducted on the other types of motors available in the industry. Then, a general concepts which agrees with all the machines should produce the desired performance and reliability levitated to a much satisfactory level in cost effective wise. However, all the methods and techniques discussed above have been proven to be worthy in terms of achieving its objectives in improving reliability and performance for specific faults, with minimization of cost in installation of extra devices. Most of the setbacks found are backed up with scientific reasonings which promise more suggestion and doubts to be investigated and penetrated in the near future. Also the down time of the machines is greatly reduced compared to the expensive conservative and redudancy methods as in [2-3]. But these advantages might

978-1-4577-0007-1/11 $26.00 © 2011 IEEE

not be favourable to the operational time of the machines in the long run. As in most cases, the fault is located when the machine is already in bad shape and cannot be repaired permanently after that. Replacing a new machine is very costly and not always inline with the budget procedures of an organisation [37-38]. What is needed at this moment is the proper diagnostic techniques to be employed and studied on the machines to detect any incipient fault way before it occurs especially in critical load systems. Rehabilitation of the incipient fault can be done while the machine is still running by means of minimum fault tolerant drives and other preventative maintenace as per the standard operating procedures. This could lengthen the lifetime of an in-service machine and directly increasing its reliability status to a much higher level. The bar chart in Fig. 5 shows the importance of the term reliability as seen from the general bar chart of selection criteria for an electric machine.

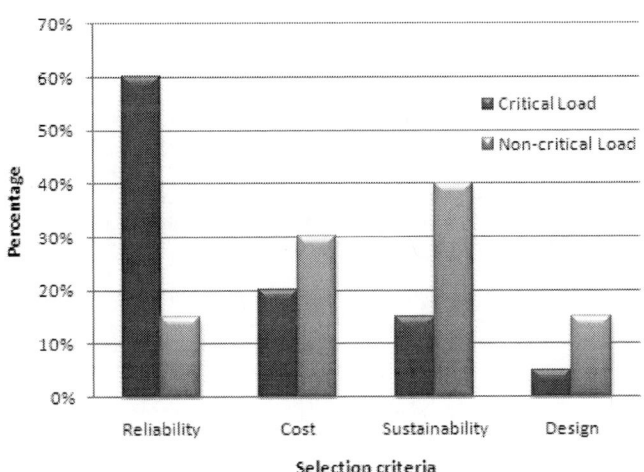

Fig. 5 Bar chart of selection criteria of an electrical machine.

Most of the selection of an electric machine for an operation in an electrical system is done based on the general selection criteria as shown above. However, sometimes the selectivity may vary based upon the mode of application and the suitability to the current system. For an example, in non-critical load system, a chain saw motor is mainly considered for its sustainability and load capacity when cutting down trees in timber logging industry. Reliability, cost, design and manufacturer of the saw are optional criteria. Contrastly, in a critical load system like in the aviation system,the reliability and sustainability of the turbine motors of aeroplanes are taken into account with highest creditability as it involves transportation of lives everyday. As the bar chart above is constructed merely on assumption and observation, more and proper surveys for this criteria could provide accurate statistics for this kind of reviews in the future. Although the priority of selection criteria varies, oftenly the term reliability may not be neglected or forgotten at any stage in purchasing a reliable electric machine.

As noticed overall from the review, basic flow in detecting a fault and compensating it [39] by any means to make sure the machine is operating in the safe mode if any fault were to occur is shown in the Fig. 6. At this point, the machine must be able to maintain most of its characteristics to sustain the load it is connected to, an important criteria taken into account in any system, be it in manufacturing or other important applications. However, a new algorithm is added to the commonly practiced one, as shown by the dotted part in the flowchart. By this way, after the fault is diagnosed its persistence is checked. Under normal practice, if the fault persists, then a normal control algorithm is performed by the technical personnel attending to it and later compensated owing to the actual situation [40]. In the new proposed method, the status of the load is checked after the fault found to be persisting. For critical load, a new critical control

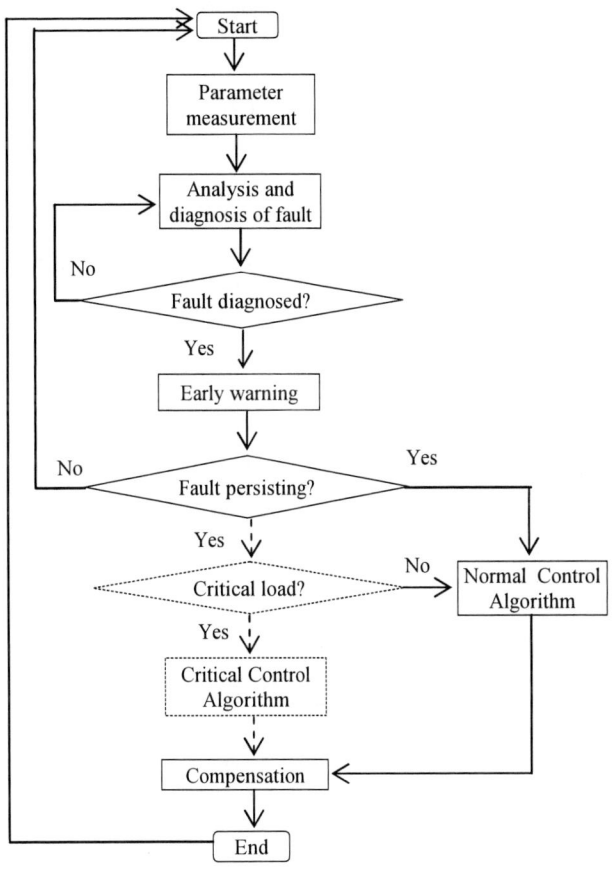

Fig. 6 Flowchart of fault detection for an electrical machine.

algorithm is employed in short duration of time to compensate and restore the machine to optimum operating status [39] or working on safe mode. Here, the purpose of including the critical control algorithm is to react faster to any signal obtained from the fault analysis in real time operation for a critical load. A following simple example could distinguish better between both the methods. A major power plant is experiencing some abnormalities in one of its generators due

978-1-4577-0007-1/11 $26.00 © 2011 IEEE

to unexpected overloading at peak hours. As the plant is supplying a large network, the generator cannot be shut down and the other generators could not provide the demanded power supply at that time. So, the engineers anticipated that the generator might fail based on the available data obtained from the fault analyse system. A major blackout is expected throughout the network. In this situation, the critical control algoritm could provide the probability of and when the fault will occur with the help of a fault detection system. This could, for some time, with a proper control system direct the generator to operate at safe mode, where some of the load actually will be despatched accordingly to the other generators in the meantime until off-peak time. This could save and increase the lifetime of the generator and the plant would be providing undisrupted quality power supply to its connected network. Where else, in a normal control algorithm of a system, the available data from the fault analyser could persuade the engineers to shut down the generator and loading all the other loads to the other generators knowing the risk that the system might go hay wire. So the proposed method is vital, in any machine whether it is operating in critical or non-critical load system.

IV. CONCLUSION

Talking about reliability, it is not only dependable on the performance of the machines, but on the service of the entire system connected to it electrically and mechanically as the drives, power supply, sensors and load. Any malfunction or fault in the system also contribute directly to the performance and reliabiltity of the machines. These setbacks are not fully discussed in literatures and are always scrapped at the initial stage of the discussion. The AI tools currently play an important role in condition monitoring and diagnostic of faults in electrical machines. The tools are used, tested and implemented at their highest potential to improve the lifetime, performance and mainly the reliability of machines. Most of the techniques reviewed here are discussed only for non-critical load systems. As reliability is the major concern in the critical load systems, priority should be given to explore other new techniques from different angles and perspectives to avoid scarcity of current ones and increase the available options in the coming years. This is to be highly anticipated by researchers in the area of electrical machines.

V. REFERENCES

[1] P.J. Tavner, "Review of condition monitoring of rotating electrical machines," *IET Electr. Power Appl.*, vol. 2, no. 4, pp. 215–247, 2008.

[2] Thomas G. Habetler, "Current-based condition monitoring and fault tolerant operation for electric machines in automotive applications," in *International Conf. on Elect. Mach. and Syst*, Seoul, Korea, Oct.8-11, 2007, pp. 2011-2016.

[3] Thomas G. Habetler, "Current-based condition monitoring of electrical machines in safety critical applications," *13th International Power Electronics and Motion Control Conf.*, Poznan, Sept. 1-3, 2008, pp. 21-26.

[4] Youngkook Lee and Thomas G. Habetler, "An on-line stator turn fault detection method for interior PM synchronous motor drives," *Applied*

Power Electronics Conf., APEC - Twenty Second Annual IEEE, Anaheim, CA, USA , Feb. 25- March 11, 2007, pp. 825-831.

[5] M. Hasanuzzaman, N.A. Rahim, R. Saidur, S.N. Kazi, "Energy savings and emissions reductions for rewinding and replacement of industrial motor," Energy 36, 24 November 2010, pp. 233-240.

[6] R. Saidur, M. Hasanuzzaman, S. Yogeswaran, H.A. Mohammed, M.S. Hossain, "An end-use energy analysis in a Malaysian public hospital," Energy 35, 9 October 2010, pp. 4780-4785.

[7] Integrated Publishing's Educational Archive, Integrated Publishing, Inc, *http://www.tpub-products.com/archive/* , Sept. 21, 2010.

[8] L.Romeral, J.A. Rosero, A.G. Espinosa, J. Cusido, and J.A. Ortega, "Electrical monitoring for fault detection in an EMA," *Aerospace and Electronic Systems Magazine, IEEE* , vol. 25, no.4, pp. 4 – 9, May 2010.

[9] P.J. Tavner, and A.F Anderson, "Core faults in large generators", *Electric Power Applications, IEE Proc.*, vol. 152 , no. 6, pp.1427 – 1439, Nov. 2005.

[10] Mohamed A. Awadallah and Medhat M. Morcos, "Application of AI Tools in fault diagnosis of electrical machines and drives – an overview," *IEEE Trans on Energy Conversion*, vol. 18, no. 2, pp. 245-251, June 2003.

[11] M. A. Awadallah and M. M. Morcos, "Switch fault diagnosis of PM brushless DC motor drive using adaptive fuzzy techniques," *IEEE Trans. on Energy Conversion*, vol. 19, no. 1, pp. 226-227, March 2004.

[12] M. A. Awadallah and M. M. Morcos, "ANFIS-based diagnosis and location of stator interturn faults in PM brushless DC motors," *IEEE Trans. on Energy Conversion*, vol. 19, no. 4, pp. 795-796, Dec. 2004.

[13] Mohamed A. Awadallah, Medhat M. Morcos, Suresh Gopalakrishnan, and Thomas W. Nehl, "A neuro-fuzzy approach to automatic diagnosis and location of stator inter-turn faults in CSI-fed PM brushless DC motors," *IEEE Trans. on Energy Conversion*, vol. 20, no. 2, pp. 253-259, June 2005.

[14] Mohammad Awadallah and M.M. Morcos, "Detection of insulation failure in BLDC motors using neuro-fuzzy systems," *Annual Report Conference on Electrical Insulation and Dielectric Phenomena*, Oct. 16-19, 2005, pp. 18-21.

[15] M. A. Awadallah, and M. M. Morcos, "Diagnosis of stator short circuits in brushless DC motors by monitoring phase voltages," *IEEE Trans. on Energy Conversion*, vol. 20, no. 1, pp. 246-247, March 2005.

[16] Mohamed A. Awadallah, Medhat M. Morcos, Suresh Gopalakrishnan, and Thomas W. Nehl, "Detection of stator short circuits in VSI-fed brushless DC motors using wavelet transform," *IEEE Trans. on Energy Conversion*, vol. 21, no. 1, pp. 1-8, March 2006.

[17] Mohamed A. Awadallah and Medhat M. Morcos, "Automatic diagnosis and location of open-switch fault in brushless DC motor drives using wavelets and neuro-fuzzy systems," *IEEE Trans. on Energy Conversion*, vol. 21, no. 1, pp. 104-111, March 2006.

[18] F.F. Costa, L.A.L. de Almeida, S.R. Naidu, E. R. Braga-Filho, and R. N. C. Alves, "Improving the signal data acquisition in condition monitoring of electrical machines," *IEEE Instrumentation and Measurement Technology Conf.*, Vail, CO, USA., May 20-22, 2003, pp. 562-566.

[19] F. F. Costa, L. A. L. de Almeida, S. R. Naidu, and E. R. Braga-Filho "Improving the signal data acquisition in condition monitoring of electrical machines," *IEEE Trans. on Instrumentation And Measurement*, vol. 53, no. 4, pp.1015-1019, August 2004.

[20] G.G. Acosta, C.J. Verucchi, E.R. Gelso, A current monitoring system for diagnosing electrical failures in induction motors, Mechanical Systems and Signal Processing, Volume 20, Issue 4, May 2006, Pages 953-965.

[21] F. Gu, Y. Shao, N. Hu, A. Naid, A.D. Ball ,Electrical motor current signal analysis using a modified bispectrum for fault diagnosis of downstream mechanical equipment, Mechanical Systems and Signal Processing, Volume 25, Issue 1, January 2011, Pages 360-372.

[22] Subhasis Nandi and Hamid A. Toliyat, "Condition monitoring and fault diagnosis of electrical motors- a review," *Ind. Applications Conf., 1999. Thirty-Fourth IAS Annual Meeting*, Phoenix, AZ , Oct. 3-7, 1999, pp. 197-204.

[23] Subhasis Nandi, Hamid A. Toliyat, and Xiaodong Li, "Monitoring and fault diagnosis of electrical motors- a review," *IEEE Trans. on Energy Conversion*, vol. 20, no. 4, pp. 719-729, Dec. 2005.

[24] D.B. Durocher, and G.R Feldmeier, "Predictive versus preventive maintenance," *Industry Applications Magazine, IEEE*, vol. 10, no. 5, pp. 12-21, Sept.-Oct. 2004.

[25] P.J. Rodriguez, A. Belahcen, and A. Arkkio, "Signatures of electrical faults in the force distribution and vibration pattern of induction motors," *Electric Power Applications, IEE Proc.*, vol. 153, no. 4, pp. 523 – 529, July 2006.

[26] A. Yazidi, H. Henao, G.A. Capolino, M. Artioli, F. Filippetti, and D. Casadei, "flux signature analysis: an alternative method for the fault diagnosis of induction machines, " *Power Tech. IEEE Russia*, St. Petersburg, 27-30 June, 2005, pp. 1-6.

[27] A. Yazidi, H. Henao, G.A. Capolino, M. Artioli, F. Filippetti, "Improvement of frequency resolution for three-phase induction machine fault diagnosis", *Ind. Appl. Conf. at 40th IAS Annual Meeting*, vol. 1, 2005, pp. 20-25.

[28] S.H. Kia, H. Henao, and G. Capolino "Analytical and experimental study of gearbox mechanical effect on the induction machine stator current signature," *IEEE Transactions on Industry Applications*, vol. 45, no. 45, pp. 1405 – 1415, July-Aug. 2009.

[29] S.H. Kia, H. Henao, and G. Capolino, "Torsional vibration effects on induction machine current and torque signatures in gearbox-based electromechanical system," *IEEE Transactions on Industrial Electronics*, vol. 56, no. 11, pp. 4689 – 4699, Nov. 2009.

[30] S.H. Kia, H. Henao, and G. Capolino "Torsional vibration assessment using induction machine electromagnetic torque estimation," *IEEE Transactions on Industrial. Electronics,* vol. 57, no. 1, pp. 209 – 219, Jan. 2010

[31] S.H. Kia, A. M. Mabwe, H. Henao, and G. Capolino, "Wavelet based instantaneous power analysis for induction machine fault diagnosis," *32nd Annual Conference on IEEE Industrial Electronics*, Paris, Nov. 6-10, 2006, pp.1229 – 1234.

[32] S.H. Kia, H. Henao, and G. Capolino, "Diagnosis of broken-bar fault in induction machines using discrete wavelet transform without slip estimation," *Industry Applications Conference - 42nd IAS Annual Meeting*, New Orleans, LA, Sept. 23-27, 2007, pp. 1917-1922.

[33] S.H. Kia, H. Henao, and G. Capolino, "Diagnosis of broken-bar fault in induction machines using discrete wavelet transform without slip estimation," *IEEE Transactions on Industry Applications*, vol. 45, no.4, pp. 1395 – 1404, Jul- Aug 2009.

[34] Basel Isayed, Lahouari Cheded and Fadi Al-Badour "Vibration monitoring and faults detection using wavelet techniques," *9th International Symposium on Signal Processing and Its Applications*, Sharjah, Feb 12-15, 2007, pp. 1-4.

[35] P. S. Barendse, B. Herndler, M. A. Khan, and P. Pillay, "The application of wavelets for the detection of inter-turn faults in induction machines" *IEEE International Electric Machines and Drives Conference*, Miami, FL, May 3-6, 2009, pp. 1401 – 1407.

[36] Werynski, P. Roger, D. Corton, R. and Brudny, J.F., "Proposition of a new method for in-service monitoring of the aging of stator winding insulation in AC motors," *IEEE Transactions on Energy Conversion*, vol. 21, no. 3, pp. 673-681, Sept. 2006.

[37] Penrose, H., "***Simple time-to-failure estimation techniques for reliability and maintenance of equipment,***" *Electrical Insulation Magazine, IEEE* , vol. 25, no. 4, pp. 14-18, July-Aug. 2009.

[38] Lorenz, R.D., "The future of electric drives: where are we headed?" - *Eighth International Conference on Power Electronics and Variable Speed Drives)*, London, 2000, pp. 1-6.

[39] Farshad Fahimi, David Brown and Marzuki Khalid, " Feature set evaluation and fusion for motor fault diagnosis," *Symposium on Inds. Electronics and Applications*, Penang, Malaysia, Oct. 3-5, 2010, pp. 617-622.

[40] Xin Wen, David Brown and Qizheng Liao, "Online motor fault diagnosis using hybrid intelligence techniques," *Symposium on Inds. Electronics and Applications*, Penang, Malaysia, Oct. 3-5, 2010, pp. 343-348.

A Wavelet-Based Technique for Discrimination of Inrush Currents from Faults in Transformers Coupled with Finite Element Method

M. Jamali, M. Mirzaie, S. Asghar Gholamian and S. Mahmodi Cherati
Babol University of Technology, Babol, Iran

*Abstract-**The phenomenon of magnetizing inrush is a transient condition, which occurs primarily when a transformer is energized. The magnitude of inrush current may be as high as ten times or more times of transformer rated current that causes malfunction of protection system. So, for safe running of a transformer, it is necessary to distinguish inrush current from fault currents .In this paper, a wavelet-based technique has been used to discriminate inrush current from faults. For this purpose, a wavelet transform (WT) concept is presented and a criterion based on different behaviors of the differential currents under fault and inrush conditions have been derived. For checking the validity of the wavelet-based technique in discriminating of inrush currents from faults a three-phase 20/63 kV transformer has been simulated in Maxwell software that is based on finite element method (FEM). Then, obtained data from different situations have been used in MATALB software for implementation of the WT technique.**

I. INTRODUCTION

Power transformers are a class of very expensive and vital component of electric power systems. The cost associated with unplanned outage of a power transformer is very high. So, it is important to minimize the frequency and duration of unwanted outages in this component. One of the main reasons for wrong operation of protective system for the transformer is inrush current. Inrush current is a transient current that occurs in a transformer due to flux saturation in the core and its magnitude can be as high as fault currents. Because inrush current is not a fault current, it is important to develop a technique to distinguish this current from fault currents. In this regard, some techniques have been proposed in the literature. Since a magnetizing inrush current generally contains a larger second harmonic component, conventional transformer protection systems are designed to restrain during inrush transient phenomenon by sensing second harmonics [1-3]. However, the second harmonic component may also be generated during faults due to CT saturation or the existence of a shunt capacitor or the distributive capacitance in a long EHV transmission line to which the transformer may be connected. In certain cases, the magnitude of the second harmonic in the fault current can be close to or greater than that present in the inrush current. Consequently, the methods based on measurement of the second harmonic are not sufficiently effective for differential protective relays. So, in order to improve the accuracy and speed of transformer protection some alternative methods have been proposed in

the literature. In [4], the sum of active power flowing into transformer from each terminal has been considered as a criterion. Because the average power in the inrush current is almost zero, fault current can be discriminated from inrush by large power consumption. The equivalent instantaneous inductance-based technique has been proposed in [5]. In this method, it has been shown that the inrush current can be characterized by the drastic variation of the equivalent instantaneous inductance, but this criterion for the fault current is almost constant and therefore, it can be used to distinguish inrush current from faults. Identification of inrush current in transformer using error estimation technique is discussed in [6]. Based on error estimation technique, first, the dead angles are extracted from the differential current waves. Then, with comparing this wave with two reference waves, inrush current has been discriminated from fault currents. In [7], by examination of the main flux variation that is constructed by the voltages and currents of transformer windings, inrush current is identified. In [8-10], the fuzzy logic concept has been used to discriminate inrush currents from faults. In [11-12], using S transform, different features are extracted for inrush current and fault currents. For discrimination of inrush current from fault currents, improved correlation algorithm is used in [13]. In this technique, by examination of correlation coefficients between two successive half cycles, inrush current can be distinguished from fault currents.

In this paper, a wavelet-based method is used for discriminating inrush currents from faults. This method has been developed by considering different behaviors of the differential current under fault and inrush current conditions. The discrete wavelet transform (DWT) has been applied on the differential currents and different features for inrush currents and faults have been extracted based on wavelet components. Also, the finite element method (FEM) has been used to simulate a three-phase power transformer under different conditions in a much realistic way.

II. FUNDAMENTALS of INRUSH CURRENT

It is very well known that a transformer will experience magnetizing inrush current during energization. Inrush current occurs in a transformer whenever the residual flux does not match the instantaneous value of the steady-state flux which would normally be required for the particular point on the voltage waveform at which the circuit is closed [14].

978-1-4577-0007-1/11 $26.00 © 2011 IEEE

For the explanation of the mechanism causing inrush current in a transformer's primary winding when connected to an AC voltage source, we consider (1), where λ and v are the instantaneous flux linkage in a transformer core and voltage drop across the primary winding, respectively.

$$v = \frac{d\lambda}{dt} \qquad (1)$$

As we see from (1), the rate of change of instantaneous flux in a transformer core is proportional to the instantaneous voltage drop in the primary winding or on the other hand, the flux waveform is the integral of the voltage waveform. In continuously-operating transformer, these two waveforms are shifted by 90°. But a significant difference exists between continuous-mode operation and energization of a transformer. During continuous operation, the flux level is at its negative peak when voltage is at its zero point, but during energization the flux has to start at zero. So, for a rising voltage just started from zero, the magnetic flux will reach approximately twice its normal peak as it integrates the area under the voltage waveform's first half-cycle. This amount of flux, because of the nonlinear characteristic of the magnetization curve, causes saturation of the transformer. During saturation, disproportionate amounts of mmf are needed to generate magnetic flux. This means the winding current, which creates the mmf to cause flux in the core, will disproportionately rise to a value easily exceeding twice its normal peak. Fig. 1 shows the generation of inrush current in a transformer. As seen from the figure, exceeding flux from the knee point, results in large magnetizing current that in some circumstances can be ten times of the rated current in a transformer.

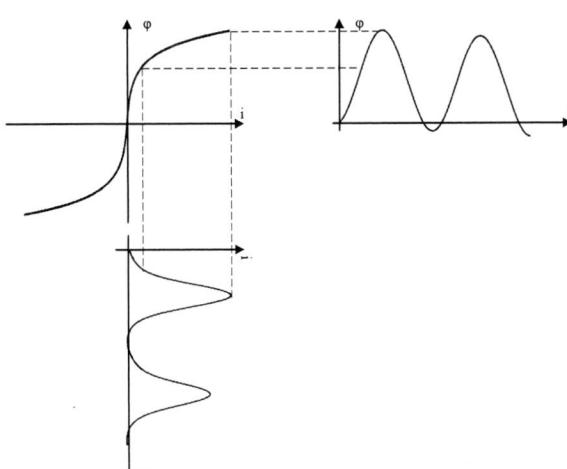

Fig. 1. Generation of inrush current in a transformer

III. WAVELET TRANSFORM

The waveforms associated with fast electromagnetic transients typically are nonperiodic signals which contain both high-frequency oscillations and localized impulses superimposed on the power frequency and its harmonics. These characteristics present a problem for traditional discrete Fourier transform (DFT) because its use assumes a periodic

signal and that the representation of a signal by the DFT is best reserved for periodic signals [15].

In order to overcome these problems, the WT has been used as a powerful tool in the analysis of transient phenomena. The ability of WT to focus on short time intervals for high-frequency components and long intervals for low-frequency components improves the analysis of signals with localized impulses and oscillations. For this reason, wavelet decomposition is ideal for studying transient signals and obtaining a much better current characterization and a more reliable discrimination.

Fig. 2 shows the implementation procedure of a DWT where S is original signal, HPF and LPF are the high-pass and low-pass filters, respectively.

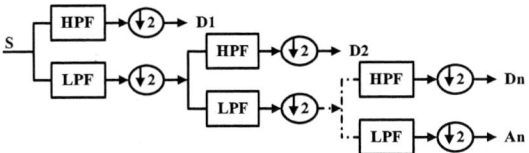

Fig. 2. Implementation procedure of DWT

As seen from Fig. 2, at the first stage, the original signal (S) is divided into two halves of the frequency bandwidth and sent to both HPF and LPF. Then the output of LPF is further cut in half of the frequency bandwidth and sent to second stage. This procedure is repeated until the original signal is decomposed to a pre-defined certain level. If the original signal is being sampled at F Hz, the highest frequency that the signal could contain, from Nyquist's theorem, would be F/2 Hz. This frequency would be seen at the output of the high frequency filter, which is the first detail (D1). Thus, the band of frequencies between F/4 and F/2 would be captured in D1. Similarly, the band of frequencies between F/8 and F/2 would be captured in D2, and so on [16]. In this paper, the sampling frequency is taken to be 2 kHz and TABLE I shows the frequency levels of the wavelet function coefficients.

TABLE I.
FREQUENCY LEVEL of WAVELET FUNCTION COFICIENTS

Wavelet analysis	Frequency Components Hz
D1	500-1000
D2	250-500
D3	125-250
D4	62.5-125
D5	31.25-62.6
A5	0-31.25

IV. DISCRIMINATION METHOD BASED on WAVELET TRANSFORM

The method for discriminating of inrush currents from faults is based on different behaviors of these waveforms following the disturbance. Since the magnetizing inrush current corresponds to the transformer core saturation, its waveform has a conical shape. One of the characteristics of this waveform is that its slope at the switching time is small and as time passes the slope increases. However, for a fault current, the waveform has higher slope compared with the starting of the inrush current and its slope decreases as time passes [17].

These features are shown in Fig. 3. It should be mentioned that a larger slope in the time domain shows that there are higher frequencies in the frequency domain. So, it can be concluded that for a fault current the amplitude of the high frequencies at the initial instants has decreasing trend, while for an inrush current an increasing trend is expected.

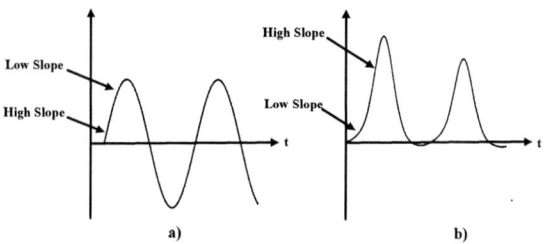

Fig. 3. Different behaviors of a) fault current, b) inrush current

These trends have been shown in Fig. 4 and Fig. 5 for line-to-ground fault and inrush currents for a three-phase 220/110 kV, 200MVA transformer that has been simulated in MATLAB SIMULINK. As seen from these figures, the aforementioned featured are clearly visible in the frequency level D3. So, this frequency level has been used as a criterion in the simulations based on finite element method that will be presented. Also, it should be mentioned, with using the absolute value of D3, the aforementioned trends can be seen in a better way. So, this value can be used in the algorithms for discrimination of inrush current and faults.

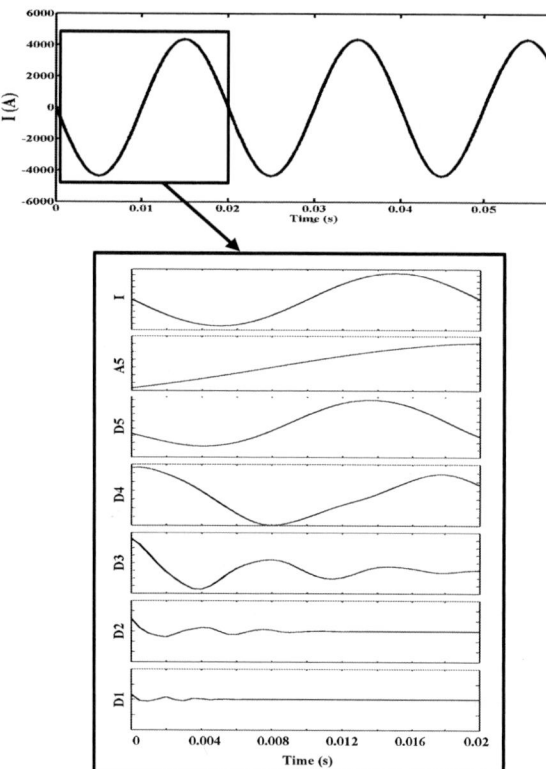

Fig. 4. Differential current of phase A as a result of line-to-ground fault and frequency levels obtained from WT

V. SIMULATION RESULTS

In order to have a more realistic simulation to verify the validity of the mentioned method in the discrimination of inrush current from fault currents, a three-phase power transformer has been simulated in Maxwell software that is based on FEM.

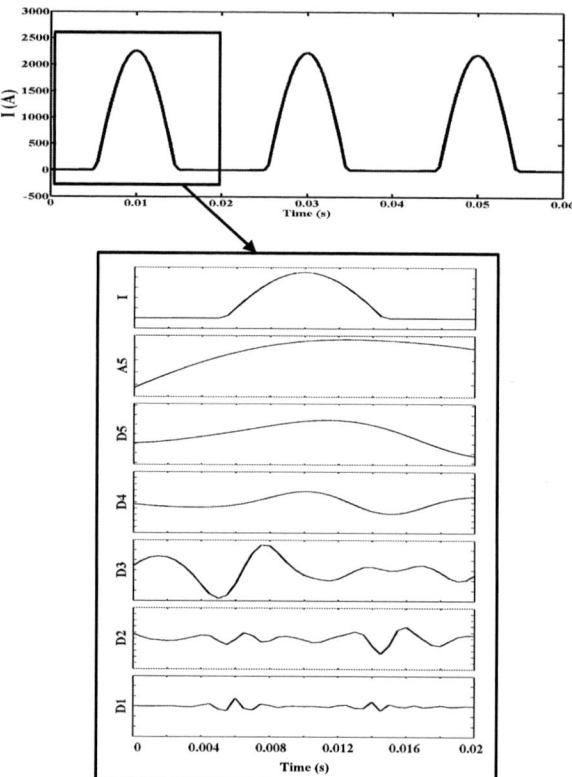

Fig. 5. Differential current of phase A as result of inrush current and frequency level obtained from WT

In this software, simulation with a lot of triangles in mesh operation is selected to increase accuracy, although it will take longer time for analysis. Also, Simplorer software has been used to create transformer connections and different situations including magnetizing inrush and fault conditions. The parameters of the simulated transformer are given in TABLE II. Also, some geometry of the employed transformer is drawn in Fig. 6.

TABLE II
PARAMETERS of the SIMULATED TRANSFORMER

Transformer connection	Δ/Y (LV/HV)
Rated apparent power	30 MVA
Voltage ratio	20/63 kV
Rated frequency	50 Hz
Number of LV turn (each limb)	232
Number of HV1 turn (each limb)	352
Number of HV2 turn (each limb)	70
Number of HV3 turn (each limb)	63
Core steel type	M5

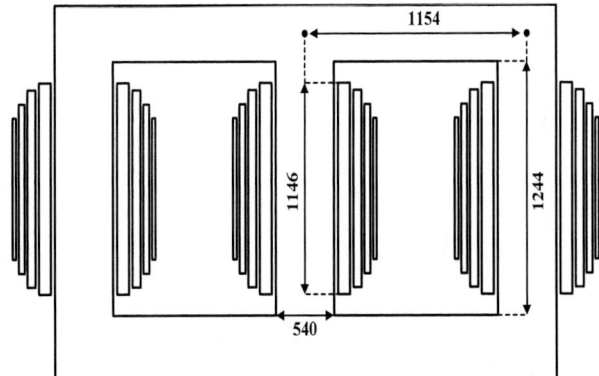

Fig. 6. Primitive geometry of the simulated transformer in TABLE II
(dimensions are in mm)

It should be mentioned that the analysis has been investigated on the lower tap that is including HV1 and the power source is applied to the LV windings. The following sections present the results of implementation of WT on the obtained data from different finite element-based simulations.

A. Fault Current

In this case, short-circuit of 30 turns in the phase A for the on-load transformer has been considered. Fig. 7 shows the differential currents of all phases in this situation. As seen from the figure, the differential currents of B and C phases are within the range of nominal magnetizing current, but A phase current is larger than the nominal magnetizing current and it must be investigated using WT.

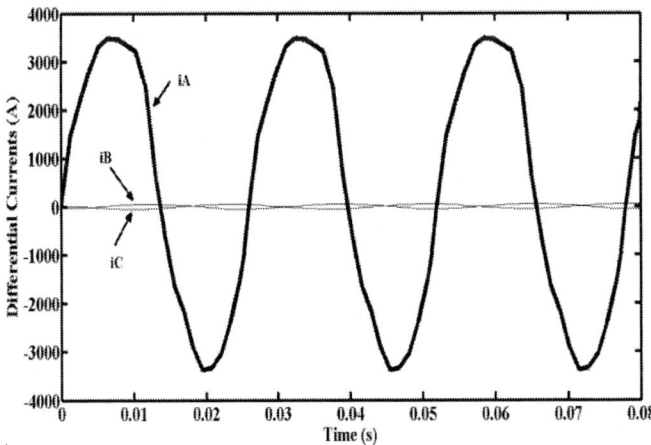

Fig. 7. Differential currents for the case of short-circuit of 30 turns in the phase A

Fig. 8 shows the absolute value of the frequency level D3 for phase A due to short-circuit of 30 turns in the phase A. As seen from this figure, the frequency level D3 has a decreasing trend that is correspond on the aforementioned criterion.

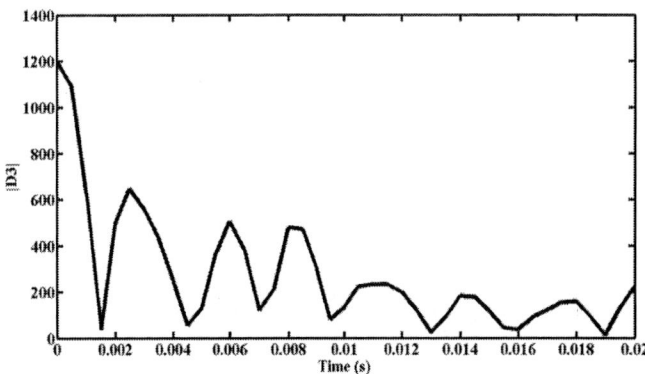

Fig. 8. The absolute value of the frequency level D3 for phase A due to short-circuit of 30 turns in the phase A

B. Inrush Current

Fig. 9 shows the inrush currents of transformer due to energization under no-load condition. As seen from the figures, all phases have shown inrush current situation that must be analyzed with WT criterion.

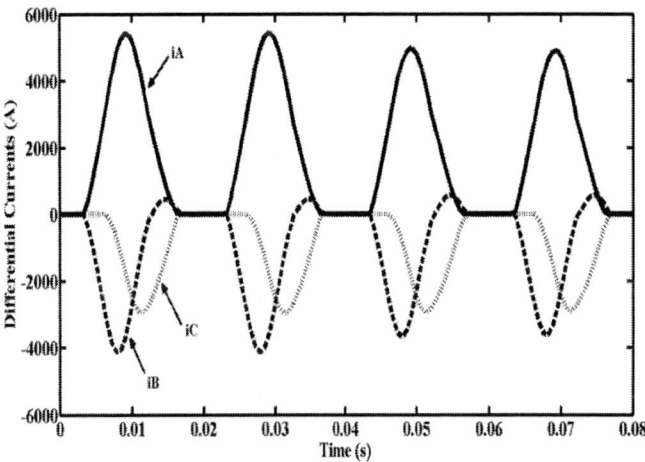

Fig. 9. Differential currents for the case of unloaded inrush current

Fig. 10 shows the absolute value of the frequency level D3 for phase A due to inrush current. As seen from this figure, unlike the fault situation, the frequency level D3 has an increasing trend from the switching time that is characteristic of an inrush situation.

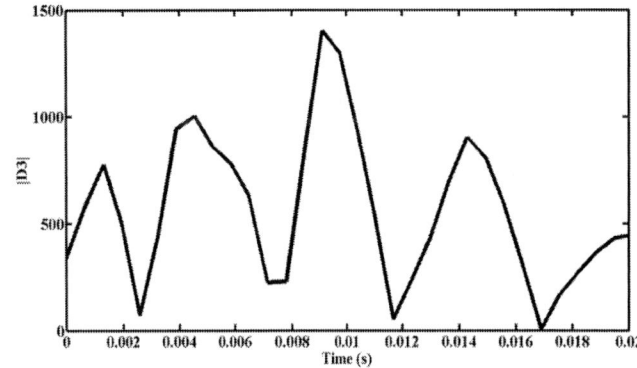

Fig. 10. The absolute value of the frequency level D3 for phase A due to inrush current

As the results show, the mentioned method is accurate for building a reliable differential protecting system for power transformers. Also, it should be noted that using WT criterion fault currents can be identified in less than a quarter of cycle based on 50 Hz supply.

VI. CONCLUSION

In this paper, a method based on different behaviors of the differential currents under fault and inrush current conditions has been developed for discriminating of inrush current from fault current. The different behaviors of inrush and fault currents have been characterized by the wavelet coefficients over a specific frequency range. In order to have a more realistic simulation to verify the validity of the mentioned method in the discrimination of inrush current from fault currents, the finite element method that is a rapid and effective way in simulation and modeling of advanced engineering systems has been used to simulate the three-phase power transformer. The results show that the method can discriminate inrush current from fault current in less than a quarter a cycle based on 50 Hz supply.

REFERENCES

[1] R. L. Sharp and W. E. Glassburn, "A transformer differential relay with second harmonic restraint," Trans. AIEE, vol. 77, pp. 884–892, 1958.

[2] J. A. Sykes and I. F. Morrison, "A proposed method of harmonic restraint differential protection of transformers by digital computer," IEEE Trans. Power App. Syst. vol. PAS-91, pp. 1266-1272, May 1972.

[3] B. Kasztenny and A. Kulidjian, "An improved transformer inrush restraint algorithm increases security while maintaining fault response performance," 53rd Annual Conference for Protective Relay Engineers, pp. 1-27, Apr. 2000.

[4] K. Yabe, "Power differential method for discrimination between fault and magnetizing inrush current in transformers," IEEE Trans. Power Del., vol. 12, no. 3, pp. 1109-1118, July 1997.

[5] G. Baoming, A. T. Almeida, Z. Qionglin and W. Xiangheng, "An equivalent instantaneous inductance-based technique for discrimination between inrush current and internal faults in power transformers," IEEE Trans. Power Del., vol. 20, no. 4, pp. 2473-2482, Oct. 2005.

[6] B. He, X. Zhang and Z. Q. Bo, "A new method to identify inrush current based on error estimation," IEEE Trans. Power Del., vol. 21, no. 3, pp. 1163-1168, July 2006.

[7] X. Zhao, J. Chai and P. Su, "Identification of magnetizing inrush currents of power transformers based on features of flux locus," Sixth International Conference on Electrical Machines and Systems, vol. 1, pp. 317-320, Nov. 2003.

[8] A. Wiszniewski and B. Kasztenny, "A multi-criteria differential transformer relay based on fuzzy logic," IEEE Trans. Power Del., vol. 10, no. 4, pp. 1786–1792, Oct. 1995.

[9] M. Delshad, S. M. Mosavian and B. Fani, "A new method for improved of power transformer protection using fuzzy logic," International Conference on Electrical Machines and Systems, pp. 1279-1282, Oct. 2007.

[10] H. Khorashadi, "Fuzzy-neuro approach to investigating transformer inrush current," Transmission and Distribution Conference and Exhibition, pp. 1302-1306, May 2006.

[11] S. Jiao, S. Wang and G. Zheng, "A new approach to identify current based on generalized S-transform," International Conference on Electrical Machines and Systems, pp. 4317-4322, Oct. 2008.

[12] Q. Zhang, S. Jiao and S. Wang, "Identification inrush current and internal faults of transformer based hyperbolic S-transform," Conference on Industrial Electronics and applications, pp. 258-263, May 2009.

[13] X. Lin, P. Liu and O. P. Malik, "Studies for identification of the inrush based on improved correlation algorithm," IEEE Trans. Power Del., vol. 17, no. 4, pp. 901-907, Oct. 2002.

[14] Sonnemann, W.K., Wagner, C.L., Rockefeller, G.D., Magnetizing Inrush Phenomena in Transformer Banks, AIEE Trans. Part III, vol. 77, pp. 884-892, Oct. 1958.

[15] O. A. S. Youssef, "A wavelet-based technique for discrimination between faults and magnetizing inrush currents in transformers," IEEE Trans. Power Del., vol. 18, no. 1, pp. 170–176, Jan. 2003.

[16] S.G. Mallat, "A Theory for Multiresolution Signal Decomposition: The Wavelet Representation", IEEE Trans. on Pattern Anal. Machine Intel., pp. 674-693, vol. 11, Apr. 1989.

[17] A. Guzman, S. Zocholl, G. Benmouryal, and H. J. Altuve, "A currentbased solution for transformer differential protection—Part I: Problem statement," IEEE Trans. Power Del., vol. 16, no. 4, pp. 485–491, Oct. 2001.

A New Schematic for Hybrid Active Power Filter controller

Emad Samadaei*, S. Lesan**, S. Mahmodi Cherati***

*, ** Department of Electrical and Computer Engineering, Babol University of Technology, Babol, IRAN
*** University Technology Malaysia, Malaysia
E-mail: *e.samadaei@stu.nit.ac.ir, ** s_lesan@nit.ac.ir, *** sam.mcherati@fkegraduate.utm.my

Abstract— nowadays power quality in power network requests using from existing equipments as high-performance and low cost such as improvement of compensator equipment and ways of efficiency increasing can be reduced destroyer effect on network. Active power filters is more importance and finance in network and industrial. It has depended detector algorithm and switching technique. This paper presents a novel Schematic for active power filter algorithm. This algorithm based on harmonic extraction is divided into two parts as feedback loop and feedforward loop. Also it increases accuracy of harmonic extraction by reducing of isolator calculation due to elimination of voltage calculation and changing of reference signal from voltage reference to current reference than similar algorithms. Also this changing cause that it can be used with switching techniques that work base on current as they are simpler and efficacy. In this algorithm, switching losses are considered that it helps to increase accuracy. Thus using the combination of this algorithm and current switching technique increases accuracy and decreases complication of previous structures that consequently, it increases controller efficacy of active power filter. Simulation result shows performance this model clearly.

Keywords: hybrid active power filters, hysteresis current control, Matlab/Simulink, Self tuning filter

Nomenclature

SRF	Synchronous Reference Frame
LPF	Low Pass Filter
HPF	High Pass Filter
STF	Self Tuning Filter
PWM	Pulse Wide Modulation Switching
HYS	Hysteresis Switching
HB	Hysteresis Bandwidth
PI	Proportional Integral
VSI	Voltage Source Inverter
THD	Total Harmonics Distortion
PCC	Point of Coupling of Circuit
P	Active power drawn from mains
L	Inductor
C	Capacitor
IL(a,b,c)	Load Currents
Is(a,b,c)	Source(Network) Currents
Vs(a,b,c)	Source(Network) Voltages
dq	PARK Reference Frame
$\alpha\beta$	Constant PARK Reference Frame
ωn	Arbitrary Frequency (n)
I,c	Feedback Current Signals
I*c	Reference Current Signals
E	Error

I. INTRODUCTION

Harmonic currents are the source of adverse effects for many types of equipments such as heating in transformers, perturbation of sensitive control equipments and resonances with the grid. These harmonic currents are mostly generated by the power conversion units and the power electronic equipments [1]. In the last few years, many different topologies of hybrid active filters with various control strategies have been proposed in the literature as lower cost alternatives to active filtering for harmonic compensation. Nowadays, hybrid active filters are considered as one of the best solutions for improving power quality [2]-[4]. Fig.1 shows the hybrid active filter studied in this paper.

As well known, active filter system performances mostly depend on accurate of the harmonic isolation and current control technique used to generate the switching patterns for the inverter.

In the literature, we can find many methods for active filter control based on the SRF and implementing a low-pass or high-pass filters to produce the harmonic references [5]-[7]. By replacement of STF with LPF and HPF, calculation is reduced and efficiency is increased also [8]. In [9] by using STF an algorithm based on voltage reference is presented.

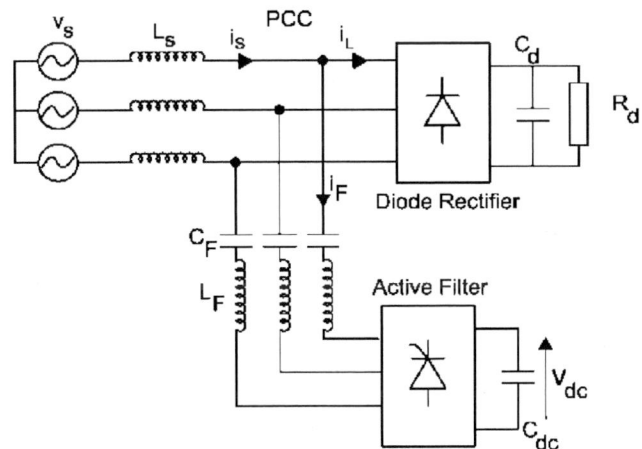

Fig.1 Parallel hybrid active power filter configuration

978-1-4577-0007-1/11 $26.00 © 2011 IEEE

Many PWM techniques have been introduced to produce the reference current. They are divided into two main groups: current control and voltage control. Current control techniques are widely used due to high accuracy and considering peak current [10]. Among the various current control techniques, fixed-band hysteresis current control technique is the simplest control approach. However, the fixed band hysteresis current control has some drawback such as switching frequency is not constant but its simplicity and fast dynamic response and no sensitivity to load parameters, makes it acceptable method for pulse generation [11], [12]. In [13] active filter controller with adjustable-band hysteresis current control technique are investigated that results show stabilizing of switching frequency.

This paper presents an improved approach of algorithm and novel algorithm "STF-CPC". Then, in next section, fixed-band hysteresis current controller has been introduced to produce and inject the reference current made by STF-CPC algorithm. At last a comparison has been performed. The effectiveness of the proposed control scheme is verified by numerical simulation using Matlab/Simulink.

II. Algorithm of harmonic elimination (STF-CPC)

According to Fig.1 the hybrid active filter algorithm studied in this paper consists in a three-phase LC filter that according to equation.1 tuned to the 7th harmonic frequency, connected in series with an active filter without any transformer. The passive filter absorbs the 7th harmonic currents generated by the load whereas the active filter improves filtering performances of the passive filter.

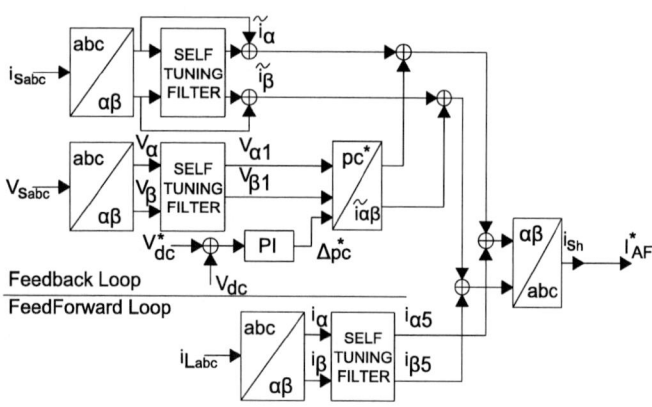

Fig.2 algorithm of STF-CPC

$$f = \frac{1}{2\pi\sqrt{LC}} \qquad (1)$$

This collection connected to a non-linear load as parallel that they are supplied by the network. The associated control algorithm combines a feedback and a feedforward loop. The feedforward loop is dedicated to the most dominant 5th

harmonic current component to improve filtering characteristics of the hybrid filter and the feedback loop is dedicated remained harmonics; in other hand harmonics that weren't dedicated in passive filter and feedforward loop (Fig.2).

III. Block of STF

Before description the feedback and feedforward, it is necessary to introduce block of STF that it has an important role in both loop. Performance of STF is extraction of arbitrary frequency (ωn) from input of this block. In old algorithm (SRF) is used LPF and HPF [14], [15].

STF has more accuracy and less calculation. Fig.3 shows this block.

From Fig.3, following expressions can be obtained [16]:

$$\hat{i}_a = \left(\frac{k}{s}[i_a(s) - \hat{i}_a(s)] - \frac{\omega_n}{s}.\hat{i}_\beta(s)\right) \qquad (2)$$

$$\hat{i}_\beta = \left(\frac{k}{s}[i_\beta(s) - \hat{i}_\beta(s)] + \frac{\omega_n}{s}.\hat{i}_\alpha(s)\right) \qquad (3)$$

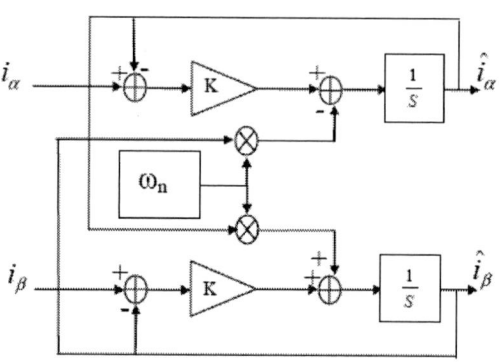

Fig.3 Self-tuning filter tuned to the pulsation ωn

Where, ωn is arbitrary frequency that is appeared in output. K is gain of block. One can see that small value of K increases filter selectivity and accurate. Also, the transient time is increased when K is decreased.

Fig.3 shows STF block at dq axis (αβ) that it causes increasing of responsibility and reducing of calculation.

IV. Algorithm

According to Fig.2, this algorithm is divided into two parts. A part is dedicated to the most dominant 5th harmonic current component. This part, the measured three phase load currents are as input signal and it is named feedforward loop. Either is dedicated remained harmonics (11th, 13th, 17th,

19th, etc.) and in this part the measured three phase supply currents are as input signals and it is named feedback loop.

A. Feedforward loop

The three phase load currents, iLa, iLb, and iLc are measured and transformed into α-β reference frame by:

$$\begin{bmatrix} i_{l\alpha} \\ i_{l\beta} \end{bmatrix} = \sqrt{\frac{2}{3}} \begin{bmatrix} 1 & -\frac{1}{2} & -\frac{1}{2} \\ 0 & \frac{\sqrt{3}}{2} & -\frac{\sqrt{3}}{2} \end{bmatrix} \begin{bmatrix} i_{la} \\ i_{lb} \\ i_{lc} \end{bmatrix} \quad (4)$$

We tuned the STF at the 5h harmonic frequency by ωn= ω5 in order to compute fifth components i_{5L} ($i_{\alpha5}$, $i_{\beta5}$) at the output of the self tuning filter.

Those references are added to the output current references established by the feedback loop to define the total current references for the active filter. Then total current reference is transferred to abc axis by αβ/abc block. Equation.5 shows αβ/abc block calculation.

$$\begin{bmatrix} i_a \\ i_b \\ i_c \end{bmatrix} = \sqrt{\frac{2}{3}} \begin{bmatrix} 1 & 0 \\ -\frac{1}{2} & \frac{\sqrt{3}}{2} \\ -\frac{1}{2} & -\frac{\sqrt{3}}{2} \end{bmatrix} \begin{bmatrix} i_\alpha \\ i_\beta \end{bmatrix} \quad (5)$$

In this loop, in comparison between new algorithm and previous algorithms ([5]-[7],[9]), the calculation is based on current reference and prevents from any complicated and severe calculation. The voltage calculation needs parameters of passive filter that it can become imprecise and changed by lapse. Thus algorithm based on current is simpler, more accuracy and fast dynamic response.

B. Feedback loop

In feedback loop, with extraction of remained harmonics will be increased efficacy of algorithm.

The three phase supply current, are measured and transformed into α-β reference frame by:

$$\begin{bmatrix} i_{s\alpha} \\ i_{s\beta} \end{bmatrix} = \sqrt{\frac{2}{3}} \begin{bmatrix} 1 & -\frac{1}{2} & -\frac{1}{2} \\ 0 & \frac{\sqrt{3}}{2} & -\frac{\sqrt{3}}{2} \end{bmatrix} \begin{bmatrix} i_{sa} \\ i_{sb} \\ i_{sc} \end{bmatrix} \quad (6)$$

Then this current is entered to STF block that tuned on original frequency whereas STF block output is supply fundamental component. Extraction of remained harmonics is achieved by subtracting the self tuning filter input signals from the corresponding outputs. The resulting signals are the harmonics components in the stationary reference frame.

To considering of switching loss, variation of capacitor voltage is a good index. It can be injected to dq axis by Proportional integral (PI) as current. This current must be injected in fundamental axis.

We know power equation in Park axis is as follows:

$$p = i_\alpha \hat{v}_\alpha + i_\beta \hat{v}_\beta \quad (7)$$
$$q = i_\beta \hat{v}_\alpha - i_\alpha \hat{v}_\beta \quad (8)$$

It can be rewritten:

$$i_\alpha = \frac{\hat{v}_\alpha}{\hat{v}^2{}_\alpha + \hat{v}^2{}_\beta} p - \frac{\hat{v}_\beta}{\hat{v}^2{}_\alpha + \hat{v}^2{}_\beta} q \quad (9)$$

$$i_\beta = \frac{\hat{v}_\beta}{\hat{v}^2{}_\alpha + \hat{v}^2{}_\beta} p + \frac{\hat{v}_\alpha}{\hat{v}^2{}_\alpha + \hat{v}^2{}_\beta} q \quad (10)$$

Thus according upon equations, loss index should be injected as $\dfrac{\hat{v}_\alpha}{\hat{v}^2{}_\alpha + \hat{v}^2{}_\beta} P_C$ in i_α and $\dfrac{\hat{v}_\beta}{\hat{v}^2{}_\alpha + \hat{v}^2{}_\beta} P_C$ in i_β. By using simple block of **pc*/iαβ** that is shown in Fig.4, the algorithm's performance will be more accurate and it does not need any external supply to compensation of switching loss.

According to upper explanation, three phase supply voltages, are measured and are transferred to α-β reference frame.

$$\begin{bmatrix} V_\alpha \\ V_\beta \end{bmatrix} = \sqrt{\frac{2}{3}} \begin{bmatrix} 1 & -\frac{1}{2} & -\frac{1}{2} \\ 0 & \frac{\sqrt{3}}{2} & -\frac{\sqrt{3}}{2} \end{bmatrix} \begin{bmatrix} V_a \\ V_b \\ V_c \end{bmatrix} \quad (11)$$

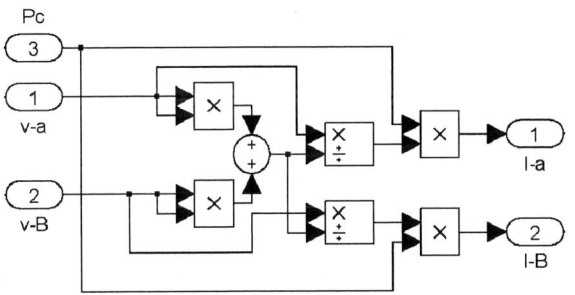

Fig.4 **pc*/iαβ** block

Then this voltage is entered to STF block that tuned on original frequency, as STF block output is supply fundamental component.

Then, after computation based on d-q transformation, current reference of feedback loop are added to the current reference of feedforward loop and we obtain the three-phase harmonic reference by equation.5

For DC voltage regulator, voltage of DC link is compared with reference voltage, and then is injected to algorithm by PI and it doesn't need any external supply.

We have:

$$\Delta E = \frac{1}{2} C_{DC} (V^*{}_{DC}{}^2 - V_{DC}{}^2) \quad (12)$$

Where $V^*{}_{DC}$, V_{DC} and C_{DC} are reference voltage, DC link voltage and capacitor respectively. That is, $\Delta E = P \times T$, P= active power drawn from mains , T= period time $= \dfrac{1}{f}$. Either:

$$p = \frac{3}{2} V_s I_c \qquad (13)$$

Where, V_s is supply voltage and I_c is compensator current. Thus:

$$I_C = \frac{C_{DC}(V^*{}_{DC}{}^2 - V{}_{DC}{}^2)}{3.V_S.T} \qquad (14)$$

V. Hysteresis Current control

Hysteresis current control is used for the generation of switching pulses. Among the various current control techniques, hysteresis current control is the most extensively used technique because of the noncomplex implementation, outstanding stability, absence of any tracking error, very fast transient response, inherent limited maximum current, and intrinsic robustness to load parameters variations. As indicated in [11] a review of used current control techniques for PWM converters reveals that hysteresis control shows certain superiority for active power filter applications. Hysteresis control provides a better low-order harmonic suppression than PWM control, which is the main target of the active power filter. It is easier to realize with high accuracy and fast response. However, as a disadvantage its switching frequency might fluctuate.

In the hysteresis control technique the error function is centred in a preset hysteresis band. When the error exceeds the upper or lower hysteresis limit the hysteretic controller makes an appropriate switching decision to control the error within the preset band and send these pulses to VSI to produce the reference current as shown in Fig.5.

Ic* is a vector of the desired compensation current reference signals. Ic is a vector of the feedback actual voltage source inverter output currents. Ic* and Ic signals are each demultiplexed to 3 signals phase A, B and C reference current signals and phase A, B, and C actual feedback current signals. Reference and actual signals are compared and fed into a hysteresis block and the output of the hysteresis block is the firing pulse.

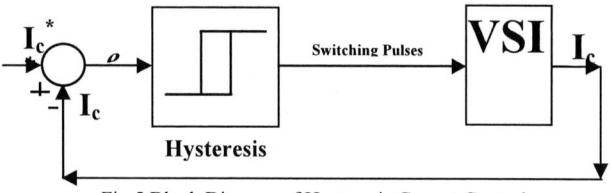

Fig.5 Block Diagram of Hysteresis Current Control

The outputs of the hysteresis blocks are directly fed as the firing pulse of upper bridge device of each leg of the inverter and NOT of that signal is fed as the firing pulse of lower bridge device of each leg. This is necessary for operation and avoiding the conduction of same leg switches simultaneously.

When Ic* is greater than Ic, the resultant difference signal e, is positive. If the magnitude of e is bigger than the upper boundary of the specified hysteresis band, the hysteresis block

output goes high, firing the upper bridge device of the leg and making the leg current increase.

When Ic becomes greater than Ic*, e becomes negative. If the magnitude of e is smaller than the lower boundary of the hysteresis band, the hysteresis block goes low, firing the lower bridge device of the leg and making the leg current decrease. If e is within the limits of upper and lower boundaries of hysteresis band, hysteresis block keeps its current state. Fig.6 shows this act. Note that less wide band causes more accurate, either more switching frequency that switch has to capacitate.

In hysteresis current controller, the hysteresis bandwidth (HB) has been taken as a small amount of system current and in many researches it has been taken as 5% of main current which will be HB=1.5A, here.

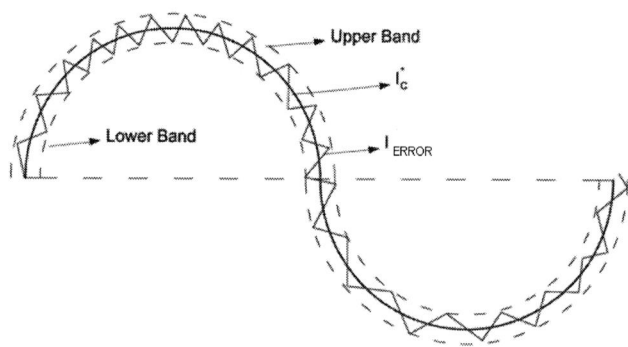

Fig.6 pulsing of hysteresis band controller

VI. Simulation result of STF-CPC by HYS switching technique

Corresponding to above analysis of three phase hybrid filter, the simulation has been performed with specified parameters shown in Table I.

Table I
Simulation parameters

preamble	Parameters	Value
Network voltage	V	400
Network frequency	HZ	60
Network inductor	mH	16.7
Network resistance	R	0.8229
Passive filter inductor	mH	2.5
Passive filter capacitor	μF	57.6
Passive filter resistance	R	0.001
DC load capacitor	μF	1500
DC load resistance	R	21
DC link capacitor	μF	1500
Factor K in STF block	K	60
Factor KV in feedback loop	KV	6

Fig.7 shows load current (IL) with its harmonic Bar graph. In Table II is shown that THD of current is 44.06% before using filter. Also Values of each harmonic and permitted value (Europe standard) are presented.

978-1-4577-0007-1/11 $26.00 © 2011 IEEE

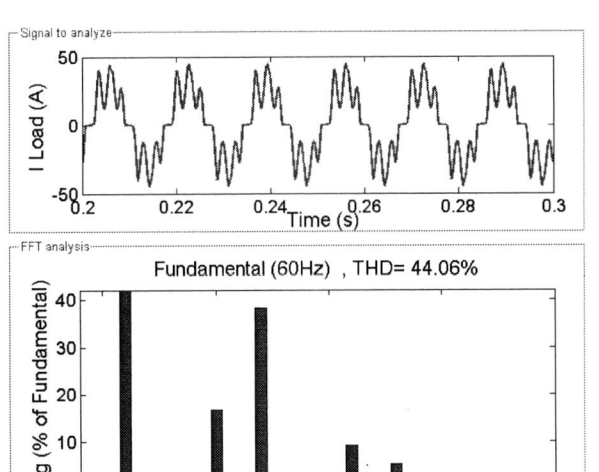

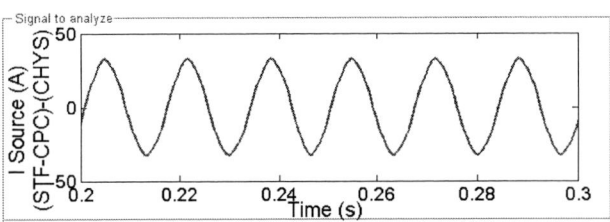

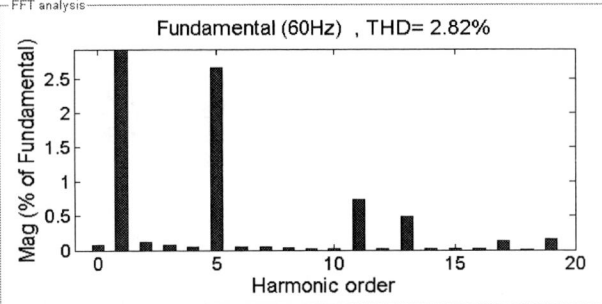

Fig.7 wave form of network current and it's harmonics bar graph before using active filter

Fig.9 wave form of network current and it's harmonics bar graph by use STF-CPC algorithm and HYS switching technique

Table II
Comparison between value of load current harmonics and permitted value base on Europe standard

Order harmonic	THD	5	7	11	13	17
Load current (%)	44.06	17.13	39.96	9.73	5.33	1.6
Permit value (%)	8	4	4	2.5	2	1

Table III
Value of network current harmonics by use STF-CPC algorithm and HYS switching technique

Order harmonic	THD	5	7	11	13	17
Network current (%)	2.82	2.66	0.04	0.73	0.48	0.14

As we know for correction of source current, filter current is added to load current sake to make a sinusoidal wave of network current. Fig.8 shows this act. Vector adding of load current and filter current is equal to network current.

So Fig.9 shows network current form and it's THD by use STF-CPC algorithm and HYS switching technique.

In presented algorithm, reference signal was current type. Either in hysteresis switching technique, kind of input must be current. Thus, this approach has more efficiency, more responsibility, low complex and independent from other reference signals (for example: triangle wave in PWM).

Simulation result shows that this proposed algorithm with hysteresis current controls has high accuracy and efficiency.

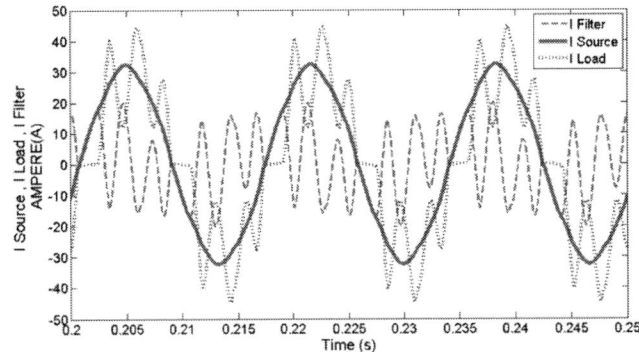

Fig.8 wave form of network current, filter current and load current

Table III shows value of each harmonic value after using power filter. Comparison of Table II and Table III shows performance of active power filter with using from STF-CPC algorithm. It reduces harmonics value under permitted value of Europe standard. THD of network current is 2.82% after using filter.

This results presents the hybrid active filter by using STF-CPC algorithm and hysteresis switching technique can removes harmonics as efficacy, and reduce until under of permit limit. This algorithm has more performance than other algorithm (for example: SRF). Using hysteresis current control due its simplicity and fast dynamic response will be reasonable than other switching technique (for example: PWM).

VII. Conclusion

According to development of power systems in industrial such as UPS, speed control of electrical machine, electrical furnace, computers and non-linear load that cause increasing of harmonic in network, Undesirable effect of harmonic is one of power transfer problem. This is why of standard codifying on THD limitation. Thus, it is necessary to detect and remove it until under permit limit. In this paper the novel algorithm of "STF-CPC" has been proposed to detect harmonics in power system that is based on current reference. Then hysteresis current control has been used to switching the reference current due to its simplicity and high accuracy. Simulation results had shown the efficiency of this active power filter in harmonic elimination by using the STF-CPC algorithm and hysteresis switching technique.

Reference

[1] J. C. Das," Passive filters- Potentialities and limitations" IEEE-Transactions on industry applications, vol. 40, pp. 345-362, (2004).

[2] Park, J-h. Sung and K. Nam," A New parallel hybrid filter configuration minimizing active filter size" IEEE/PESC Ann. Meeting Conf, vol. 1, pp.400-405 (1999)

[3] B. N. Singh, Bhim Singh, A. Chandra and K. Al-Haddad," Digital implementation of new type of hybrid filter with simplified control

strategy" Conference Proceeding IEEE-APEC 99., vol 1, pp. 642-648 (1999)

[4] H. Fujita, and H. Akagi," A practical approach to harmonic compensatreion in power systems-Series connection of passive and shunt active filters," IEEE Trans. Ind. Appl, vol 27, pp. 1020-1025 (1991)

[5] Michael John Newman, Daniel Nahum Zmood , Donald Grahame Holmes," Stationary Frame Harmonic Reference Generation for Active Filter Systems", IEEE TRANSACTIONS ON INDUSTRY APPLICATIONS, VOL. 38, NO. 6, NOVEMBER/DECEMBER 2002

[6] S. Sujitjorn, T. Kulworawanichpong," The DQ Axis With Fourier (DQF) Method for Harmonic Identification", IEEE TRANSACTIONS ON POWER DELIVERY, VOL. 22, NO. 1, JANUARY 2007

[7] Bhim Singh, Vishal Verma," An Indirect Current Control of Hybrid Power Filter for Varying Loads ", IEEE TRANSACTIONS ON POWER DELIVERY, VOL. 21, NO. 1, JANUARY 2006

[8] Mohamed Abdusalam, Philippe Poure and Shahrokh Saadate,"A New Control Scheme of Hybrid Active Filter Using Self-Tuning-Filter", POWERENG 2007, April 12-14, 2007, International Conference on SetiTbal, Portugal

[9] E. Samedaei, H. Vahedi, A. Sheikholeslami, S. Lesan" Using "STF-PQ" Algorithm and Hysteresis Current Control in Hybrid Active Power Filter to Eliminate Source Current Harmonic", IEEE Conf, PQC2010, , September 21-22, 2010

[10] Marian P. Kaźmierkowski, Ramu Krishnan, Frede Blaabjerg " Control in power electronics", Academic Press, 2001

[11] M. P. Kazmierkowski and L. Malesani, 'Current control techniques for three-phase voltage source PWM converters: a survey', IEEE Trans. Industrial Electronics, vol. 45(5), pp. 691–703, 1998.

[12] M. S. Dawande, V. R. Kanetkar, and G. K. Dubey, "Three-phase switch mode rectifier with hysteresis current control," IEEE Transactions on Power Electronics, vol. 11(3), pp. 466–471, May 1996.

[13] H. Vahedi, A. Sheikholeslami, E. Samedaei "variable hysteresis current control applied in an shunt active filter with constant switching frequency ", IEEE Conf, PQC2010, , September 21-22, 2010

[14] H. Akagi, "Active and hybrid filters for power conditioning" IEEE/ISIE Proceedings, vol 1, pp.TU26-TU36 (2000)

[15] J. Skramlik, V. Valouch,"Combined feedback and feedforwxard and control strategy of parallel hybrid filter" IEEE Inter. Conf Ind. Tech, vol 1, pp 449-453 (2004)

[16] S. Hong-Seok, P. Hyun-Gyu, N. Kwanghee, An instantaneous phase angle detection algorithm under unbalanced line voltage condition, in: IEEE 30th Annual Power Electronics Specialist Conference PESC'99, vol. 1, August, 1999, pp. 533–537.

Harmonic Study for MDF Industries: A Case Study

Mohammad Yazdani-Asrami[1], Sayyed Mohammad Bagher Sadati[2], Emad Samadaei[1]

1Department of Electrical and Computer Engineering, Babol University of Technology, Babol, Iran
[2]SAMA Technical and Vocational Training College, Islamic Azad University, Sari Branch, Sari, Iran

Abstract- **In recent years, with the increase of nonlinear loads drawing Nonsinusoidal currents, power quality distortion has become a serious problem in electrical power systems. Distortions in voltage and current wave-shapes can upset end-use equipment and cause other problems. Harmonics are a particularly common type of distortion that repeats every cycle. Poor power quality may cause many problems, such as the nonlinearity, instability and unbalance of electric network. Also, power quality problem will decrease life time of electrical instrumentation. In this paper, effects of harmonic distortion on the performance of the transformers, cables and induction motors have been studied. Then after theoretical study, harmonic and several power quality indices have been measured in a Medium Density Fiberboard (MDF) factory. The measurements show that harmonic distortion will decrease the life time of transformer and motors and also, will increase the losses and temperature of these equipments.**

I. INTRODUCTION

Power quality is a term that means different things depending on one's frame of reference. For instance, a utility may define power quality as reliability, whereas a manufacturer of load equipment may define power quality as those characteristics of the power supply that enable the equipment to work properly. These characteristics can be very different for different criteria. Power quality is ultimately a consumer-driven issue, and the end user's point of reference takes precedence. Therefore, the following definition of a power quality problem is sometimes used: any power problem manifested in voltage, current, or frequency deviations those results in failure or mis-operation of customer equipment. Institute of Electrical and Electronic Engineers (IEEE) defines power quality as the concept of powering and grounding sensitive electronic equipment in a manner suitable for the equipment. Lack of power quality may cause many problems, such as the nonlinearity, instability and unbalance of electric network. Understanding power quality issues is a good starting point for solving any harmonic problems. The significant reason that electrical engineers are interested in power quality is economic value. There are economic impacts on utilities, their customers, and suppliers of load equipment. The quality of power can have a direct economic impact on many industrial consumers. There has recently been a great emphasis on revitalizing industry with more automation and more modern equipments. This usually means electronically controlled, energy-efficient equipment that is often much sensitive to deviations in the supply voltage [1, 2].
Many events and disturbances included in power quality issue. Some of these disturbances and events are such as; voltage sag and swells, over-voltage, under-voltage, harmonics, inter-

harmonics and etc. It should be mentioned that, between these types of disturbances, harmonic studies have a great concern in power quality study [1-6].
In this paper, first theoretical aspects of effects of harmonic distortions on the behavior and performance of several important electrical equipments such as transformers, induction motor drives and power cables have been described, punctually and then, harmonic analysis has been done on a MDF factory, as a case study. For the sake of this analysis, several measurements have been performed on some of installed electrical instruments in this factory. All measurements have been done using a C.A8310 power analyzer. After measurements, results have been analyzed. The C.A8310 power analyzer has been shown in Fig. 1. It should be mentioned that, some of various abilities of this device are as follow:

1) Store the three phase current and voltage waveforms
2) Measure the rms value of phase and line voltage and current
3) Measure the active and reactive and apparent power of each phase
4) Measure the voltage and current harmonic orders up to 25th order
5) Measure the THD-V and THD-I
6) Measure the frequency, power factor and displacement factor

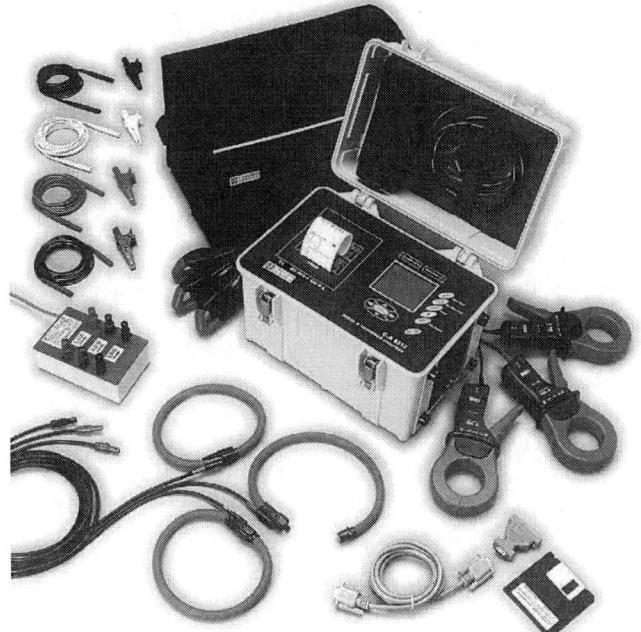

Fig.1. The C.A8310 power analyzer

II. Harmonic Definition and Its Impacts on Electrical Instruments

Harmonic is defined as a sinusoidal component of a periodic wave or quantity having a frequency that is an integral multiple of the fundamental frequency. Therefore, harmonic is the presence of voltage/current with the frequency of a multiple of fundamental voltage/current in the voltage/current of the system [5].

The common power quality problem in power and distribution networks and also, in industrial factories is harmonic distortion particularly current harmonic distortion. More recently the range of types and the number of units of equipment causing harmonics have risen sharply, and will continue to rise, so designers must now consider harmonics and their side effects very carefully. The most important source of voltage and current harmonics in electrical networks are nonlinear loads. Nonlinear loads, which were only fifteen percent of total loads in 1987, have been increased to seventy percent in 2000 [7]. So, ever-increasing of nonlinear load causes inordinate increase of harmonics in network. A nonlinear load is created when the load current is not proportional to the instantaneous voltage. Nonlinear currents can be no sinusoidal, even when the source voltage is a clean sinusoidal wave.

Harmonic studies play an important role in characterizing and understanding the extent of harmonic problems. Harmonic studies provide a means to evaluate various possible solutions and their effectiveness under a wide range of conditions before implementing a final solution. Harmonic studies are often performed when [2]:
1) Finding a solution to an existing harmonic problem
2) Installing large nonlinear devices or loads
3) Designing a harmonic filter
4) Installing large capacitor banks on utility distribution systems or industrial systems
5) Converting a power factor capacitor bank to a harmonic filter

Generally, the procedure of the harmonic study in industry is shown in Fig.2.

The harmonic sources can classified into two major groups; commercial and industrial loads. Commercial facilities such as office complexes, department stores, hospitals, universities, and Internet data centers are dominated with high-efficiency fluorescent lighting with electronic ballast, adjustable-speed drives for the heating, ventilation, and air conditioning (HVAC) loads, elevator drives, and sensitive electronic equipment supplied by single-phase switch-mode power supplies. Beside, modern industrial facilities are characterized by the widespread application of nonlinear loads. These loads can make up a significant portion of the total facility loads and inject harmonic currents into the power system, causing harmonic distortion in the voltage. Some of these industrial loads that inject harmonic distortion into electrical networks are such as; three-phase power converters (DC and AC drives), arcing devices (includes arc furnaces, arc welders, and discharge-type lighting such as fluorescent, sodium vapor, mercury vapor with magnetic ballasts) and saturable devices (includes power and distribution transformers and other electromagnetic devices with a steel core, like induction motors) [1-5 books].

Most of the electrical instruments and loads are sensitive to harmonics. In fact, harmonics may lead to their improper operation. Also, the general effects of harmonics on electrical instruments can be classified into [5]:

1) Disturbance to electric and electronic devices
2) Higher losses
3) Extra neutral current
4) Improper working of metering devices
5) Resonance problems

Beside, the main effects of voltage and current harmonics within the power system are, as follow [3]:

1) The possibility of amplification of harmonic levels resulting from series and parallel resonances.
2) A reduction in the efficiency of the generation, transmission and utilization of electric energy.
3) Ageing of the insulation of electrical plant components with consequent shortening of their useful life.
4) Malfunctioning of system or plant components.

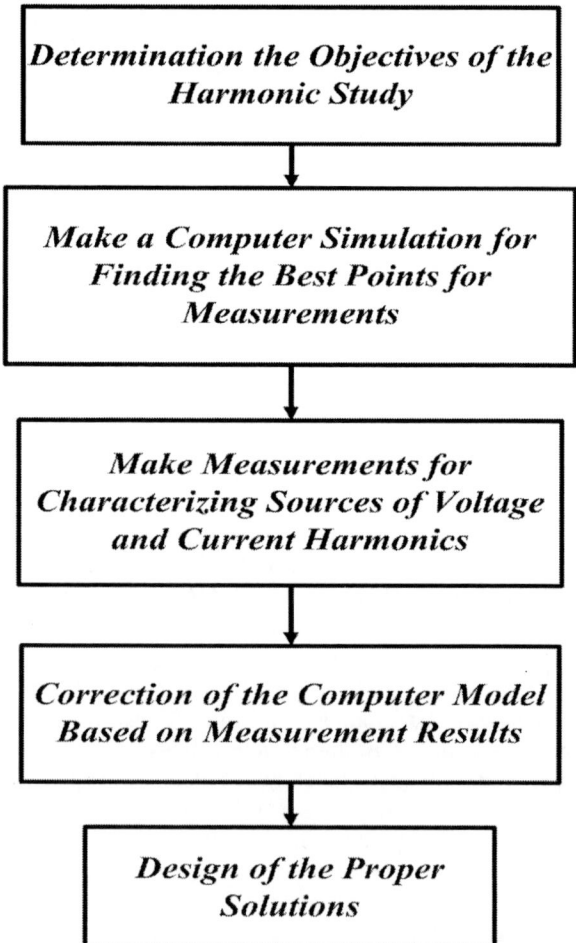

Fig. 2. Description of the harmonic study procedure

On the other hand, the effects of harmonics on the electrical components of power and distribution networks can be described as follow:

A. Power Cables and Conductors:
Resistance increases with frequency due to skin effect and proximity effect, particularly on larger conductors. As an example, at the seventh harmonic, a 500-kcmil cable has a resistance that is 2.36 times its dc resistance. Neutral conductors in facilities are especially prone to problems; third harmonic currents from single-phase loads add in the neutral, which can actually increase neutral current above that in the phase conductors [6].

B. Electrical Motors:
Voltage distortion induces heating on the stator and on the rotor in motors and other rotating machinery. The rotor presents relatively low impedance to harmonics, like a shorted transformer winding. The effects are similar to motors operating with voltage unbalance, although not quite as severe since the impedance to the harmonics increases with frequency.
Application of distorted voltage to a motor results in additional losses in the magnetic core of the motor. Hysteresis and eddy current losses in the core increase as higher frequency harmonic voltages are impressed on the motor windings. Besides, harmonic currents produce additional ohmic loss in the motor windings. Another effect, and perhaps a more serious one, is torsional oscillations due to harmonics. Two of the more prominent harmonics found in a typical networks are the fifth and seventh harmonics. It should be mentioned that, large motors supplied from ASDs are usually provided with harmonic filters to prevent motor damage due to harmonics [6, 8-10].

C. Power and Distribution Transformers:
Transformers are usually designed for utilizing at the rated frequency and linear load. Nowadays, present of nonlinear load, leads to higher losses and reduction of useful life in transformer. If the transformer cannot be operated up to its standard lifetime, there will be an economic loss.
The primary effect of harmonics on transformers is the additional heat generated by the losses caused by the harmonic content of the load current. Other problems include possible resonances between the transformer inductance and system capacitance, mechanical insulation stress (winding and lamination) due to temperature cycling and possible small core vibrations. The flow of harmonic currents increases the copper losses; this effect is more important in the case of converter transformers because they do not benefit from the presence of filters, which are normally connected on the AC system side. Apart from the extra rating required, converter transformers often develop unexpected hot spots in the tank.
The harmonic currents increase the rms current flowing in the transformer windings which results in additional ohmic loss. Also, winding eddy current loss will increase. Eddy current concentrations are higher at the ends of the windings due to

the crowding effect of the leakage magnetic field at the coil extremities. The winding eddy current losses increase as the square of the harmonic current and the square of the frequency of the current. Eddy currents due to harmonics can significantly increase the transformer winding temperature. The formulations of these increases in different losses of transformers under harmonic current according to the IEEE C. 57-110 standard are, as follow [11-17]:

$$P_{dc} = R_{dc} \times I^2 = R_{dc} \times \sum_{h=1}^{h=h_{max}} I^2_{h,max} \qquad (1)$$

$$P_{EC} = P_{EC-R} \times \sum_{h=1}^{h=h_{max}} h^2 [\frac{I_h}{I_R}]^2 \qquad (2)$$

$$P_{OSL} = P_{OSL-R} \times \sum_{h=1}^{h=h_{max}} h^{0.8} [\frac{I_h}{I_R}]^2 \qquad (3)$$

The presence of harmonic voltages increases the hysteresis and eddy current losses in the laminations and stresses the insulation. The increase in core losses due to harmonics depends on the effect that the harmonics have on the supply voltage and on the design of the transformer core.
It should be mentioned that, measurement will play an important role for solving any power quality problem. This is the primary method of characterizing the problem or the existing system that is being evaluated. When performing the measurements, it is important to record impacts of the power quality variations at the same time so that problems can be correlated with possible causes. It should be mentioned that there are many devices for measuring power quality in different cases, such as harmonic analyzers, transient disturbance analyzers, oscilloscopes, data loggers and chart recorders, true rms meters, flicker meters and etc [1, 2].
Basic categories of instruments that may be applicable include:

1) Wiring and grounding test devices
2) Multi-meters
3) Oscilloscopes
4) Disturbance analyzers
5) Harmonic analyzers and spectrum analyzers
6) Combination disturbance and harmonic analyzers
7) Flicker meters
8) Energy monitors

Instruments in the disturbance analyzer category have very limited harmonic analysis capabilities. Some of the more powerful analyzers have add-on modules that can be used for computing Fast Fourier transform (FFT) calculations to determine the lower-order harmonics. However, any significant harmonic measurement requirements will demand an instrument that is designed for spectral analysis or harmonic analysis. Important capabilities for useful harmonic measurements include:

1) Capability to measure both voltage and current simultaneously so that harmonic power flow information can be obtained.
2) Capability to measure both magnitude and phase angle of individual harmonic components
3) Synchronization and a sampling rate fast enough to obtain accurate measurement of harmonic components
4) Capability to characterize the statistical nature of harmonic distortion levels

III. HARMONIC STUDY OF A TYPICAL MDF FACTORY

For the sake of harmonic analysis, a MDF factory has been selected as a case study. This factory has many dc and induction motors, electrical drives with switching devices and transformers. The used ac to dc drives in this factory, produce and inject a significant value of harmonic distortion in voltage and current waveform. The harmonic content of voltage or current can cause a big set of problems in terms of loss increasing, increasing of insulation stress and eventually, life reduction in transformers and motors. Therefore, the authors admit the effectiveness of the impacts of harmonic study on the power quality indexes for electrical instruments. Hence, three dc induction motor and a three-phase transformer have been selected as case study subjects. It should be mentioned that, all measurements have been accomplished using a C. A8310 power analyzer that shown in Fig. 1.

A. Drive of a 250 kW DC Motor:
For harmonic analysis of this equipment, the power analyzer has been put on output of drives for sampling and recording duration of three days. The data have been recorded with time interval of 15 minutes with sampling frequency of 256 sample per cycle. For this case, voltage and current qualities and variation of power factor have been measured and analyzed. So, the Total Harmonic Distortion (THD) of voltage and Total Demand Distortion (TDD) of current of each phase have been shown in Table I and Table II, respectively.
In addition, the magnitude of voltage harmonic per harmonic order has been shown in Table III.
It should be mentioned that, the first disadvantage of harmonic distortion is extra losses in electrical machines, especially motors and transformers. So, for this motor the amount of extra losses for each phase have been shown in Table IV for time duration of one month. These losses measured using above mentioned power analyzer, too.

B. Drive of a 30 kW DC Motor:
For harmonic analysis of this equipment, the power analyzer has been put on output of drives for sampling and recording duration of two days. The data have been recorded with time interval of 15 minutes with sampling frequency of 256 sample per cycle. For this case, voltage and current qualities and variation of power factor have been measured and analyzed. So, the voltage variation, THD of voltage and TDD of current

of each phase have been shown in Table V, Table VI, and Table VII, respectively. The amount of extra losses for each phase of this motor, have been shown in Table VIII for time duration of one month. A typical waveform of voltage and current and their harmonic content of this motor have been shown in Fig. 3.

TABLE I
THE THD VALUES OF 250 KW MOTOR VOLTAGES

VTHD	Maximum (%)
THD Vab	5.6
THD Vbc	4.9
THD Vca	4.7

TABLE II
THE TDD VALUES OF 250 KW MOTOR CURRENTS

ITDD	Maximum (%)
TDD Phase A	42.6
TDD Phase B	40.9
TDD Phase C	38.7

TABLE III
THE MAGNITUDE OF THE VOLTAGE HARMONIC FOR 250 KW MOTOR

Harmonic Order	Phase A (%)	Phase B (%)	Phase C (%)
1	100	100	100
2	0.38	0.27	0.29
3	0.39	0.36	0.31
4	0.27	0.23	0.22
5	4.36	3.88	3.78
6	0.22	0.22	0.19
7	0.99	1.09	1.01
8	0.22	0.23	0.20
9	0.33	0.33	0.29
10	0.21	0.21	0.19
11	0.8	0.69	0.62
12	0.21	0.22	0.18
13	0.43	0.43	0.34
14	0.21	0.21	0.19
15	0.33	0.32	0.28
16	0.22	0.22	0.19
17	0.67	0.62	0.47
18	0.21	0.22	0.18
19	0.38	0.39	0.31
20	0.21	0.21	0.19
21	0.33	0.32	0.28
22	0.21	0.22	0.19
23	0.59	0.56	0.43
24	0.21	0.22	0.18
25	0.36	0.36	0.29
THD	5.6	4.9	4.7

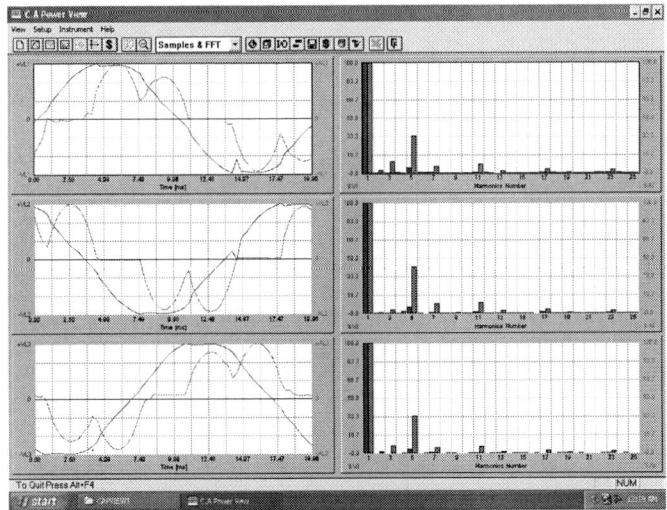

Fig. 3. Typical waveform of voltage and current and their harmonic content for drive of a 30kW Motor

TABLE IV
THE VALUES OF EXTRA LOSSES FOR 250 kW MOTOR IN ONE MONTH

Phase	Extra Value (kW)
Phase A	1260
Phase B	1188
Phase C	1195
Total Value	3643

TABLE V
VOLTAGE VARIATION OF EACH PHASE FOR 30 kW MOTOR

Phase Voltages	Va	Vb	Vc
Maximum Voltage	220.44	222.71	221.2
Minimum Voltage	214.71	216.89	215.99

TABLE VI
THE THD VALUES OF 30 kW MOTOR VOLTAGES

VTHD	Maximum (%)
THD Vab	5.8
THD Vbc	5.3
THD Vca	4.6

TABLE VII
THE TDD VALUES OF 30 kW MOTOR CURRENTS

ITDD	Maximum (%)
TDD Phase A	36.7
TDD Phase B	43.5
TDD Phase C	36.5

TABLE VIII
THE VALUES OF EXTRA LOSSES FOR 30 kW MOTOR IN ONE MONTH

Phase	Extra Value (kW)
Phase A	77
Phase B	100
Phase C	67
Total Value	244

TABLE VIIII
THE THD VALUES OF 15 kW MOTOR VOLTAGES

VTHD	Maximum (%)
THD Vab	1.7
THD Vbc	2.2
THD Vca	1.8

TABLE X
THE TDD VALUES OF 15 kW MOTOR CURRENTS

ITDD	Maximum (%)
TDD Phase A	37.8
TDD Phase B	45.7
TDD Phase C	37.8

TABLE XI
THE THD AND TDD VALUE OF PHASE A FOR 1250 VA TRANSFORMER

Parameter	Maximum (%)
THD Va	2.5
TDD Phase A	8.7

TABLE XII
FREQUENCY VARIATION FOR 1250 VA TRANSFORMER

Variation	Maximum	Mean	Minimum
Frequency	50.2	50.05	49.9

C. Drive of a 15 kW DC Motor:

For harmonic analysis of 15 kW motor, the measurement device has been put on output of drives for sampling and recording duration of three days. The data have been recorded with time interval of 15 minutes with sampling frequency of 256 sample per cycle. For this case, voltage and current qualities and variation of power factor have been measured and analyzed. So, the THD of voltage and TDD of current of each phase have been shown in Table VIIII, and Table X, respectively.

D. A 1250 VA Transformer Supplying AC Drives:

For harmonic analysis of this transformer, the measurement device has been put on output of it for sampling and recording duration of four days. The data have been recorded with time interval of 15 minutes with sampling frequency of 256 sample per cycle. This transformer supplies all of AC drives. So, it will have a great amount of harmonic distortion.

The phase A voltage THD and current TDD with their harmonic spectrum have been shown in Table XI, Fig. 4 and Fig. 5, respectively.

It should be mentioned that, the frequency variation on this transformer has been recorded. The variation of frequency has been shown in Table XII.

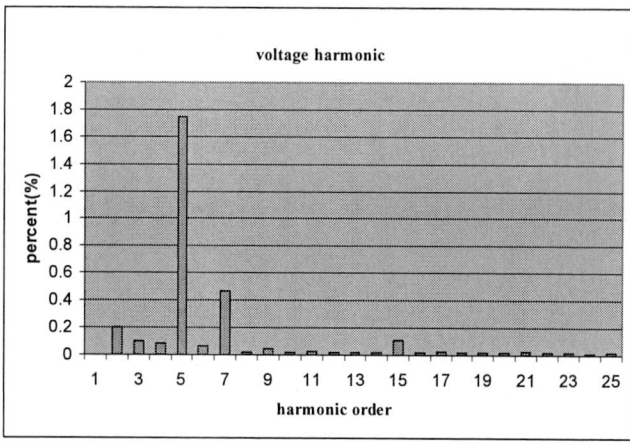

Fig. 4. Harmonic spectrum of phase A voltage for 1250 VA transformer

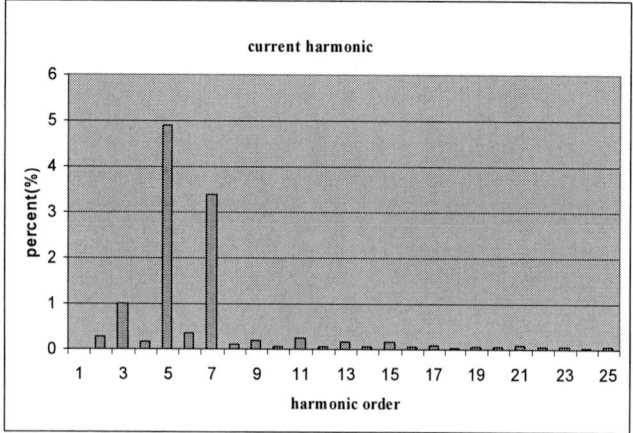

Fig. 5. Harmonic content of phase A current for 1250 VA transformer

IV. CONCLUSION

In this paper, some power quality indices especially harmonic problems have been investigated based on study of the behavior of transformer and induction motors under harmonic distortion. Therefore, a medium density fiberboard factory has been considered for case study. For this purpose, effect of harmonic on the electrical instruments has been evaluated using a harmonic analyzer. Therefore, under this condition, voltage THD and current TDD and harmonic spectrum of this quantities has been measured. It is shown that using dc drives, the voltage peak and its harmonic spectrum will be distorted. This condition is very dangerous for life time of motor and the transformer that supply the drives.

ACKNOWLEDGMENT

The authors would like to thanks Mr. Behrouz Darvishi (Department of Engineering and Planning, Mazandaran Electric Power Distribution Company, Sari, Mazandaran, Islamic Republic of IRAN) for his technical support and discussion during the experimental procedure of this paper.

REFERENCES

[1] C. Sankaran, *Power Quality*, CRC Press, 2002.
[2] R. C. Dugan, M. F. Mc Granaghan, S. Santoso, and H. W. Beaty, Electrical Power Systems Quality, Second Edition, McGraw-Hill, 2004.
[3] J. Arrillaga, and N. R. Watson, Power System Harmonics, Second Edition, John Wiley & Sons, 2003.
[4] M. H. J. Bollen, and I. Y. H. Gu, Signal Processing of Power Quality Disturbances, IEEE Power Engineering Society, John Wiley & Sons, 2006.
[5] A. Emadi, A. Nasiri, and S. B. Bekiarov, Uninterruptible Power Supplies and Active Filters, CRC Press, 2005.
[6] T. A. Short, Distribution Reliability and Power Quality, CRC Taylor & Francis Group, 2006.
[7] M. Radmehr, S. Farhangi, A. Nasiri, "Effect of Power Quality Distortion on Electrical Drives and Transformer Life in Paper Industries: Simulation and Real Time Measurements", *IEEE Pulp and Paper Industry Technical Conference*, 18-23 June 2006, Iran, pp.1-9.
[8] S. X. Duarte, and N. Kagan, "A Power-Quality Index to Assess the Impact of Voltage Harmonic Distortions and Unbalance to Three-Phase Induction Motors," IEEE Transactions on Power Delivery, vol. 25, no. 3, pp. 1846-1854, July 2010.
[9] M. Anwari, and A. Hiendro, "New Unbalance Factor for Estimating Performance of a Three-Phase Induction Motor With Under- and Over-voltage Unbalance," IEEE Transactions on Energy Conversion, vol. 25, no. 3, pp. 619-625, September 2010.
[10] M. Yazdani-Asrami, S. B. Sadati, M. Jamali, and B. Darvishi, Power Quality Evaluation for Industrial appliances of a Medium Density Fiberboard Factory, Accepted for publication in IEEE 3rd International Conference on Power Electronics and Intelligent Transportation System (PEITS2010), China, 2010.
[11] M. Yazdani-Asrami, M. Mirzaie, and A. Shayegani Akmal, "Calculation of Transformer Losses under Non-Sinusoidal Currents Using: Two Analytic Methods and Finite Element Analysis," World Applied Science Journal, vol. 9, no. 8, pp. 889-897, 2010.
[12] S. B. Sadati, M. Yazdani-Asrami and M. Taghipour, "Effects of Harmonic Current Content and Ambient Temperature on Load Ability and Life Time of Distribution Transformers," International Review on Electrical Engineering (I. R. E. E.), vol. 25, no. 3, pp. 1444-1451, July-August 2010.
[13] S. B. Sadati, A. Tahani, M. Jafari, and M. Dargahi, "Derating of transformers under Non-sinusoidal Loads," 11th IEEE International Conference on Optimization of Electrical and Electronic Equipment (OPTIM 2008), pp. 263-268, 2008.
[14] M. Yazdani-Asrami, M. Mirzaie and A. Shayegani Akmal, Investigation on Impact of Current Harmonic Contents On the Distribution Transformer Losses and Remaining Life, IEEE 2nd Power and Energy Conference 2010 (PECon 2010), Malaysia, pp. 689-694, 2010.
[15] M. Yazdani-Asrami, M. Mirzaie, A. Shayegani Akmal and S. Asghar Gholamian, "Life Estimation of Distribution Transformers Under Non-Linear Loads Using Calculated Loss by 2D-FEM," Journal of Electrical Systems (JES), vol. 7, no. 1, pp. 1444-1451, March 2010.
[16] S. B. Sadati, B. Darvish Motevali, M. Dargahi and M. Yazdani-Asrami, "Evaluation of Distribution Transformer Losses and Remaining Life considering Network Harmonic, Based on Analytical and Simulation Methods," Australian Journal of Basic and Applied Sciences (AJBAS), vol. 4, no. 10, pp. 5291-5299, 2010.
[17] IEEE Std C57.110-1998, "IEEE Recommended Practice for Establishing Transformer Capability when Supplying Non sinusoidal Load Currents", IEEE Publication, Piscataway, NJ, USA, 1998.

978-1-4577-0007-1/11 $26.00 © 2011 IEEE

A Novel Hysteresis Bandwidth (NHB) Calculation To Fix the Switching Frequency Employed In Active Power Filter

Hani Vahedi

Department of Electrical Engineering
Islamic Azad University, Sari Branch
Sari, Iran

hvahedi@ieee.org,

Abdolreza Sheikholeslami

Department of Electrical Engineering
Babol University of Technology
Babol, Iran

asheikh@nit.ac.ir

Mohammad Tavakoli Bina

Department of Electrical Engineering
K.N.Toosi University of Technology
Tehran, Iran

tavakoli@eetd.kntu.ac.ir

Abstract - Variable switching process is the main issue in practical implementation of fixed band hysteresis current controller in active power filters (APF) that increases the switching frequency and switching losses in power systems. Preventing this case, the Adaptive Hysteresis Current Control (AHCC) has been introduced and developed by many researchers. By this way, The Hysteresis Band (HB) will change adaptively by system parameters in order to control the switching speed and fix the switching frequency.

In this paper a revisory method for the hysteresis bandwidth calculation procedure is proposed that makes switching frequency invariable significantly. The APF Simulations using proposed method (NHB) in Matlab/Simulink environment has been done. Results including switching frequency and current source TDH have been given to prove the efficiency of this method.

Index Terms - Active Power Filter, Hysteresis Current Control, Instantaneous Power Theory, Power System Harmonic, Voltage Source Inverter (VSI).

I. INTRODUCTION

In recent years, shunt active power filters have being applied by many industries and researchers to remove the current harmonics caused by nonlinear loads [1-3]. An APF as can be seen in Figure 1 is a parallel power inverter with loads that can remove large amounts of current harmonics through the injection of reference current to the power system that contains harmonic components of the source current. Complete compensation occurs when the APF produces a same current as harmonic current with the same amplitude and opposite in sign.

Hysteresis current control is one of the most appropriate PWM switching methods to produce reference current in APFs [4]. Hysteresis current control has desirable characteristics such as high stability, fast and accurate dynamic behavior. On the other hand, conventional hysteresis method includes some undesirable results, such as variable switching frequency that causes audio noises, high switching losses and injection of high frequency current components to the source current that makes it difficult to design suitable filters to remove these high- frequency harmonics [5].

AHCC was presented to solve this problem [6-8]. According to this method, a variable band is defined for reference current in each phase so that the switching frequency remains constant. Since the bandwidth is changed every time the switching pattern becomes symmetrical and so switching

speed is limited to a fixed amount and switching frequency will be invariable and its range will be reduced.

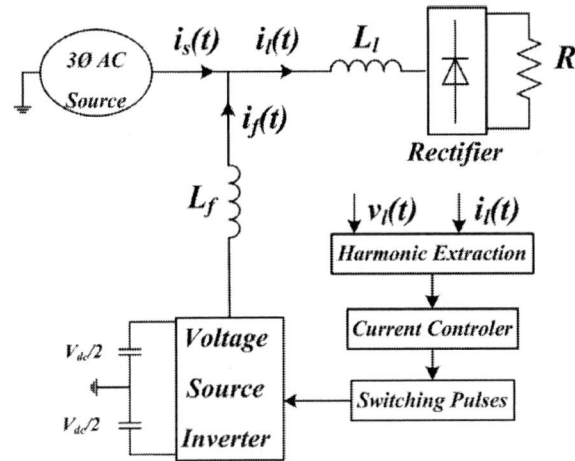

Fig. 1. Overview of APF connected to the power network

In this paper a new method (NHBC) has been suggested for calculating the variable bandwidth that has great effect on the switching frequency and reduces its range more than usual modes leads to lower switching losses and audio noises. Bandwidth, in the new formula will be slightly smaller but the switching frequency will be more constant and the source current THD is improved. In Section 2, the Instantaneous power theory is explained to extract harmonic components of load current, due to its effective performance. In section 3 the NHB method is presented and the necessary equations to calculate the bandwidth are given in this way. Then in Section 4, simulation of an APF in power network with nonlinear load in Matlab/Simulink environment has been done. The results of simulation that include instantaneous switching frequency and source current TDH show that the NHB has significant effect on reducing the switching frequency changes and source current TDH.

II. EXTRACTION OF THE COMPENSATION REFERENCE CURRENTS

One of the popular compensation reference current extraction methods is the instantaneous reactive power theory (p-q theory). Although there are some problems with this theory, it is well-established and simple in implementation. The p-q theory could be briefly reviewed as follow [9]:

Assume a three-phase load with the instantaneous voltages as $\mathbf{v}(t)=[v_a(t)\ v_b(t)\ v_c(t)]^t$ and the instantaneous currents as $\mathbf{i_l}(t)=[i_{la}(t)\ i_{lb}(t)\ i_{lc}(t)]^t$ (Fig. 1). Using (1), $\mathbf{v}(t)$ and $\mathbf{i_l}(t)$ can be converted to o-α-β coordination where C is the matrix (2):

$$\left[v_{0\alpha\beta}(t)\right]^t = C\left[v_{abc}(t)\right]^t, \left[i_{0\alpha\beta}(t)\right]^t = C\left[i_{abc}(t)\right]^t \quad (1)$$

$$C = \sqrt{\frac{2}{3}}\begin{bmatrix} \frac{1}{\sqrt{2}} & \frac{1}{\sqrt{2}} & \frac{1}{\sqrt{2}} \\ 1 & \frac{-1}{2} & \frac{-1}{2} \\ 0 & \frac{\sqrt{3}}{2} & \frac{-\sqrt{3}}{2} \end{bmatrix} \quad (2)$$

Let's assume that the zero sequence current ($i_{l0}(t)$) is null. Thus, the instantaneous active ($p(t)$) and reactive ($q(t)$) powers can be calculated as:

$$\begin{bmatrix} p(t) \\ q(t) \end{bmatrix} = \begin{bmatrix} v_\alpha(t) & v_\beta(t) \\ -v_\beta(t) & v_\alpha(t) \end{bmatrix}\begin{bmatrix} i_{l\alpha}(t) \\ i_{l\beta}(t) \end{bmatrix} \quad (3)$$

$p(t)$ and $q(t)$ can be decomposed to the average parts ($\bar{p}(t),\bar{q}(t)$) and the oscillating parts ($\tilde{p}(t),\tilde{q}(t)$). It is notable that $\bar{p}(t)$ is produced by the fundamental harmonic of the positive sequence component of the load current. Therefore, in order to compensate the harmonics and the instantaneous reactive power, compensation reference currents can be extracted as follow:

$$\begin{bmatrix} i_{f\alpha}^*(t) \\ i_{f\beta}^*(t) \end{bmatrix} = \begin{bmatrix} v_\alpha(t) & v_\beta(t) \\ -v_\beta(t) & v_\alpha(t) \end{bmatrix}^{-1}\begin{bmatrix} -\tilde{p}(t) \\ -q(t) \end{bmatrix} \quad (4)$$

$$\begin{bmatrix} i_{fa}^*(t) & i_{fb}^*(t) & i_{fc}^*(t) \end{bmatrix}^t = C^{-1}\begin{bmatrix} 0 & i_{f\alpha}^*(t) & i_{f\beta}^*(t) \end{bmatrix}^t \quad (5)$$

III. THE PROPOSED NHBC METHOD

The basic implementation of hysteresis current controller derives the switching signals by comparing the current error signal with a fixed hysteresis band (Fig. 2). In case the error signal touches the upper band, inverter voltage decreases to reduce the filter current and if the error signal violates the lower band, the inverter voltage increases to raise the filter current.

The hysteresis current control technique is the most suitable method for current control of Voltage Source Inverters (VSIs) in APFs due to its very fast response and good accuracy. On the other hand, the conventional hysteresis technique exhibits several undesirable features, such as uneven switching frequency that causes acoustic noise and difficulty in the designing input filters. The switching frequency of the hysteresis current control method depends on how fast the current changes from the upper limit to the lower limit of the hysteresis band, or vice versa [6, 8].

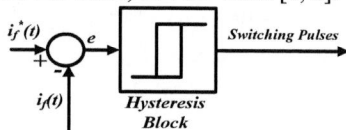

Fig. 2 hysteresis current control loop.

As above-mentioned, the crucial concern with the fixed band hysteresis current control is producing a varying modulation frequency of the power converter which, in turn, results in increasing the risk of resonance in power system. To avoid this situation, adaptive hysteresis current controller methods with the variable hysteresis band have been recommended in literature [6, 7]. Hence, a variable hysteresis band is defined for each phase so that the switching frequency remains almost constant.

In circuit instances of power system, the current's path can be selected desirably. If the current path is opposite selected path, the circuit analysis results will differ only in signs that are negative and won't make a difference in their values. If we consider the current path in calculating HB, we will achieve the new formula (NHB). In this case, we consider the current path from the network to the inverter side as it is observed in figure 2. It should be mentioned that analyzing the circuit by choosing reverse path for the current, makes no different in result.

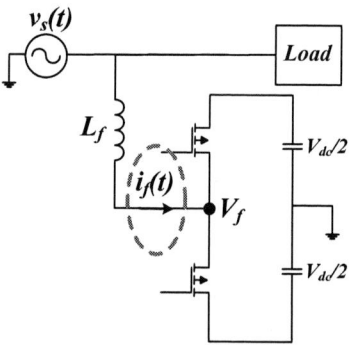

Figure 3: One phase from 3-phase VSI by reserved current assumption

The NHB formula can be calculated based on Fig. 1, also it can be achieved from Fig. 3 easier. According to Fig. 3, the following KVL equation can be easily achieved:

$$\frac{di_f(t)}{dt} = \frac{1}{L_f}\left(v_s(t) - V_f\right) \quad (6)$$

Where V_f is the inverter-side voltage and can be elaborated as below:

$$V_f = \begin{cases} \dfrac{V_{dc}}{2} & \text{the upper switch is ON} \\ \dfrac{-V_{dc}}{2} & \text{the lower switch is ON} \end{cases} \quad (7)$$

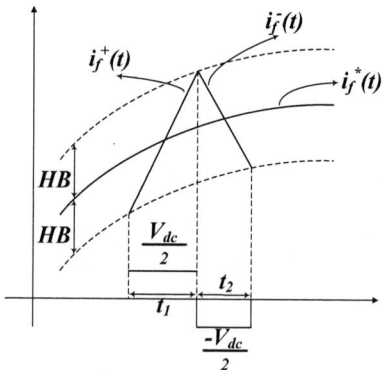

Fig. 4 The upper and lower bands of the reference compensation current.

Having paid attention to Fig. 4, the below relations can be obtained:

978-1-4577-0007-1/11 $26.00 © 2011 IEEE

$$\frac{di_f^+(t)}{dt} = \frac{1}{L_f}\left(v_s(t) - V_f\right) \qquad (8)$$

$$\frac{di_f^-(t)}{dt} = \frac{1}{L_f}\left(v_s(t) + V_f\right) \qquad (9)$$

Where $i_f^+(t)$ and $i_f^-(t)$ are the rising current and the falling current, respectively. Furthermore, the following relations can be extracted:

$$\begin{cases} \dfrac{di_f^+(t)}{dt} \times t_1 - \dfrac{di_f^*(t)}{dt} \times t_1 = 2HB \\[2mm] \dfrac{di_f^-(t)}{dt} \times t_2 - \dfrac{di_f^*(t)}{dt} \times t_2 = -2HB \end{cases} \qquad (10)$$

$$f = \frac{1}{t_1 + t_2} \qquad (11)$$

Where t_1 and t_2 are switching intervals and f is the switching frequency.

By substituting (8), (9) and (11) in (10), the novel hysteresis bandwidth (NHB) can be derived as follow:

$$NHB = \frac{-V_{dc}}{8fL_f} + \frac{L_f}{2fV_{dc}}\left(\frac{v_s(t)}{L_f} - \frac{di_f^*(t)}{dt}\right)^2 \qquad (12)$$

The adaptive HB should be derived instantaneously during each sample time to keep the switching frequency constant.
The conventional HB formula which is calculated with normal direction of the filter current is [6, 7]:

$$HB = \frac{V_{dc}}{8fL_f} - \frac{L_f}{2fV_{dc}}\left(\frac{v_s(t)}{L_f} + \frac{di_f^*(t)}{dt}\right)^2 \qquad (13)$$

As it can be seen, the new formula for the bandwidth has certain differences with the present formula regarding to its initial terms. Proposed formula will result smaller band width, but reduces the frequency changes sensibly that will be proved by simulation in next section.

IV. SIMULATION RESULTS

In this part, simulation of an APF connected to the power network has been done by three methods of fixed band, AHCC and the proposed NHB with a balanced nonlinear load, including three phase rectifier and 20Ω resistor in Matlab/Simulink software.

TABLE 1
APF SIMULATION PARAMETERS

Supply Phase Voltage	200 V
Grid Frequency	60 Hz
Load Resistanc R_l	20 Ω
Inverter Side Inductance L_f	5 mH
Rectifier Side Inductance L_l	2 mH
APF dc-link Voltage V_{dc}	500 V
Fixed Hysteresis Bandwidth	0.67 A

Then the simulation results including switching frequency and source, load and filter currents are given below. In Fig. 5, load, filter and source currents can be seen for all three simulations. As this figure implies, the first column shows the fixed band results, the second column represents the conventional AHCC and the third column is related to the proposed NHB method. The top, middle and bottom rows show the load current, filter current and source current respectively.

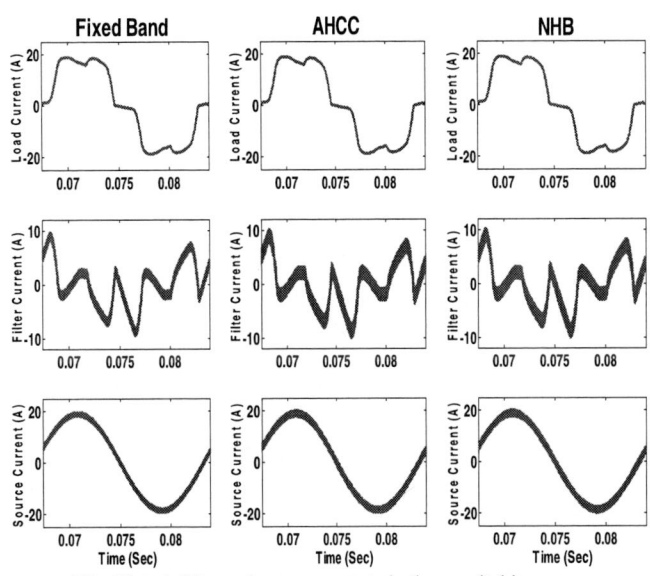

Fig. 5 Load, filter and source currents in three switching cases

Table 2 shows the load and source current RMS value that shows load current supplied properly by the source and filter.

TABLE 2
RMS VALUE OF LOAD AND SOURCE CURRENTS

	Source Current RMS (A)			Load Current RMS (A)		
	Phase a	Phase b	Phase c	Phase a	Phase b	Phase c
Fixed-band	13.40	13.40	13.39	13.86	13.86	13.86
AHCC	13.43	13.44	13.44	13.86	13.86	13.86
NHB	13.44	13.43	13.44	13.86	13.86	13.86

The THD% values of the source and load currents have been shown in Table 3. These numerical results show the appropriate performance of the proposed method.

TABLE 3
THD% VALUE OF LOAD AND SOURCE CURRENTS

	Source Current THD%			Load Current THD%		
	Phase a	Phase b	Phase c	Phase a	Phase b	Phase c
Fixed-band	3.01	2.72	2.38	22.01	22.01	22.01
AHCC	3.50	3.42	3.28	21.96	21.99	22.01
NHB	3.18	3.45	3.96	21.96	21.99	22.01

Important issue in switching with hysteresis method is its switching frequency changes. Fig. 6 shows instantaneous switching frequency for all three methods in this survey. Maximum and average and minimum numerical values of the switching frequency can be seen in Table 4.

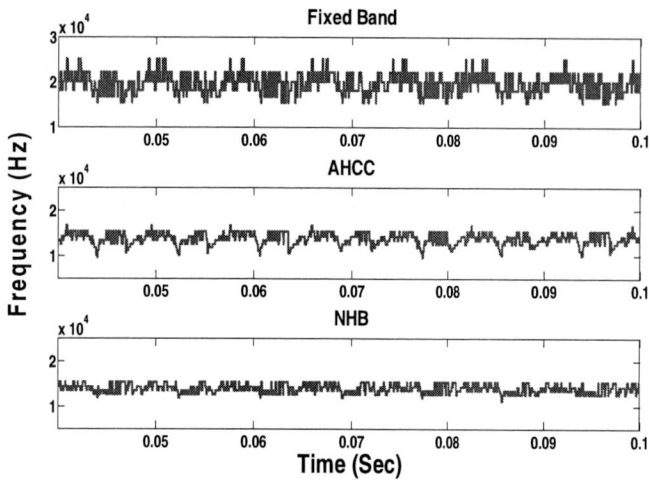

Fig. 6 Instantaneous switching frequency for 3 switching methods

TABLE 4
NUMERICAL SWITCHING FREQUENCY RESULTS FOR 3 SWITCHING METHODS

	SwitchingFrequency (KHz)		
	Min.	Average	Max.
Fixed-band	15.38	19.70	25.00
AHCC	9.52	13.72	16.66
NHB	11.11	13.85	15.38

Figure 6 and Table 4 indicate that in fixed band method, switching frequency increases and its variation is also higher. By definition of AHCC method, switching frequency and its range has been significantly reduced and the simulation continues with the proposed NHB method, the switching frequency range is limited to a smaller one than the two other cases.

As noted, changing in switching frequency not only causes audio noise and increases the switching losses but also injects high frequency current components to the source current. Therefore, by stabilizing switching frequency we can eliminate large amounts of current harmonics in high frequency.

The results of the proposed method including switching frequency and source current THD indicates suitable performance of the NHB formula in producing HB leads to switching pulses and making reference current , while reducing the source current THD optimally, switching frequency stabilization is done.

IV. CONCLUSION

Many studies have been performed to fix the switching frequency in hysteresis current control method using adaptive HB. This paper shows an efficient attempt in introducing a new formula for HB calculation (NHB) that produces more appropriate bandwidth leads to more smooth switching and more constant switching frequency. The simulations results for three switching methods prove the fact that the NHB formula introduces superior results in calculating the proper hysteresis bandwidth results o fix the switching frequency and solve the related problems like switching losses and audio noises.

REFERENCES

[1] Bhim Singh, Jitendra Solanki, "An Implementation of an Adaptive Control Algorithm for a Three-Phase Shunt Active Filter", IEEE Transaction, 2009, Vol. 56, Iss. 8, pp. 2811-2820.

[2] O. Vodyakho, T. Kim, "Shunt active filter based on three-level inverter for three-phase four-wire systems", IET Power Electronics, 2009, Vol. 2, Iss. 3, pp. 216-226.

[3] Oleg Vodyakho, Chris C. Mi, "Three-Level Inverter-Based Shunt Active Power Filter in Three-Phase Three-Wire and Four-Wire Systems", IEEE Transaction, 2009, Vol. 24, Iss. 5, pp. 1350-1363.

[4] Marian P. Kaźmierkowski, Ramu Krishnan, Frede Blaabjerg, "Control in power electronics", Academic Press, 2002.

[5] M. Tavakoli Bina, E. Pashajavid, "An efficient procedure to design passive LCL-filters for active power filters", Elsevier Journal Electric Power Systems Research, 2009, Vol. 79, Iss. 4, pp. 606-614.

[6] B.K.Bose, "An Adaptive Hysteresis Band Current Control Technique of a Voltage Feed PWM Inverter for Machine Drive System", IEEE Trans. On Ind. Elec., 1990, pp. 402- 406

[7] Hani Vahedi, Abdolreza Sheikholeslami, "Variable hysteresis current control applied in a shunt active filter with constant switching frequency", IEEE Conf. on Power Quality, PQC, 2010, pp.1-5

[8] Hani Vahedi, Abdolreza Sheikholeslami, "The Source-Side Inductance Based Adaptive Hysteresis Band Current Control to be Employed in Active Power Filters", International Review on Modeling and Simulation Journal, Vol. 3, No. 5, October 2010, pp. 840-845.

[9] Wenjin Dai, Yongtao DaiTingjian Zhong, "A New Method for Harmonic and Reactive Power Compensation", IEEE Int. Conf. on Industrial Technology, ICIT, 2008, pp.1-5.

Simulation Single Phase Shunt Active Filter Based on p-q technique using MATLAB/Simulink Development Tools Environment

Musa Yusup Lada[*], Othman Mohindo[**], Aziah Khamis[**], Jurifa Mat Lazi[**], Irma Wani Jamaludin[**]

[*] Universiti Teknikal Malaysia Melaka/FKE, Durian Tunggal, Melaka. Email: musayl@utem.edu.my
[**] Universiti Teknikal Malaysia Melaka//FKE, Durian Tunggal, Melaka, Malaysia.

Abstract – **This paper presents a single phase shunt active power filter based on instantaneous power theory. The active filter will be connected directly to utility in order to reduce THD of load current, in this case the utility is TNB. The instantaneous power theory also known as p-q theory is used for three phase active filter and this paper proves that the p-q theory can also be implemented for single phase active filter. Since the system has only single phase signal for both voltage and current, thus the dummy signal with 120 ° different angels must be generated for input of the p-q theory. The p-q technique will generate six signals PWM for switching IGBT, but only two of the signals will be used to control the switching IGBT. The simulation results are on MATLAB/Simulink environment tools presented in order to demonstrate the performance of the current load on single phase shunt active power filter.**

Keywords - Shunt Active Power Filter, Total Harmonic Distortion, Instantaneous Power Theory,

I. Introduction

Increasing demand on power converter or others non-linear load will cause usage of active power filter which widely applied eliminates the total harmonic distortion of load current. By generating harmonic that came from non-linear load, will facing a serious problem in the power system such as low power factor, increases losses, reduces the efficiency and increase the total harmonic distortion. The instantaneous power theory or p-q theory was introduced by Akagi, Kanazawa and Nabae in 1983 [1], [2]. The p-q theory was introduced and implemented only for three phase power system as shows in Fig. 1. Based on the term of p and q, the p-q theory will manipulate the

active and reactive power in order to maintain the purely sinusoidal current waveform at three phase power supply.

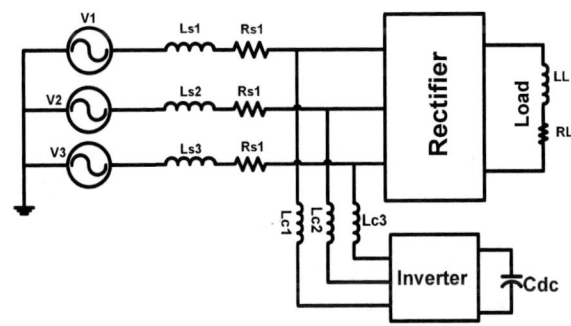

Fig. 1: Three phase active filter

There are a few techniques which can be used to eliminate harmonic others than active filter namely: L-C filter and Zig-Zag transformer. These techniques facing many disadvantage either the controller or the system such as fixed compensation, possible resonance, bulkiness, electromagnetic interference, voltage sag and flicker [1-6].

There are some advantages of implementing shunt active filter on grid power system since it can be installed at housing estate or others system that using single phase grid power system. The aim of this paper is to implement the p-q theory in single phase shunt active filter connected directly to gird power system. The technique is simulated by using MATLAB/Simulink simulation development tools environment.

978-1-4577-0007-1/11 $26.00 © 2011 IEEE

II. Mathematical Model

The p-q theory also known as instantaneous power theory is widely used for three wires three phase power system and also extended to four wires three phase power system. Although this theory using three current and three voltage signals, it also can be used for single phase active filter by duplicating two more current and voltage signal with 120° angel shifting. This theory based on separation power component separation in mean and oscillating values. Consider load current of single phase load as phase "*a*" and others phase (phase "*b*" and phase "*c*") are generated by duplicating technique. The load current can be assumed as phase "*a*" current and with be expressed mathematically as shows in eq. (1). By assuming that eq. (1) as phase "*a*" load current, load current for phase "*b*" and *c* can be represented as eq. (2) and eq. (3).

$$i_a = \sum_{i=0}^{n} \sqrt{2} I_i \sin(w_i + \theta_i) \tag{1}$$

$$i_b = \sum_{i=0}^{n} \sqrt{2} I_i \sin(w_i + \theta_i - 120°) \tag{2}$$

$$i_c = \sum_{i=0}^{n} \sqrt{2} I_i \sin(w_i + \theta_i + 120°) \tag{3}$$

Equation (1), (2) and (3) can be transformed in matrix form as shown in (4) and (5) for load current and load voltage respectively:

$$\begin{bmatrix} i_a \\ i_b \\ i_c \end{bmatrix} = \begin{bmatrix} 1 \\ 1\angle 120° \\ 1\angle 240° \end{bmatrix} [i_a] \tag{4}$$

$$\begin{bmatrix} v_a \\ v_b \\ v_c \end{bmatrix} = \begin{bmatrix} 1 \\ 1\angle 120° \\ 1\angle 240° \end{bmatrix} [v_a] \tag{5}$$

Determine the α and β reference current by using Clarke transformation as shown in (6) for load current and in (7) for load voltage.

$$\begin{bmatrix} i_\alpha \\ i_\beta \\ i_o \end{bmatrix} = \sqrt{\frac{2}{3}} \begin{bmatrix} 1 & -\frac{1}{2} & -\frac{1}{2} \\ 0 & \frac{\sqrt{3}}{2} & -\frac{\sqrt{3}}{2} \\ \frac{1}{\sqrt{2}} & \frac{1}{\sqrt{2}} & \frac{1}{\sqrt{2}} \end{bmatrix} \begin{bmatrix} i_a \\ i_b \\ i_c \end{bmatrix} \tag{6}$$

$$\begin{bmatrix} v_\alpha \\ v_\beta \\ v_o \end{bmatrix} = \sqrt{\frac{2}{3}} \begin{bmatrix} 1 & -\frac{1}{2} & -\frac{1}{2} \\ 0 & \frac{\sqrt{3}}{2} & -\frac{\sqrt{3}}{2} \\ \frac{1}{\sqrt{2}} & \frac{1}{\sqrt{2}} & \frac{1}{\sqrt{2}} \end{bmatrix} \begin{bmatrix} v_a \\ v_b \\ v_c \end{bmatrix} \tag{7}$$

The active and reactive power is written as:

$$p = v_\alpha i_\alpha + v_\beta i_\beta + v_o i_o \tag{8}$$

$$q = v_\alpha i_\alpha - v_\beta i_\beta \tag{9}$$

$$\begin{bmatrix} p \\ q \end{bmatrix} = \begin{bmatrix} v_\alpha & v_\beta \\ -v_\beta & v_\alpha \end{bmatrix} \begin{bmatrix} i_\alpha \\ i_\beta \end{bmatrix} \tag{10}$$

Active power and reactive power consist of two part which are mean part and oscillating part also known as DC part and AC part. The equations of active power and reactive power can be given as:

$$p = \bar{p} + \tilde{p} \tag{11}$$

$$q = \bar{q} + \tilde{q} \tag{12}$$

The DC part can be calculated by using low-pass filter, which is can remove the high frequency and give the fundamental component or the DC part. From DC part active power and reactive power, the α-β reference current can be represented in (13).

$$i_{\alpha\beta} = \frac{1}{\Delta} \begin{bmatrix} v_\alpha & v_\beta \\ v_\beta & -v_\alpha \end{bmatrix} \begin{bmatrix} p \\ q \end{bmatrix} \tag{13}$$

Where $\Delta = v_\alpha^2 + v_\beta^2$

The three phase current reference of active power filter is given in (14) before the signal will subtracted to load current. The subtracted three phase current will be used to generated PWM signal using hysteresis band. Hysteresis band will produce six PWM signals and for single phase active filter it is only two are used as input of hysteresis band.

$$i_{abc}{}^* = \sqrt{\frac{2}{3}} \begin{bmatrix} 1 & 0 \\ -\frac{1}{2} & \frac{\sqrt{3}}{2} \\ -\frac{1}{2} & -\frac{\sqrt{3}}{2} \end{bmatrix} i_{\alpha\beta}{}^* \qquad (14)$$

III. Single Phase Shunt Active Filter

Single phase shunt active filter consists of supply utility single phase, single phase rectifier, single phase active filter, controller and load. Schematic of single phase shunt active filter is shown in Fig. 2. They are two kinds of active power filter such as current source active filter and voltage source active filter. The different between these two topologies is the storage element. Current source active filter will use inductance as the storage element mean while voltage source active filter use capacitance as the storage element.

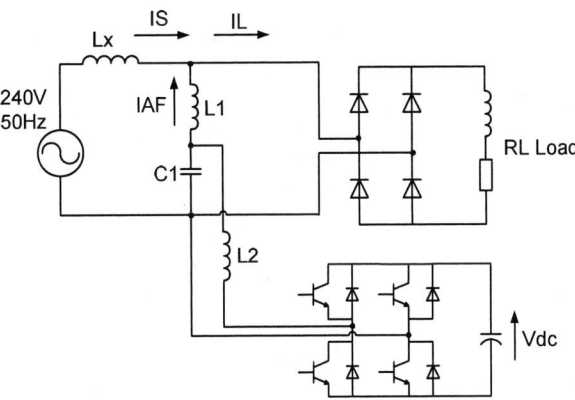

Fig. 2: Schematic diagram of single phase shunt active filter

Fig. 3 shows the control strategy based on p-q theory that is used to generate PWM signal for single phase shunt active filter. The simulation of single phase shunt active

filter uses this control strategy on MATLAB/ Simulink software.

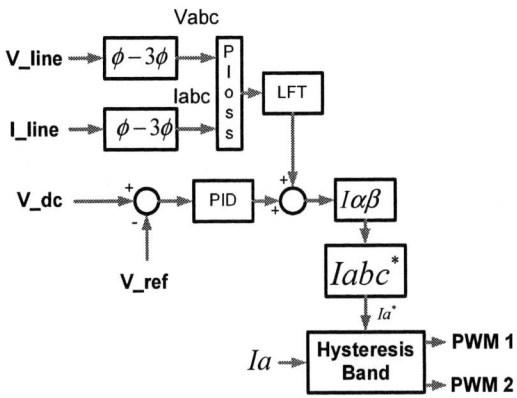

Fig. 3: Control strategy

IV. Simulation Result

A simulation of single phase shunt active filter is simulated using MATLAB/Simulink. The simulation use single phase system 240V 50Hz directly from TNB as shows in Fig. 4. The non-linear load with 3KVA for compensation is connected before single phase diode rectifier.

978-1-4577-0007-1/11 $26.00 © 2011 IEEE 161

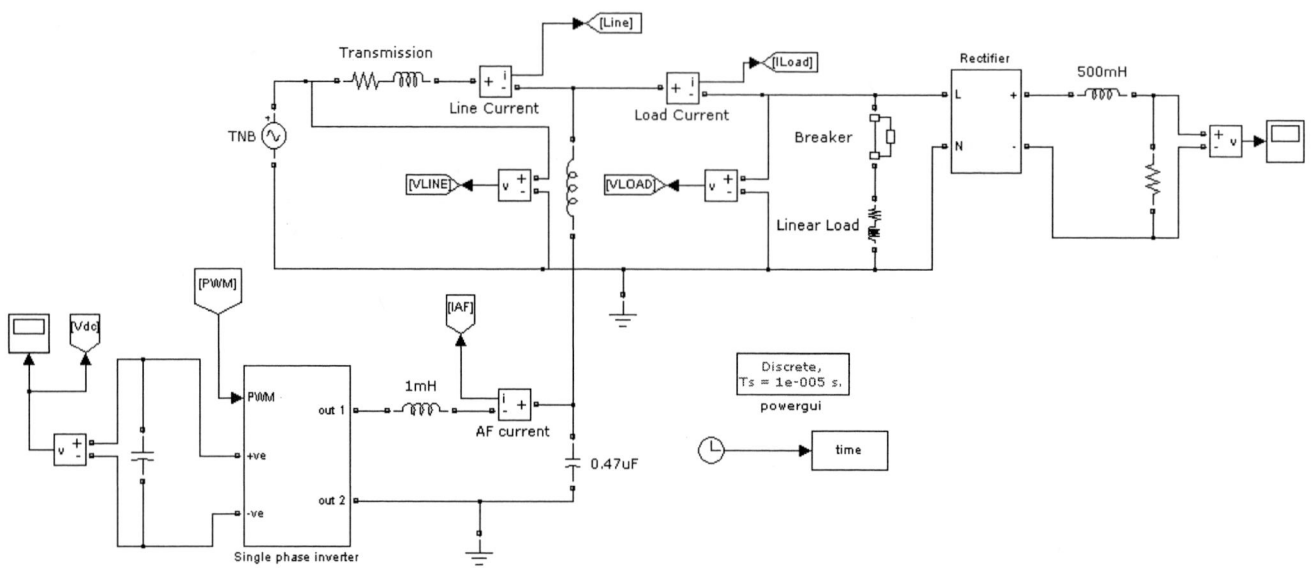

Fig. 4: Modelling of single phase active filter

Fig. 5 shows the modelling of p-q theory which consists of single to three phase block, algebra transformation of p-q theory three phase to two phase, two phase to three phase transformation and hysteresis band. Hsyteresis band will produce six signals PWM and for single phase active filter only use two signals to control the single phase active filter.

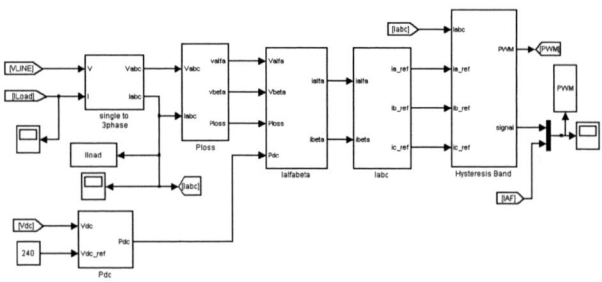

Fig. 5: Modelling of p-q theory

Current response for single phase active filter is shown in Fig. 6. The switching PWM signal and the active filter current are shown in Fig. 7. The load current in Fig. 8 will be compensated by injecting active filter current as shown in Fig. 9, so that the line current will be kept maintain in purely sinusoidal form as shown in Fig. 10. Fig. 11 shows the three phase load current that will be used for p-q theory application.

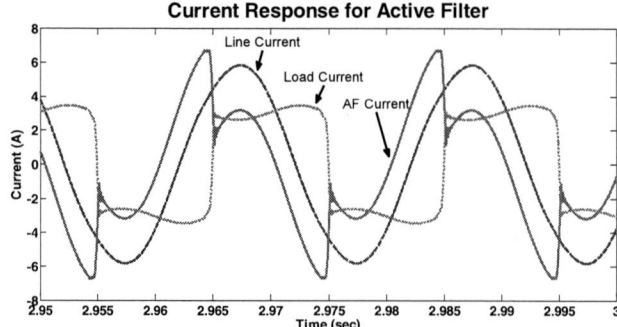

Fig. 6: Current response for single phase active filter

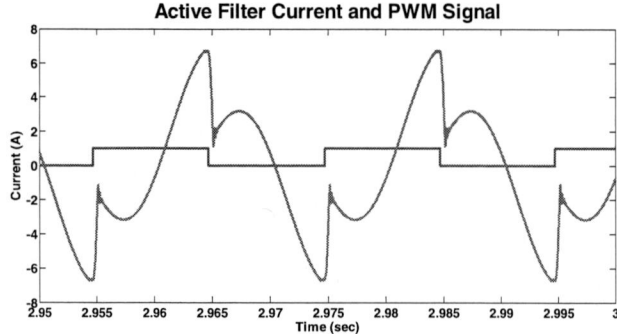

Fig. 7: Active filter current and PWM signal

978-1-4577-0007-1/11 $26.00 © 2011 IEEE   162

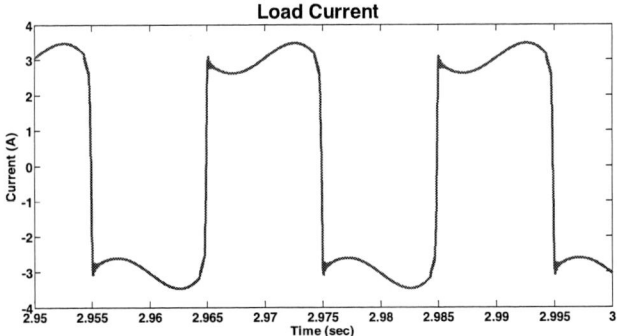

Fig. 8: Load current

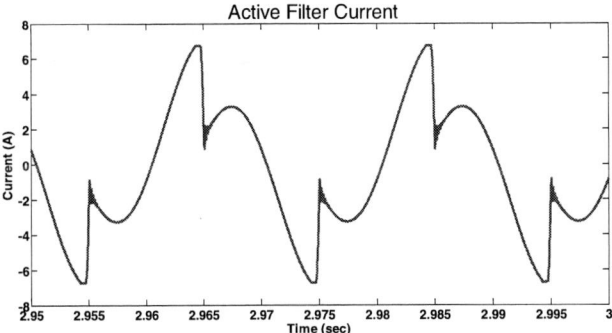

Fig. 9: Active filter current

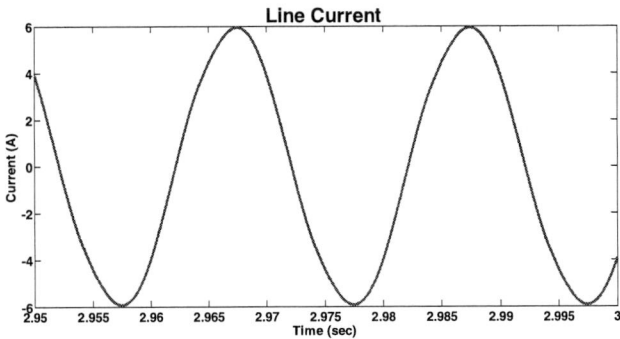

Fig. 10: Line current

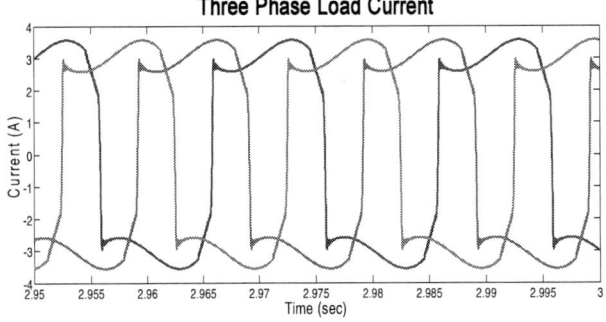

Fig. 11: Three phase load current

The effected non-linear load of the system will make the THD of load current increase up to 44.92% as shown in Fig. 12. By injecting the active filter current the THD of line current will reduce to 2.85% as shown in Fig. 13.

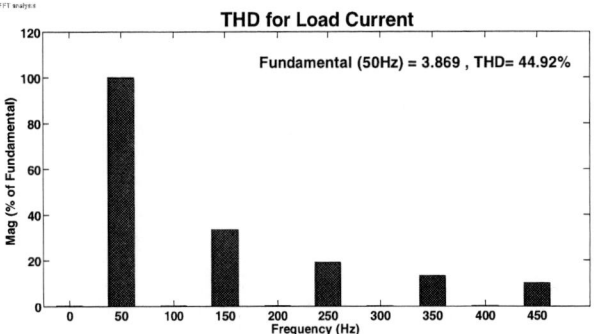

Fig. 12: THD for load current

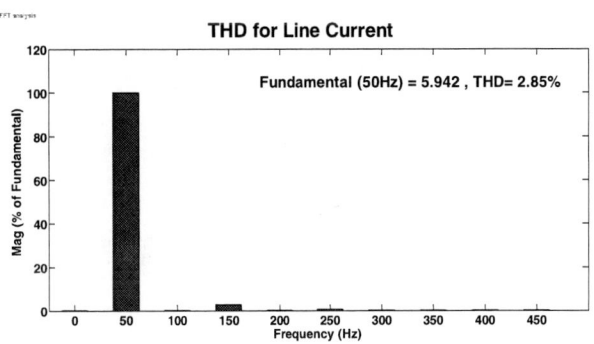

Fig. 13: THD for line current

V. Conclusion

In recent years the increasing usages of non-linear load facing of harmonic and power factor problem in power system. Many technique or topologies can be used to eliminate harmonics from power system; one of the techniques is active power filter. This paper proves that p-q theory can be implemented to control single phase active filter, which the theory widely used to control three phase active power filter. It is discovered from simulation that by implemented the p-q theory the THD of the load current can be reduced from 44.92% to 2.85%.

References

[1]   A. Emadi, A. Nasiri, and S. B. Bekiarov, *Uninterruptible power supplies and active filters*: CRC, 2005.

[2]   H. Akagi, E. H. Watanabe, and M. Aredes, *Instantaneous power theory and applications to power conditioning*: Wiley-IEEE Press, 2007.

[3]   N. A. Rahim, S. Mekhilef, and I. Zahrul, "A single-phase active power filter for harmonic compensation," *Industrial Technology. IEEE International Conference*, 2006, pp. 1075-1079.

[4]   K. Ryszard, S. Boguslaw, and K. Stanislaw, "Minimization of the source current distortion in systems with single-phase active power filters and additional passive filter designed by genetic algorithms," *Power Electronics and Applications, European Conference*, 2006, p. 10.

[5]   D. W. Hart, *Introduction to power electronics*: Prentice Hall PTR Upper Saddle River, NJ, USA, 1996.

[6]   M. McGranaghan, "Active filter design and specification for control of harmonics in industrial and commercial facilities," Knoxville TN, USA: Electrotek Concepts, Inc., 2001.

[7]   S. Round, H. Laird, R. Duke, and C. Tuck, "An improved three-level shunt active filter." vol. 1: *Power Electronic Drives and Energy Systems for Industrial Growth International Conference*, 2004, pp. 87-92.

[8]   H. Lev-Ari and A. M. Stankovic, "Hilbert space techniques for modeling and compensation of reactive power in energy processing systems." vol. 50: *IEEE Transcactions on Circuits and System Part 1: Regular Papers*, 2003, pp. 540-556.

[9]   A. Emadi, "Modeling of power electronic loads in ac distribution systems using the generalized state-space averaging method." vol. 51: *IEEE Transactions on Industrial Electronics*, 2004, pp. 992-1000.

[10]   J. Afonso, C. Couto, and J. S. Martins, "Active filters with control based on the pq theory," *IEEE Industrial Electronics Society newsletter*. ISSN 0746-1240. 47:3, 2000.

[11]   C. Cai, L. Wang, and G. Yin, "A three-phase active power filter based on park transformation," *Nanning Computer Science & Education 2009. 4th International Conference*, 2009, pp. 1221-1224.

[12]   M. George and K. P. Basu, "Three-Phase Shunt Active Power Filter." vol. 5: *American Journal of Applied Sciences*, 2008, pp. 909-916.

Designing Dynamic Controller and Passive Filter for a Grid Connected Micro-turbine

R. Rahmani, M. Tayyebi, M. S. Majid, M. Y. Hassan, H. A. Rahman
Center of Electrical Energy Systems
University Technology of Malaysia
81310 UTM Johor Baru Malaysia

Abstract- **This paper presents a simulation model of electrical components of a grid connected 50 kW single shaft micro-turbine in MATLAB-simulink GUI environment. Micro-turbine consists of a PWM inverter to control the injected power to the load. Deploying micro-turbines with Combined Heat Power (CHP) capability make it so interesting to use in distributed generation. Employing DG creates an un-sinusoidal voltage waveform supplied to the load. Hence, a band pass filter is installed to provide acceptable sinusoidal waveform with the lowest rate of THD for 1 MW load. Simulation results and discussions show that THD confirms to the requirements of IEEE Std 519-1992.**

Key words: Micro-turbine; Dynamic controller; PQ controlling system; Distributed Generation; Passive Filters.

I. INTRODUCTION

Variety of Distributed Resources (DR) in Distributed Generation (DG) such as micro-turbine, fuel cells in non-renewable class and wind turbines with photovoltaic in renewable category are implemented in power systems. Different types of DGs have different ranges of usage which can be observed in table 1 [1]. Providing ways to by-pass problems of conventional methods of energy delivery will bring a suitable manner of more flexible power production with higher generation efficiencies and a lower rate of pollution emission to consumers. In this category, micro-turbine could be classified as a Low Voltage (LV) network. Possibility of heat and power generation put the micro-turbine in combine cooling heating power generation (CCHP) class as well. Micro-turbines are divided into two classes. Split-shaft Micro-turbines have induction or synchronous generators which are connected to turbines within gearboxes as discussed in [2].

Another category is single-shaft micro-turbine. Design of this type consists of high speed single shaft, a compressor and a turbine installed on same shaft along with a permanent magnet synchronous generator. Single-shaft micro-turbine is applied in this study. Rapid response capability and ability of peak-shaving are some of the advantages of this kind of power supply. Furthermore, benefits of DG could be classified in: improving power quality, reduction of losses, releasing the capacity of transmission lines and distribution systems and improving utility system reliability. Micro-turbine as a DG system has all of the above advantages while it has its own

TABLE I
TYPES OF DG SOURCES

Type of source	Power range
Micro-turbine	25 kW – 1 MW
Wind turbine	100 kW – 2 MW
Photovoltaic	5kW – 100 kW
Fuel-cell	100 kW – 2 MW

disadvantages and problems. The main problems of this kind of energy production are divided into two categories.

First and the more important concern is controlling a system which consists of inverter that makes micro source capable to dispatch with the grid [3].

There are two strategies for controlling the inverter. PQ inverter control and voltage source inverter (VSI) control. In PQ controlling manner, active and reactive power set point are used. Voltage waveform of DG must be synchronizing with the grid voltage. This class of controlling is suitable for connected to grid micro sources which are in interconnection with grid in dealing the active and reactive power. Also it is also name as grid parallel operation. In case of controlling the isolated systems in local loads, inverter is controlled by given voltage and frequency. By measuring the load demand in islanded mode, output active and reactive power would be injecting automatically [4].

Second problem of implementing micro-turbine is harmonics which makes the load waveforms disputed. These harmonics are the effects of switching in inverter. To make these power suppliers usable and beneficial for the system, above problems must be mitigated. There are two steps unsolved in mitigating the above problems. At first step, power injection to load and grid side of system would be controllable by designing a system for inverter switching through the PQ control manner. As second step, a band pass filter has been used which is designed especially for this system. However, different kinds of filters could be used like low pass.

978-1-4577-0007-1/11 $26.00 © 2011 IEEE

II. MODEL DESCRIPTION AND SIMULATION

A. Micro-turbine

The whole system which is known as a micro-turbine power supplier consists of three main parts. A permanent magnet synchronous machine (PMSM), air compressor, combustion chamber and turbine mounted air bearings. Single shaft micro-turbine is not suitable for being connected to grid because of high frequency AC voltage produced by it. Output generated signal is at very high frequency ranging from 1500 to 4000 Hz. High frequency voltage be rectified and then inverted to 60 Hz voltage. So power electronic devices like AC-DC rectifier, DC-AC inverter are used to attain a frequency matched with interfaced grid. The shaft in which that generator and turbine are installed on it must operate without any vibration during high speed rotation. Inlet of the air is through the generator and compressed and flows through recuperator for preheating process and increasing efficiency of system before entering the combustion chamber. Fig. 1, illustrates a schematic diagram of a typical micro-turbine system.

Excellent samples of practical micro-turbines in the market are in the range under 500kW, like 25-100 kW. These systems have high speed gas turbines (15,000-90,000 rpm). Consequently output voltage of micro-turbine would be in high frequency range which forces the consumer to use converters to reach the grid frequency.

Dynamic model of turbine is realized in [5]. Micro-turbine and rectifier is brought as a DC dependent current source here. Grid could be presented by a balanced three-phase source, 600 volts line to line, 60 Hz frequency. The total simulated model has been demonstrated in Fig. 4. In micro-turbine part, power demand has been set to 50 kW while the desired DC voltage is equal to 100 V in this case.

B. Inverter and PQ controller

According to operation mode of micro grid, there are two main control division mode consist of voltage source inverter (VSI) and PQ inverter control. In case of operation by pre-defined values of voltage and frequency in islanded mode, VSI is used. The PQ category of controlling system is used in grid connected or connection to a load Island contains of more micro sources of power generators [7]. That is the reason PQ control type has been implemented in current study.

Output voltages of PWM inverter is not in sinusoidal shape and is in one of vector part consist of six active and one zero voltage section, thus the output is discrete. Therefore, employing a dq0 stationary frame is necessary to control [8]. In Fig. 4, total simulated system is brought. As it can be seen, Inverter switching controller consists of two controlling loops. Active power and current feedback from bus number 2 are inputs of these loops.

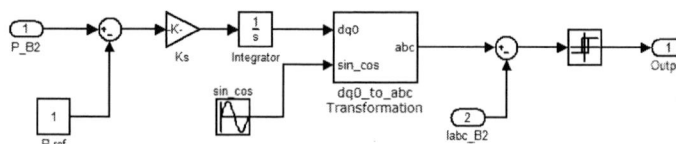

Fig. 2. Simple block diagram of the control loops.

Fig. 2, depicts block diagram of the controller which has been implemented to control the grid side inverter. For the current there are two components to be considered, I_d and I_q. I_d is the active component of current while I_q is the reactive component which has been set to zero as reference.

C. Passive Filter

Aim of implementing a filter on the grid side is to filter out the high frequencies which are generated by inverter. However, filter circuit has impact on low frequency harmonics as well. In this paper a band pass filter has been utilized to ensure that the best possible waveform has been received by load. Fig. 3, demonstrates the circuit of a band pass filter which is considered as a passive filter because there are no active elements in its structure. All the used elements are passive; resistors, inductors and capacitors.

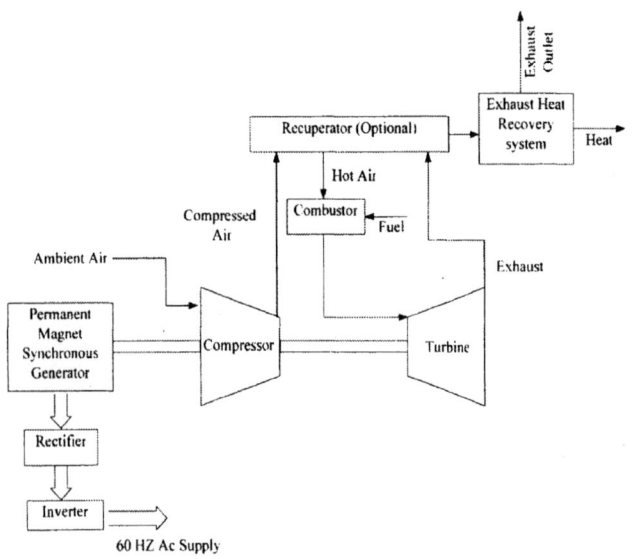

Fig. 1. Complete schematic of a micro-turbine [6].

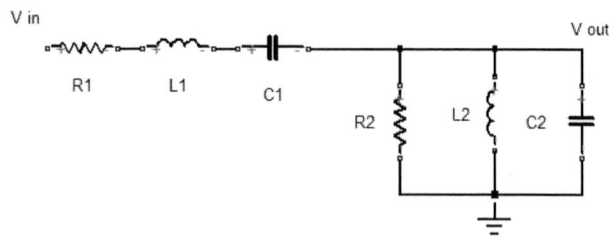

Fig. 3. Circuitry scheme of the band-pass filter.

978-1-4577-0007-1/11 $26.00 © 2011 IEEE

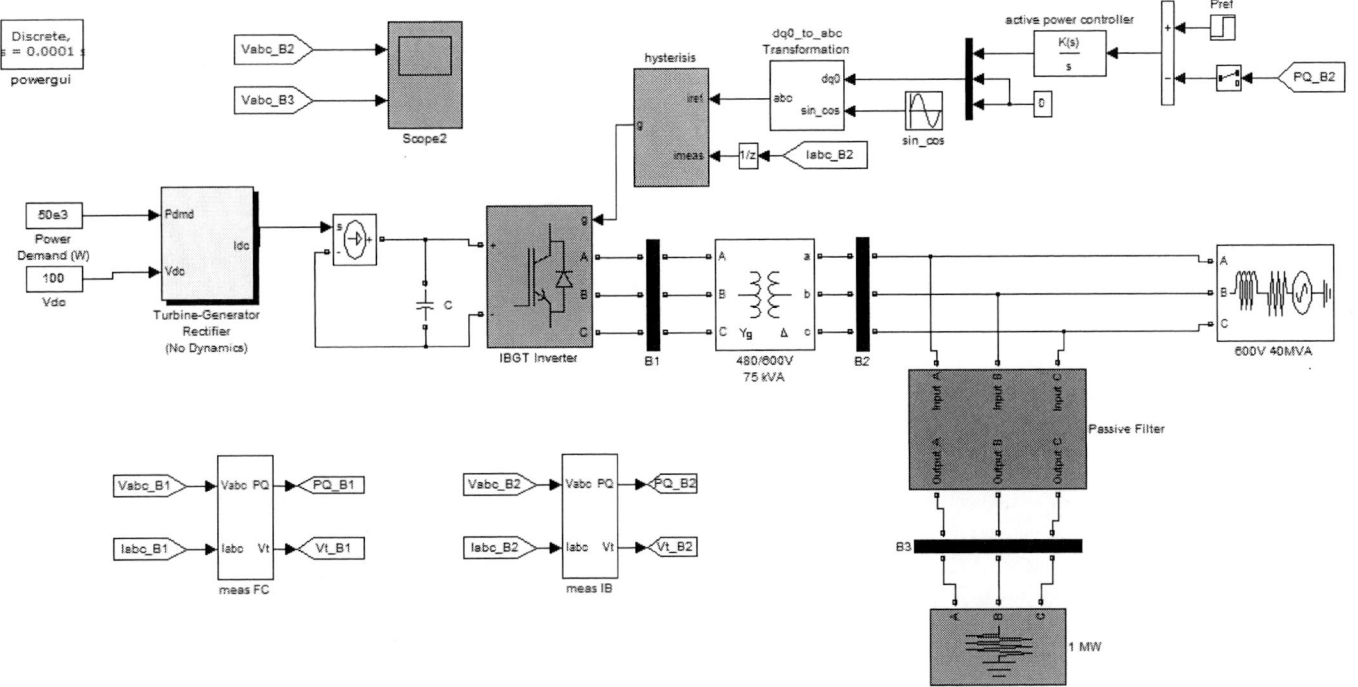

Fig. 4. Complete schematic of the simulated model of micro-turbine connected to grid.

Kirchhoff's current law for fig. 3, has a result as in (1):

$$\frac{V_{out} - V_{in}}{R_1 + SL_1 + \frac{1}{SC_1}} = V_{out}\left(\frac{1}{R_2} + \frac{1}{SL_2} + SC_2\right) \quad (1)$$

The voltages have been used here to obtain transfer function of the filter, while voltage and current also can be considered. Equation (2) is

$$V_{in} = V_{out}\left(\begin{array}{l} 1 + \dfrac{R_1}{R_2} + \dfrac{R_1}{SL_2} + SR_1C_2 + \dfrac{1}{R_2C_1S} + \dfrac{1}{S^2C_1L_2} + \\ \dfrac{C_2}{C_1} + S\dfrac{L_1}{R_2} + \dfrac{L_1}{L_2} + S^2L_1C_2 \end{array}\right) \quad (2)$$

Considering the G(s) as the transfer function of the filter the following equations will be the transfer function of the filter:

$$G(s) = \frac{V_{out}}{V_{in}}$$

$$= \frac{S^2}{S^4(L_1C_2) + S^3\left(R_1C_2 + \dfrac{L_1}{R_2}\right) + S^2\left(1 + \dfrac{R_1}{R_2} + \dfrac{C_2}{C_1} + \dfrac{L_1}{L_2}\right) + S\left(\dfrac{R_1}{L_2} + \dfrac{1}{R_2C_1}\right) + \dfrac{1}{L_2C_1}}$$

$$(3)$$

$$= \frac{S^2\left(\dfrac{1}{L_1C_2}\right)}{S^4 + S^3\left(\dfrac{R_1}{L_1} + \dfrac{1}{R_2C_2}\right) + S^2\left(\dfrac{1}{L_1C_2} + \dfrac{R_1}{R_2L_1C_2} + \dfrac{1}{L_1C_1} + \dfrac{1}{L_2C_2}\right) + S\left(\dfrac{R_1}{L_2L_1C_2} + \dfrac{1}{R_2L_1C_1C_2}\right) + \dfrac{1}{L_1L_2C_1C_2}}$$

Pairs of L_1, C_1 and L_2, C_2 will go to resonance at grid frequency upon to (4).

$$\omega_0 = \frac{1}{\sqrt{L_1C_1}} = \frac{1}{\sqrt{L_2C_2}} \quad (4)$$

In order to reduce the distortions generated by existing frequencies other than grid, having a band close to 60 Hz seems to be necessary, and then corner frequencies are given 40Hz and 80Hz. Consequently, at the resonance frequency which is 60 Hz, either series or parallel parts of the designed second order band pass passive filter would be in resonance and filter impedance got as low as possible to have the lowest value of losses on filter. Based on IEEE Std 519-1992 the filter should be designed in a manner that THD of voltage be less than 5% for low-voltage general systems.

Considering attenuation frequencies equal to 40 Hz and 80 Hz with margins and band pass equal to 40 Hz, besides assuming the series resistance equal to 0.1 Ohms, we will have the following values for the filter's elements:

$R_1 = 0.1$ ohm,

$L_1 = 0.562613$ mH,
$C_1 = 0.0125063$ F,
$R_2 = 1$ ohm,
$L_2 = 0.125063$ mH,
$C_2 = 0.0562613$ F.

The frequency response of designed filter is as shown in Fig. 5. It is obvious that filter has a pass-band of 40 Hz.

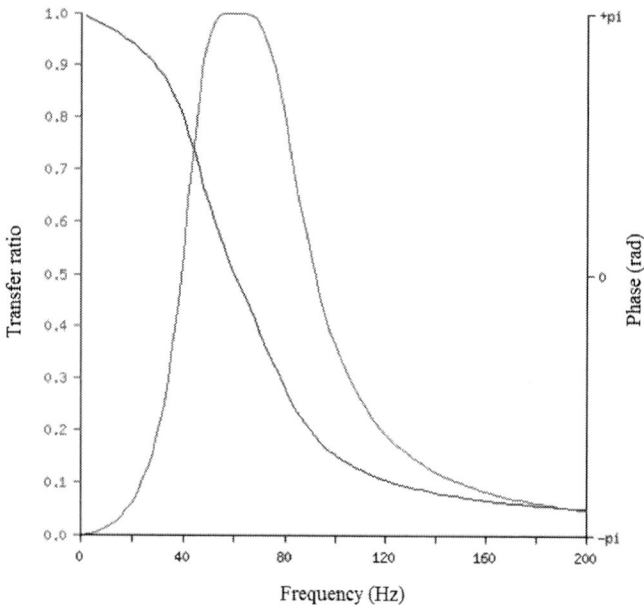

Fig. 5. Frequency response of the designed band-pass filter.

III. SIMULATION RESULTS

Based on simulated model in fig. 4, the waveforms of two buses are important and are analyzed in this paper. Bus B2 shows the waveforms before entering to the filter, while B3 has the waveform properties after filtering. The voltages of buses B2 and B3 are shown in fig. 6, and fig. 7, for first second of initiating. In order to have a comparison between the quality of waveforms of two buses before and after filtering, same short interval has been demonstrated for both of waveforms in fig. 8, and fig. 9.

Because of having a resistive 1 MW load, the waveform of voltage is the same with current, so no current waveform has been brought.

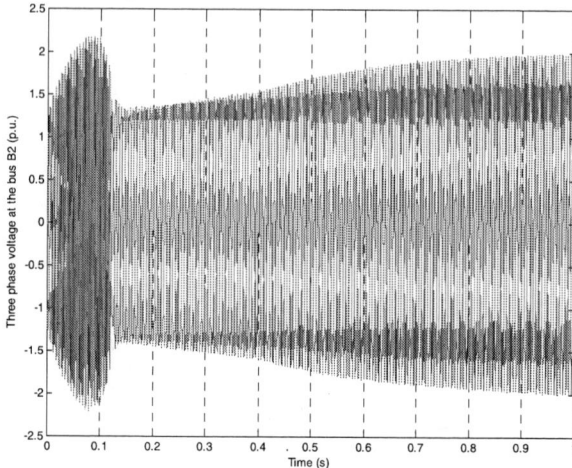

Fig. 6. Three phase voltage waveform at bus B2 before filter.

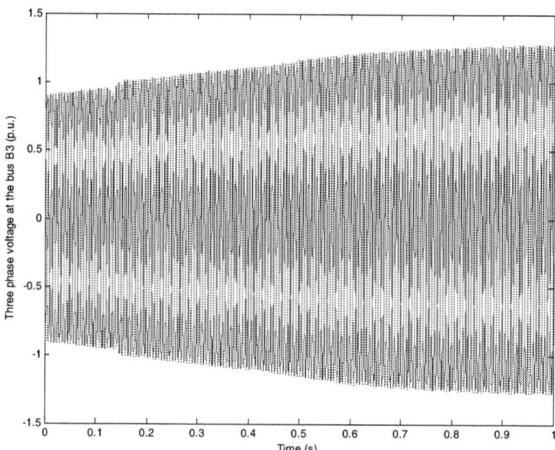

Fig. 7. Three phase voltage waveform at bus B3 after filtering.

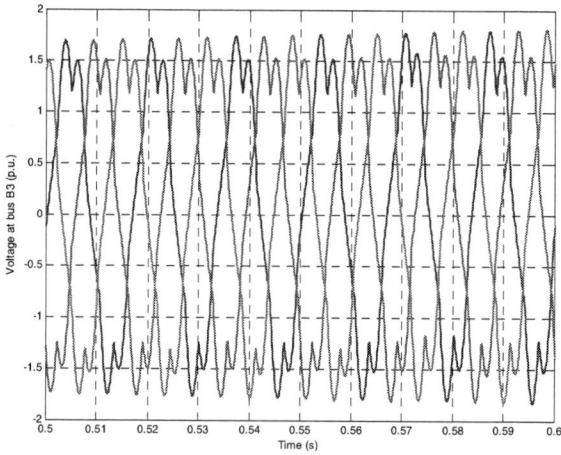

Fig. 8. Voltage waveform before being filtered out containing harmonics

978-1-4577-0007-1/11 $26.00 © 2011 IEEE

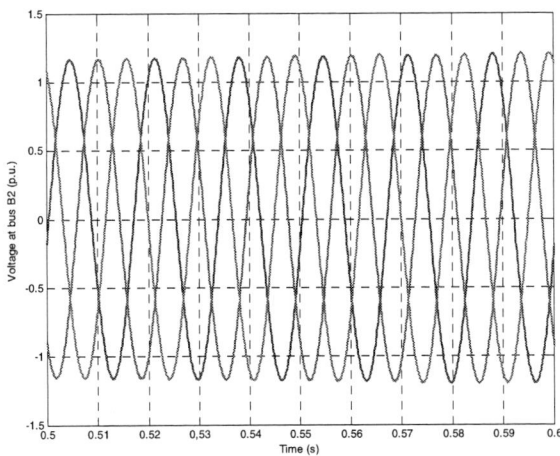

Fig. 9. Filtered waveform of voltage delivered to load.

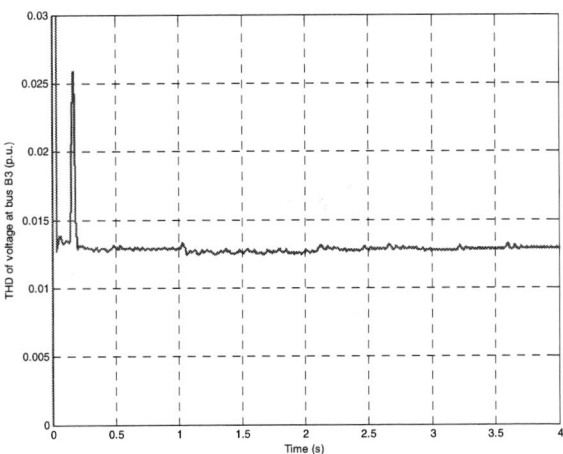

Fig. 10. THD of voltage at bus B3.

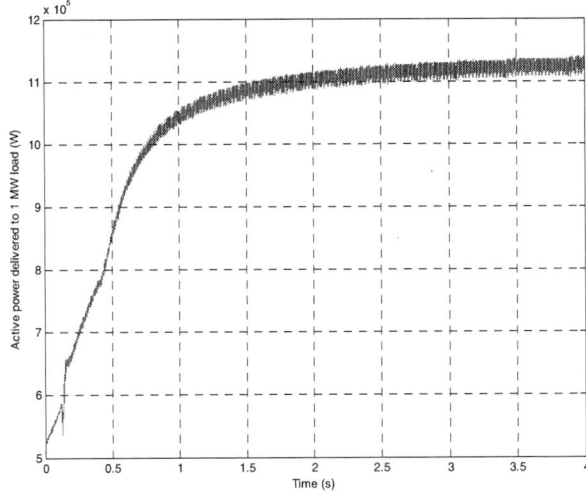

Fig. 11. Active power delivered to load at bus B3.

Fig. 10, illustrates the THD of voltage at load bus B3. The values of THD indicate that the designed and simulated system conform to the requirements of IEEE Std 519-1992 about harmonics. The active power delivered to the load has been demonstrated in fig. 11, which indicates that the proposed model has delivered the demand load.

IV. CONCLUSION

In this paper a designed and simulated model has been proposed for a micro-turbine system connected to grid. A Dynamic controller has been designed to control the power output of inverter besides a band pass filter to filter out the voltage harmonics of the 1 MW load. However, the micro-turbine model which has been brought here is not a dynamic model. The designed system has been concentrated to observe the operation of band-pass filter, and the waveforms of voltage before and after filter have been figured out. The THD of voltage delivering to the load has been illustrated as well and conformed to IEEE Std 519-1992. Considering the micro-turbine with a dynamic model will be the object of future study.

ACKNOWLEDGMENT

We thank the Ministry of Higher Education for the financial support of this work. FRGS-vot No:78355

REFERENCES

[1] G. Joos, B.T. Ooi, D. McGillis, F.D. Galiana, R. Marceau, "The potential of distributed generation to provide ancillary services," *IEEE Power Engineering Society Summer Meeting*, pp.1762-1767, 2000.
[2] A. Al-Hinai, K. Schoder, A. Feliachi, "Control of grid-connected split shaft microturbine distributed generator," Proceedings of the 35th Southeastern Symposium System Theory, pp.84-88, 2003.
[3] J.A. Pecas Lopes, C.L. Moreira, A.G. Madureira," Defining Control Strategies for MicroGrids Islanded Operation," *IEEE Transactions on power systems*, Vol 21, May, 2006.
[4] S. Barsali, M. Ceraolo, and P. Pelacchi, "Control techniques of dispersed generators to improve the continuity of electricity supply," in *Proc. PES Winter Meeting*, vol. 2, 2002, pp. 789–794.
[5] A. Al-Hinai, A. Feliachi, "Dynamic model of a micro turbine used as a distributed generator," *Proceedings of the Thirty-Fourth Southeastern Symposium System Theory*, pp.209-213, 2002.
[6] S.R. Guda, C. Wang, M.H. Nehrir, "A Simulink-based microturbine model for distributed generation studies," *Proceeding of the 37th Annual North American Power Symposium*, 2005.
[7] Huang Wei, Wu Ziping, "Dynamic Modelling and Simulation of Microturbine Generation System for the Parallel Operation of Microgrid" *IEEE conference sustainable power generation and supply*, 2009.
[8] Robert Lasseter," **Dynamic Models for Micro-Turbines and Fuel Cells,**" *IEEE conference*, **pp.7803-7173**, 2001.

Impact of Double-Loop Controller on Grid Connected Inverter Input Admittance using Virtual Resistor

Arwindra Rizqiawan[*‡], Goro Fujita[*], Toshihisa Funabashi[†] and Masakatsu Nomura[†]

[*]Department of Electrical Engineering, Shibaura Institute of Technology

[†]Meidensha Corporation

[‡]Email: m609501@shibaura-it.ac.jp

Abstract—**Grid connected inverter offers better controllability and better operation, however, there is a risk for local instabilities due to particular control scheme. As long as the input admittance is positive at particular resonance frequency, the risk of local instability can be suppressed. In this paper, the virtual resistor concept is employed to obtain the positive input admittance. Impact of double-loop controller using virtual resistor on converter passivity is analyzed by observing the input admittance of the grid connected inverter. By numerical analysis, it can be shown that employment of virtual resistor gives better input admittance of grid connected inverter, particularly in low frequency region.**

I. Introduction

Penetration of distributed generation has made the application of grid connected inverter is commonly found in current power system. Although those offer better controllability and better operation, however, there is a risk for local instabilities due to particular control scheme. HVDC transmission system has been widely reported as one of the cause of local instability. Its countermeasures usually by tuning the controller or by proposing alternative control scheme [1] [2]. Another cause is constant power operation of grid connected inverter [3]. In constant power operation, small perturbation in current or voltage may lead to negative resistance seen by the system to the converter. If a small-signal oscillation build up at this condition, it will be poorly damped due to negative input resistance.

Using resistor to provide extra damping in electrical circuit is well understood. If oscillation occurs in circuit, the power will be absorbed by resistor. However, extra resistor means extra losses in the circuit. In high power application, this method is unlikely to choose because it will degrade the efficiency. Virtual impedance concept let us to obtain the impedance effects, i.e resistor, without need to put real impedance in the circuit [4]. Thus the efficiency will not be sacrificed as we put the extra resistor into the circuit [5]. Virtual resistor is constructed from control algorithm which simulates the role of resistor in the circuit.

In this paper, the virtual resistor concept is employed to obtain the positive input admittance as previously described. The virtual resistor is employed in the current controller, which together with dc voltage controller, it forms the commonly found double-loop control. We assume that the *abc* to *dq* transformation works perfectly and the reactive power control

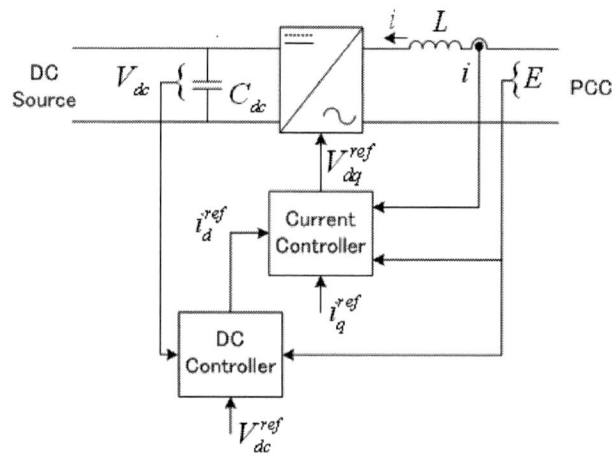

Fig. 1. Simplified control and circuit diagram of grid connected inverter.

is not considered here. We obtain the simplified control and circuit diagram of grid connected inverter as shown in Fig. 1. The analytical expression for virtual resistor employed-current controller and dc voltage controller is derived based on this figure. The input admittance of converter using virtual resistor in double-loop control is expressed based on the resulted analytical expressions. Impact of double-loop controller using virtual resistor on converter passivity is analysed by observing the input impedance of the grid connected inverter.

This paper is organized as follows: Passivity criterion which is used in this paper will briefly described in Section II. In Section III, the application of virtual resistor in current controller is explained. The dc voltage controller also explained in this section. In the Section IV, the input admittance of grid connected inverter is constructed. The impact of controller to the input admittance is studied in Section V. Finally, Section VI summarize this paper.

II. Passivity Criterion

Local passivity of any nonlinear system can be inferred from the passivity of linearized system. The passivity of linear system itself can be inferred from positive realness [6]. If a grid connected inverter is analyzed under linearization about an operating point then it can be referred as linear

978-1-4577-0007-1/11 $26.00 © 2011 IEEE

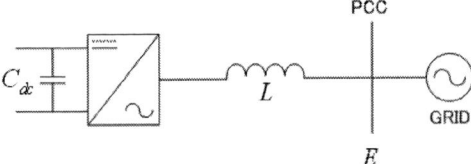

(a) Grid connected inverter: L filter.

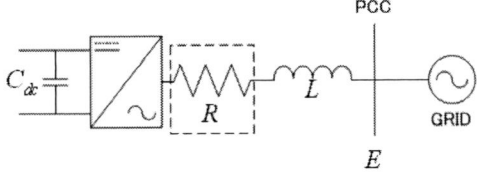

(b) Grid connected inverter: L and R filter.

Fig. 2. Single line diagram of grid connected inverter.

system. Thus, the passivity criteria of input admittance of grid connected inverter, $Y(j\omega)$, can be simplified into passivity criteria of linear system, such as

$$Re\{Y(j\omega)\} > 0, \forall\omega \qquad (1)$$

Thus as long as the input admittance is positive at particular resonance frequency, the local instability risk can be suppressed [7] [8]. In other word, the passivity is guaranted.

III. APPLICATION OF VIRTUAL RESISTOR ON INVERTER CURRENT CONTROL

In power system, oscillation will not occur as long as the impedance seen by generator set is passive [7] [8]. In order to ensure the passive impedance as seen by the rest of power system, a series resistor can be placed in the system. This section describes the application of virtual resistor concept in the converter controller to replace the extra physical series resistor in order to improve the passivity of the system.

Figure 2(a) shows typical grid connected inverter interfaced with L filter. In the previous section, it has been described that by putting an additional resistor will increase the damping in the system.

Figure 2(b) shows grid connected inverter with L and extra additional R filter. This additional extra R filter may be a physical or virtual, as will be described in this section. Figure 3 shows $F(s)$ and $H(s)$ which are standard PI current controller and low-pass filter with bandwidth of α_f for PCC feedforward voltage, respectively. Let us assume $F(s)$ and $H(s)$ are represented by those basic standard form as

$$F(s) = K_p + \frac{K_i}{s} \qquad (2)$$

$$H(s) = \frac{\alpha_f}{s + \alpha_f} \qquad (3)$$

Suppose typical grid connected inverter with L filter shown in Fig. 2(a) is controlled using current controller $F(s)$, it may be expressed in block diagram as shown in Fig. 3(a). Suppose if the resistor in Fig. 2(b) is physically exist, thus block diagram of Fig. 3(b) can be obtained.

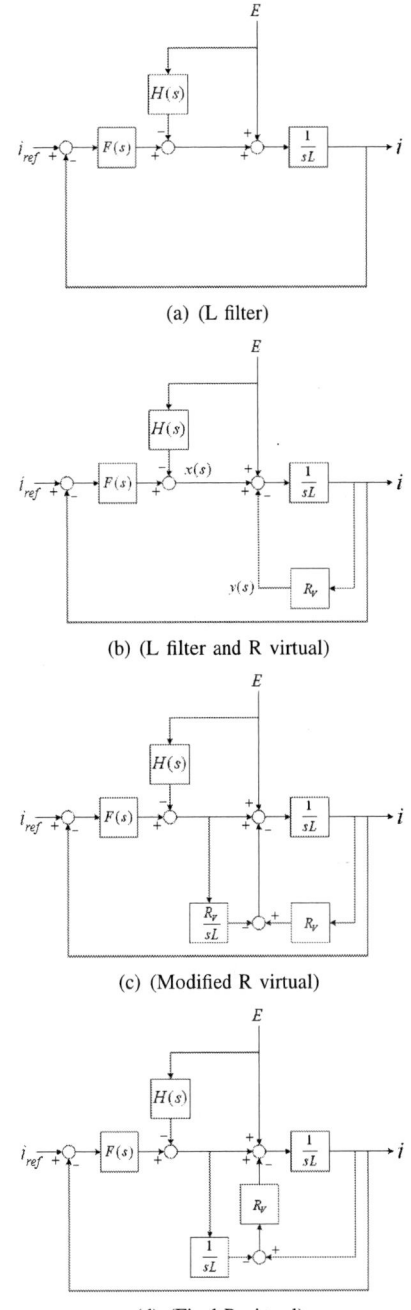

(a) (L filter)

(b) (L filter and R virtual)

(c) (Modified R virtual)

(d) (Final R virtual)

Fig. 3. Block diagram of grid connected inverter output - Virtual resistor implementation

In different circumstances, we can modify the block diagram of Fig. 3(a), by putting additional block R_v in the control domain. Then we will also obtain the block diagram in Fig. 3(b), thus we will obtain the same resistor effect as in Fig. 2(b) seen from the grid point of view. Although our grid connected inverter is physically as Fig. 2(a).

We have obtained a block diagram which acts as resistor virtually in Fig. 3(b). In here the virtual resistor will response to both the desired and disturbance in the controller, thus it may affects the actual current response of controller. In order

978-1-4577-0007-1/11 $26.00 © 2011 IEEE   171

to avoid this condition so that the actual current response is not degraded by the existence of additional block of virtual resistor, slightly modification of the block diagram is needed. Suppose the output of virtual resistor block, $y(s)$, in the Fig. 3(b) can be expressed as

$$y(s) = \frac{R_v}{sL}E + \frac{R_v}{sL}x(s) \tag{4}$$

In eq. (4), we can see that the output of virtual resistor has two components, the disturbance term, $\frac{R_v}{sL}E$, and current term, $\frac{R_v}{sL}x(s)$, the current reference and actual current are included in the $x(s)$ term. In order to make the virtual resistor only respond to the disturbance, the effect of current component should be eliminated. Thus we can subtract the current term from the output of virtual resistor block in eq. (4), such as

$$\frac{R_v}{sL}E = y(s) - \frac{R_v}{sL}x(s) \tag{5}$$

By employing the subtraction of eq. (5), Fig. 3(c) shows the modified virtual resistor block diagram which respond to disturbance term only. Fig. 3(d) shows further modification. We can derive the transfer function of current term and disturbance term based on block diagram shown in Fig. 3(d), respectively

$$G_i(s) = \frac{K_p s + K_i}{(Ls^2 + (K_p)s + K_i)} \tag{6}$$

$$G_e(s) = \frac{s^3 L + s\alpha_f R_v}{\left[\begin{array}{c}(s + \alpha_f)(s^3 L^2 + s^2 L(K_p + R_v) + \\ s(K_i L + K_p R_v) + K_i R_v\end{array}\right]} \tag{7}$$

Complete transfer function is [9]

$$i = G_i(s)i_{ref} + G_e(s)E \tag{8}$$

As shown in eq. (6) and eq. (7) the virtual resistor does not affect the current part, it only responds to disturbance part. If the virtual resistor is set to zero, eq. (7) will become normal transfer function of grid connected inverter controller.

Typical current control of grid connected inverter uses PI controller in converter dq frame with feedforwarded PCC voltage and dq cross coupling cancellation in order to control the d and q part of current independently. Thus, in converter dq frame, we can express the transfer function in matrix form as

$$
\begin{aligned}
i^{dq} &= G_I(s)i_{ref}^{dq} + G_E(s)E^{dq} \\
&= \begin{bmatrix} G_i(s) & 0 \\ 0 & G_i(s) \end{bmatrix} i_{ref}^{dq} + \begin{bmatrix} G_e(s) & 0 \\ 0 & G_e(s) \end{bmatrix} E^{dq}
\end{aligned} \tag{9}
$$

Where $i^{dq} = \begin{bmatrix} i^d & i^q \end{bmatrix}^T$, $i_{ref}^{dq} = \begin{bmatrix} i_{ref}^d & i_{ref}^q \end{bmatrix}^T$, and $E^{dq} = \begin{bmatrix} E^d & E^q \end{bmatrix}^T$.

A. DC Voltage Controller

DC voltage control loop in principle deals with the voltage level in capacitor filter due to the power flows, either inside or outside the grid connected inverter. Further analysis needs linearized expression for power flows in the converter. Linearized active and reactive powers can be expressed as

$$
\begin{aligned}
P &\approx P_0 + \Delta P \\
&\approx \left(E_0 i_0^d\right) + \left(i_0^d \Delta E^d + i_0^q \Delta E^q + E_0 \Delta i^d\right)
\end{aligned} \tag{10}
$$

$$
\begin{aligned}
Q &\approx Q_0 + \Delta Q \\
&\approx \left(-E_0 i_0^q\right) + \left(i_0^d \Delta E^q - i_0^q \Delta E^d - E_0 \Delta i^d\right)
\end{aligned} \tag{11}
$$

Where '0' sign represents steady state condition and 'Δ' represents small pertubation for any variable. If the converter is assumed as lossless converter and the current control loop works faster than dc voltage control loop, then the power into the converter will be the power to the dc link. Thus the energy stored in capacitor is the difference between the input from the converter and the energy deliver to the load. Energy stored in dc-link can be expressed in linearized form as

$$C_{dc} v_0^{dc} \frac{d\Delta v^{dc}}{dt} = \Delta P - \Delta P_{load} \tag{12}$$

As suggested by [10], to avoid the dependency on the operating point v_0^{dc}, the dc voltage controller operates on square error between dc reference and actual voltage instead of difference as commonly found. The controller output will be the active power reference

$$P_{ref} = F_{dc}(s) \frac{(v_{ref}^{dc})^2 - (v^{dc})^2}{2} \tag{13}$$

Where $F_{dc}(s) = K_p + \frac{K_i}{s}$ is common PI controller. Since in dq frame current control the d and q part can be independently controlled, thus the d part will only depend on active power delivered. From the power reference provided by eq. (13), the current reference can be determined by dividing it by filtered PCC voltage as

$$i_{ref}^d = \frac{P_{ref}}{H_{dc}(s)E} \tag{14}$$

Where $H_{dc}(s) = \frac{\alpha_d}{s + \alpha_d}$ is standard low-pass filter with bandwidth α_d.

If steady state operation has been reached, the dc voltage will be equal to its reference and the load active power will equal to steady state power, $v_{ref}^{dc} = v_0^{dc}$ and $P_{load} = P_0$. We obtain the linearized form of eq. (14) by linearizing eq. (13) and subsituting into eq. (14) as [10]

$$\Delta i_{ref}^d = -\frac{v_0^{dc}}{E_0}F_{dc}(s)\Delta v^{dc} - \frac{P_0}{E_0^2}H_{dc}(s)\Delta E^d \tag{15}$$

Since the current reference is the input for eq. (9), thus by subsituting eq. (15) into d part of eq. (9) gives us

$$
\Delta i^d = \left[-\frac{v_0^{dc}}{E_0}G_i(s)F_{dc}\Delta v^{dc} \right.
$$
$$
\left. + \left(G_E(s) - \frac{P_0}{E_0^2}G_i(s)H_{dc}(s)\right)\Delta E^d \right] \tag{16}
$$

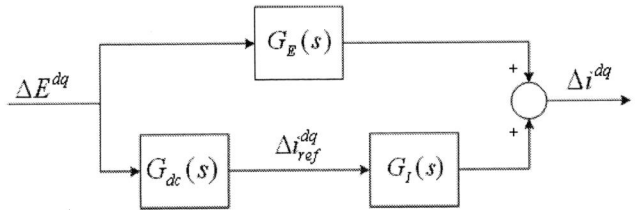

Fig. 4. Input admittance of grid connected inverter.

Given that $i_{c0}^d = \frac{P_0}{E_{c0}}$ and $i_{c0}^q = \frac{-Q_0}{E_{c0}}$, eq. (16) can be substituted into eq. (10) to obtain ΔP. The resulting expression can be substituted into eq. (12) to solve dc-link voltage as

$$\Delta v^{dc} = \frac{\left(E_0^2 G_E(s) + P_0(1 - G_i(s)H_{dc}(s))\right)\Delta E^d - Q_0 \Delta E^q}{v_0^{dc} E_0 (sC_{dc} + G_i(s)F_{dc}(s))} \tag{17}$$

From eq. (15) and eq. (17) we can obtain the relation between d current reference and the PCC voltage as [10]

$$\Delta i_{ref}^d = G_{dc}^d(s)\Delta E^d + G_{dc}^q(s)\Delta E^q \tag{18}$$

Where

$$G_{dc}^d(s) =$$
$$- \frac{\left(G_e(s) + \frac{P_0}{E_{c0}^2}(1 - G_i(s)H_{dc}(s))\right)F_{dc}(s)}{sC_{dc} + G_i(s)F_{dc}(s)} \tag{19}$$
$$- \frac{P_0 H_{dc}(s)}{E_{c0}^2}$$

$$G_{dc}^q(s) = \frac{Q_0 F_{dc}(s)}{E_{c0}^2(sC_{dc} + G_i(s)F_{dc}(s))} \tag{20}$$

Rewrite eq. (18) in matrix representation, yields

$$\Delta i_{ref}^{dq} = G_{dc}(s)\Delta E_{ref}^{dq}$$
$$= \begin{bmatrix} G_{dc}^d(s) & G_{dc}^q(s) \\ 0 & 0 \end{bmatrix}\Delta E_{ref}^{dq} \tag{21}$$

In Section I it has been clearly stated that in this paper the reactive power controller is not yet considered, thus the Δi_{ref}^q part in eq. (21) is set to be zero. This assumption holds if $\Delta i_{ref}^q = 0$ or under unity power factor operation.

IV. GRID CONNECTED INVERTER INPUT ADMITTANCE

This part is based on [10], except that reactive power controller is not considered here. From the relation between current reference and PCC voltage provided by dc voltage controller, as shown in eq. (21), we can find the relation of output current and PCC voltage of the converter in eq. (9). The relation is depicted in Fig. 4. The input admittance of the converter, $Y_i(s)$, can be expressed based on the figure as

$$i^{dq} = (G_E(s) + G_I(s)G_{dc}(s)) E^{dq} \tag{22}$$

Thus,

$$Y_i(s) = G_E(s) + G_I(s)G_{dc}(s)$$
$$= \begin{bmatrix} Y_{i11}(s) & Y_{i12}(s) \\ Y_{i21}(s) & Y_{i22}(s) \end{bmatrix} \tag{23}$$

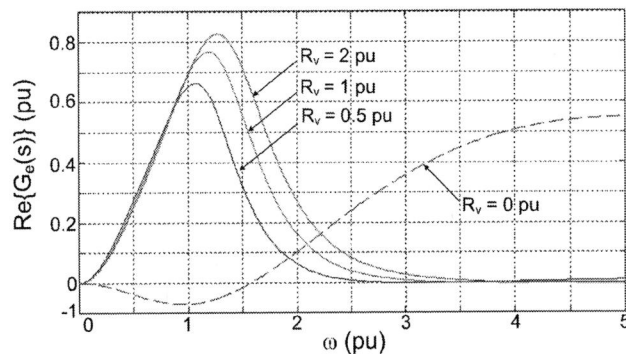

Fig. 5. Real part of converter input admittance

Where

$$Y_{i11}(s) = G_e(s) + G_i(s)G_{dc}^d(s) \tag{24}$$
$$Y_{i12}(s) = G_i(s)G_{dc}^q(s) \tag{25}$$
$$Y_{i21}(s) = 0 \tag{26}$$
$$Y_{i22}(s) = G_e(s) \tag{27}$$

We have obtained the input impedance of the grid connected inverter due to converter controller as shown in eq. (23). Based on that, we can observe the impact of virtual resistor on the input admittance of the converter.

V. IMPACT OF CONTROLLER ON INPUT ADMITTANCE

A. Impact of current controller on passivity of the converter

Suppose only current control loop is considered. The $G_E(s)$ term in eq. (8) can be seen as input admittance due to current controller from PCC node. Substituting s by $j\omega$ of the input admittance expression in eq. (7) and then take the real part give us eq. (28) in the next page.

For the purpose of numerical analysis, the parameters are shown in Table I [10] either for observing impact of current controller only or for observing impact of double-loop controller as will be described in the next section. Using the parameters as shown in Table I [10], Fig. 5 shows the real part of converter input admittance. From Fig. 5 we can see the effects of virtual resistor, it has made the real part of input admittance has positive value in low frequency. This effect is very interesting because subsynchronous oscillation frequency also lies in this range.

B. Impact of double-loop controller on 'passivity' of the converter

In Section V-A, it has been shown that the current controller itself has improved the converter passivity by providing positive input admittance value in low frequency range. However, in grid connected converter, double-loop control comprising of current controller as inner loop and dc voltage controller as outer loop is commonly found. Thus, in this section, the impact of virtual resistor in double-loop control will be observed. Since double-loop control is used, the bandwidth of inner loop control should higher than the outer control to ensure the inner

$$Re\{G_E(j\omega)\} = Re\{Y(j\omega)\} =$$

$$\frac{\omega^2(\alpha_f(K_iL + K_pR) + K_iR_v) - \omega^2L(K_p + R_v + \alpha_fL)(\alpha_fR_v - \omega^2L)}{(\omega^4L^2 - \omega^3(K_iL + K_pR_v + \alpha_fL(K_p + R_v)) + \alpha_fK_iR_v)^2 + (\omega(\alpha_f(K_iL + K_pR_v) + K_iR_v) - \omega^3L(K_p + R_v + \alpha_fL))^2} \tag{28}$$

TABLE I
CONTROLLER PARAMETERS

Parameter	Value (p.u)
α_c	4
α_f	5
α_d	$0.1\alpha_c$
K_p	α_cL
K_i	0
K_{pd}	$C_{dc}\alpha_d$
K_{id}	0
L	0.2
C_{dc}	1
E_o	1

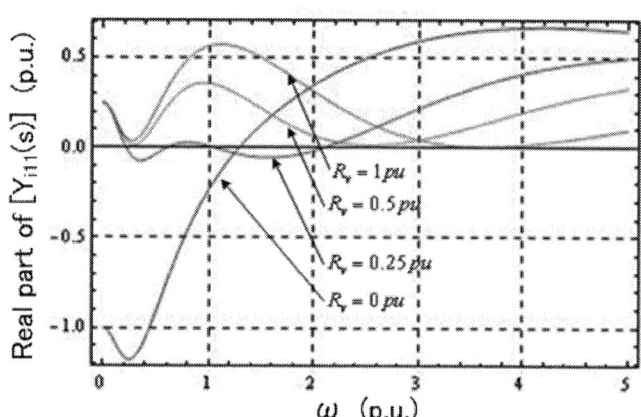

Fig. 6. Converter input admittance during rectifier operation

loop control has worked properly during transient in outer loop control. In this study, one decade ratio is considered enough to ensure the previous consideration.

We assume that the converter does not deliver any reactive power ($Q_o = 0$) in order to observe the impact of dc controller on converter passivity, because dc source is only related with active power in d component. From the matrix of input admittance in eq. (23) only one term will be considered here,

$$Y_{i11}(s) = G_e(s) + G_i(s)G_{dc}^d(s) \tag{29}$$

We can observe the 'passivity' by studying the input admittance of the converter by substituting $s = j\omega$ in eq. (29) and taking the real part. In this part, the parameters are also shown in Table I.

1) Rectifier operation (P = 1pu): For rectifier operation, the converter input admittance naturally has large negative value in low range frequency. This range of frequency has high risk of resonance and with the negative value of input admittance, the resonance will poorly damped. Application of virtual resistor will significantly increase the input admittance value especially in low range frequency. thus it will increase the damping for resonance. Figure 6 shows the relation of various values of virtual resistor to the input admittance of the converter during rectifier operation.

2) Zero power operation (P = 0pu): In zero power operation, naturally converter has small negative input impedance value. However this may still bring any risk of poorly damped resonance in low frequency range. Virtual resistor will significantly boost the input admittance value become positive in low frequency region. Figure 7 shows the relation of various values of virtual resistor to the input admittance of the converter during zero power operation.

3) Inverter operation (P = −1pu): In inverter operation, naturally converter has already had positive input impedance

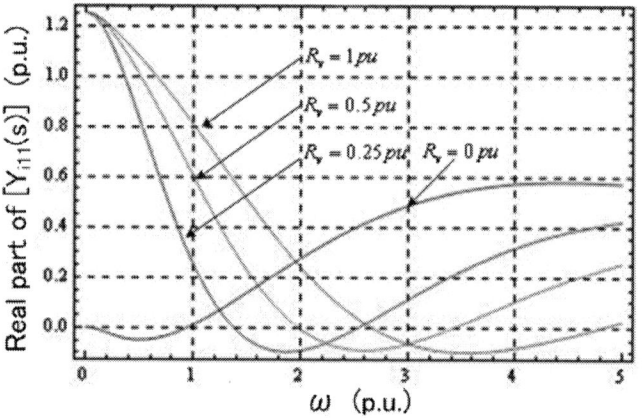

Fig. 7. Converter input admittance during zero power operation

value, thus addition of virtual resistor would not give significant effect for converter passivity. However, virtual resistor still provides higher positive input impedance in low frequency range which provides better damping in case low frequency resonance happen. Figure 8 shows the relation of various values of virtual resistor to the input admittance of the converter during inverter operation.

Figure 9 shows the input admittance as function of frequency and power delivered. We can see that in low frequency range and towards rectifier operation, the virtual resistor will gives significant impact even on double-loop controller scheme. While in inverter operation, naturally without virtual resistor, the converter input admittance has already positive. In higher frequency range, the virtual resistor seems to give

978-1-4577-0007-1/11 $26.00 © 2011 IEEE

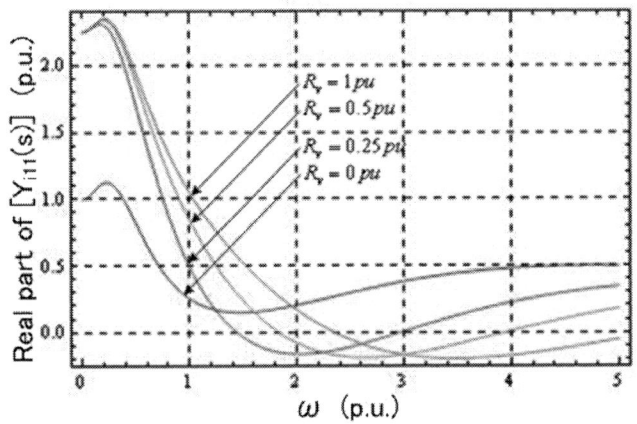

Fig. 8. Converter input admittance during inverter operation

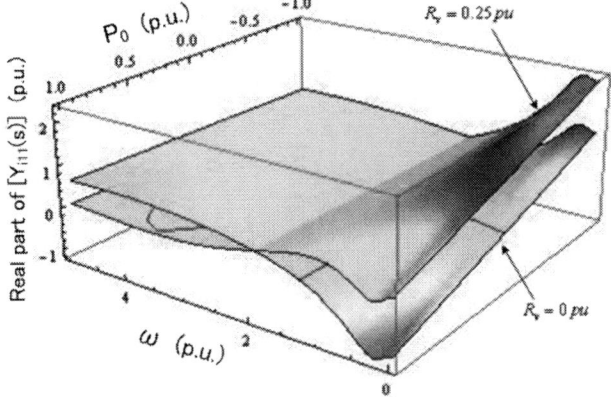

Fig. 9. Converter input admittance as a function of power and frequency.

worse input admittance condition, however the most severe resonance condition usually occurs in low frequency range.

VI. CONCLUSIONS

This paper presented the impact of dc voltage controller in converter passivity by observing the input admittance of the converter. Due to application of virtual resistor concept in the current controller, the low frequency range input admittance is significantly increased in rectifier operation. Towards inverter operation, the input admittance naturally has larger value and tends to become positive in low frequency range.

REFERENCES

[1] R. Harley and J. Balda, "Subsynchronous resonance damping by specially controlling a parallel hvdc link," *IEE Proceedings on Generation, Transmission, and Distribution*, vol. 132, no. 3, pp. 154–160, 1985.

[2] N. Kaul and R. Mathur, "Solution to the problem of low order harmonic resonance from hvdc converters," *IEEE Transactions on Power Systems*, vol. 5, no. 4, pp. 1160–1166, 1990.

[3] N. Prabhu and K. Padiyar, "Investigation of subsynchronous resonance with vsc-based hvdc transmission systems," *IEEE Transactions on Power Delivery*, vol. 24, no. 1, pp. 433–440, 2009.

[4] P. Dahono, Y. Bahar, Y. Sato, and T. Kataoka, "Damping of transient oscillation on the output lc filter of pwm inverters by using a virtual resistor," in *Proc. IEEE Power Electronics and Drive Systems Conf.*, 2001, pp. 403–407.

[5] P. Dahono, "A method to damp oscillation on the input lc filter of current-type ac-dc pwm converters by using a virtual resistor," in *Proc. IEEE Telecommunications Energy Conf.*, 2003, pp. 757–761.

[6] V. Vladimirov, "Linear passive system," *Journal of Theor. Math. Phys.*, vol. 1, no. 1, pp. 51–72, 1969.

[7] L. Harnefors, "Analysis of subsynchronous torsional interaction with power electronic converter," *IEEE Transactions on Power System*, vol. 22, no. 1, pp. 305–313, 2007.

[8] L. Harnefors, L. Zhang, and M. Bongiorno, "Frequency-domain passivity-based current controller design," *IET Power Electronics*, vol. 1, no. 4, pp. 455–465, 2007.

[9] A. Rizqiawan, G. Fujita, T. Funabashi, and M. Nomura, "Grid connected inverter input admittance improvement by using virtual resistor concept," in *Proc. The 2010 IEEJ Power and Energy Technical Meeting*, 2010.

[10] L. Harnefors, M. Bongiorno, and S. Lundberg, "Input-admittance calculation and shaping for controlled voltage-source converters," *IEEE Transactions on Industrial Electronis*, vol. 54, no. 6, pp. 3323–3334, 2007.

Design And Application Of A Novel Current Mode Controller On A Multilevel STATCOM

E.Najafi

Faculty of Electrical engineering
Universiti Technologi Malaysia (UTM)
Skudai,Johor Bahru,Malaysia
ehsan@fkegraduate.utm.my

A.H.M. Yatim

Faculty of Electrical engineering
Universiti Technologi Malaysia (UTM)
Skudai,Johor Bahru,Malaysia
halim@fke.utm.my

Abstract— **Static Compensator (STATCOM) has been widely proposed for power quality and network stability improvement. It is easily connected in parallel to the electric network and has many advantages for electrical grids. It can improve network stability; power factor, power transfer rating. It can also avoid disturbances such as voltage sag (dip) and swell which is the main concern of this paper**

This paper uses a single phase multilevel inverter for STATCOM. It also implements a new current mode controller to control a STATCOM. In order to detect voltage sag, the STATCOM utilizes Goertzel algorithm which is more efficient and practical compared to available methods.

Finally, simulations show the results of STATCOM operation for sag mitigation. The results show that the proposed controller can effectively mitigate voltage sag during disturbances.

Keywords — **STATCOM; Goertzel algorithm; multilevel inverter; current control mode; Power quality; sag.**

I.   INTRODUCTION

Sag can be defined as a decrease to between 0.1 and 0.9 pu in rms voltage or current at the power frequency for durations of several milliseconds to 1 min. [1] It is measured by the amount of the retained voltage and not by the amount of the voltage drop.

Voltage sag may be caused by switching operations associated with a temporary disconnection of supply, the flow of inrush currents associated with the starting of motor loads or the flow of fault currents. These events may emanate from the customers system or from the public supply network. Lightning strikes can also cause momentary sags [2].

STATCOM is one of the effective devices to compensate voltage sag in electric networks. It should first measure the voltage amplitude and compare it against standard levels. If the voltage is outside limits, sag is detected.

The concept of a STATCOM application and how it works is shown in Fig. 1

As can be observed from the Fig. 1 the task of the device is to inject the current to the network to correct the voltage at the Point of Common Coupling (PCC) when voltage sag or swell occurs. [3].

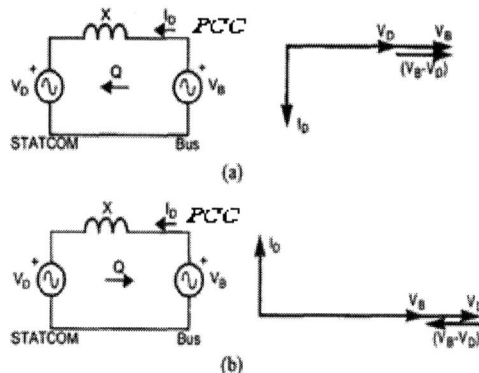

Fig. 1: Statcom operation,(a) inductive operation for swell, (b) capacitive operation for sag

According to Fig. 1 there are two modes of operation for STATCOM which may be used in different grid conditions. In order to compensate voltage sag, capacitive operation is used while for voltage swell, inductive mode is used.

In this paper Goertzel algorithm is used to detect voltage sag which is more effective and simpler compared to other available detection methods [4]. This method which is originated from FFT method is used to find limited number of frequency components within a signal. It is used to find the value of main frequency (50Hz) component of electric network. This value is compared against grid nominal value and any decrease will be considered as voltage sag. Therefore Goertzel algorithm permanently calculates the grid main component value and any decrease of voltage is considered as voltage sag which will be mitigated by the STATCOM controller [5].

In this paper, a single cascade multilevel inverter is used. Cascade is a well known multilevel inverter which is widely used in multilevel applications.

In order to control STATCOM, a new current mode controller is used. This controller is single phase and is

978-1-4577-0007-1/11 $26.00 © 2011 IEEE   176

used to control multilevel inverter to mitigate voltage sag. This controller is fixed frequency and can be easily implemented by logic gates [6].

The application of this control method on the multilevel inverter to mitigate voltage sag is described and illustrated. The practical waveforms of the STATCOM are also shown in this paper.

II. CURRENT CONTROL MODE

A new approach has been proposed to control current with Peak Current mode control (PCM) without any instability. PCM is instable for duty cycles more than half. This approach adopts PCM for duty ratios smaller than half, but it changes the control strategy for duty ratios larger than half. In traditional PCM method, the circuit current is controlled in such a way that it does not become larger than the reference. When the current in each cycle crosses the reference, the control method will change the switch status to decrease current. However, this algorithm will cause instability to the current for duty ratios greater than half. The new method will control circuit current in duty ratios larger than half in such a way that it does not become smaller than the reference. Therefore, as the current in each cycle crosses the reference, the control method will change the switch status to increase the current. So depending on the duty cycle status whether it is greater or lower than half, the controller will switch to the proposed method or the traditional PCM, respectively. This will avoid the instabilities for all duty cycles [6].

III. APPLICATION OF THE PROPOSED CURRENT METHOD ON THE STATCOM

The general schematic o the project is shown in Fig. 2. The output voltage of the inverter used as a single phase STATCOM shall be in phase with the grid voltage. At the same time, the duty cycle shall be controlled according to the new current control approach.

In order to adopt this approach, the duty cycles shall be identified whether it is greater or smaller than half.

In order to define the duty cycle for different voltage conditions, the inductor current should be calculated according to the inductor voltage drop. If the inductor voltage drop during current increase in a switching period is larger than the one for current decrease instance, the duty cycle will be less than half. It is because it takes less time for the current to rise to its final value than the time for it to decrease to its bottom value. This is well shown in voltage current equation of inductor as in (1) and (2)

$$V_{L1} = L_1 \frac{\Delta I}{\Delta t} \tag{1}$$

$$V_g - V_{inv} = L_1 \frac{\Delta I}{\Delta t} \tag{2}$$

According to Equation (2) the different voltage conditions and its related duty cycle is calculated and shown in Table I.

For example if the grid voltage is below half unit of the DC voltage supply, the voltage that is generated in the output of the converter is zero or one unit of the DC voltage. In order to increase the current, zero will be

assigned to the output of the converter. On the other hand, to decrease the STATCOM current one unit of the DC voltage supply will be assigned to output of the converter. In this case the inductor voltage drop when current increasing is equal to the grid voltage and the inductor voltage drop when current decreasing is the difference between grid voltage and one DC voltage supply. Since it was assumed that the grid voltage is lower than half of the DC voltage supply, the inductor voltage drop during current increase is lower than the voltage drop during current decrease. This means that it takes a longer time for the current to increase than for it to decrease as in fFg. 4 which leads to the duty cycle larger than half as in Table I

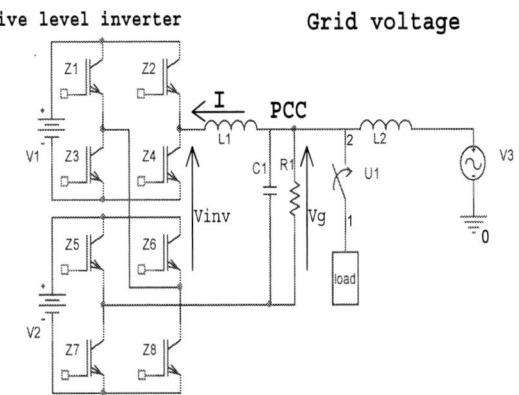

Fig. 2. General schematic of a single phase grid and the STATCOM compensator

Since the output voltage of the converter shall be in phase with the grid voltage. This is a criterion to change the switch states during voltage zero crossings. However, the main measure for changing switch states is to follow the reference current according to the mentioned control method. It means that according to the duty cycle states that can be defined by Equation (2) the inverter current is controlled. The duty cycle status for different grid voltage amplitudes is mentioned in Table I. In this table, if the absolute voltage amplitude is in certain ranges against DC voltage supplies, the duty ratio status is defined for the controller. Hence the current control strategy is based on Table I.

Table I:
Switching status and related duty cycle for different circuit conditions depending on the required voltage and current polarity.

Grid Voltage amplitude (V) compared to DC voltage supplies (V_{DC})	Duty cycle (D) status
V<0.5 V_{DC}	D>0.5
0.5 V_{DC} <V< V_{DC}	D<0.5
V_{DC} <V<1.5 V_{DC}	D>0.5

In order to manipulate the control method on the multilevel inverter, the conditions that affect the circuit current should also be identified. There should be a table to define the true voltage that should be imposed to the inverter output voltage terminals. There are some measures to select the appropriate output voltage. One of the criteria is that the grid voltage should be always between the nearest available DC-voltage magnitude that can be generated by the inverter as shown in Table II. For example, if the grid voltage amplitude is greater than V_{DC}

and lower than 2 V_{DC}, converter the output voltage will switch between V_{DC} and 2 V_{DC}.

Table II:

Appropriate inverter output voltage sets according to grid voltage and DC voltage amplitude

Grid Voltage amplitude (V) compared to DC voltage supplies (V_{DC})	Increasing current	Decreasing current
$V< V_{DC}$	0	V_{DC}
$V_{DC} <V$	V_{DC}	2 V_{DC}

In order to generate the suitable output voltage, a true combination of switches is required to turn on and off. Table III illustrates the switching combinations for the required output voltage levels.

Table III:

Switch arrangements for required inverter output

Required output voltage	Switches to be ON
-2V_{dc}	Z5,Z8,Z1,Z4
- V_{dc}	Z5,Z8,Z3,Z4
0	Z7,Z8,Z3,Z4
V_{dc}	Z7,Z8,Z3,Z2
2 V_{dc}	Z7,Z6,Z3,Z2

Table III will be used to implement the controller utilizing a DSP. Table I is used in the current control block while Table II and Table III are used to generate the appropriate PWM pulses for the switches according to the control signal.

According to the tables above and measuring the instantaneous current and voltage, the DSP decides the status of switches. The block diagram of the controller in DSP according to the tables above is in Fig. 3

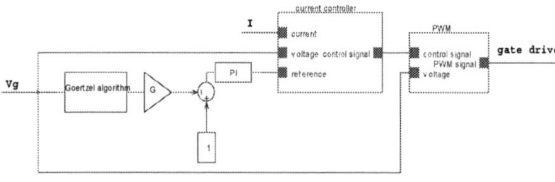

Fig. 3 current control block diagram

According to Fig. 3, the grid voltage (Vg) and inverter current (I) are measured and fed into the DSP. The DSP controls the switches according to the network parameters and tables mentioned above.

The current controller block in Fig. 3 generates the reference current by multiplying the PI output value with the sinusoidal signal which is 90 degree leading the grid voltage. This signal is considered as the reference in the current controller block. The controller will then manipulate the interfacing current (I) according to the network conditions and the tables above. The control signal will then be fed into the PWM block to activate and on the selected switches. True set of switches are selected according to Table II and Table III.

Simulation results of the proposed method are depicted in Fig. 4 and Fig. 5. In Fig. 4 the inverter current (I) in capacitive operation (inverter current leading grid voltage by 90 degree phase) and reference current, which is controlled according to the proposed criteria, is depicted. In this figure the inverter current is controlled according to

duty cycle (D) conditions in a complete sinusoidal cycle. The duty cycle (D) conditions (smaller or greater than half) are also mentioned in the figure.

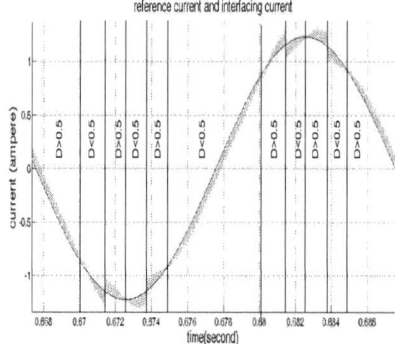

Fig. 4 duty cycle status during capacitive current complete cycle

InFig. 5, inverter output voltage (Vg) and grid voltage in point of common coupling (PCC) is illustrated. It is shown that the grid voltage is in line with inverter voltage to inject reactive power to compensate voltage amplitude.

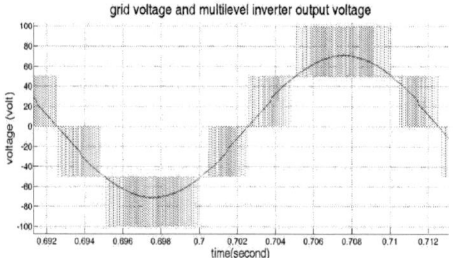

Fig. 5 STATCOM voltage and grid voltage

STATCOM is used to mitigate voltage sag in the network as shown in Fig. 2. According to the figure, the STATCOM compensate the voltage at the point of common coupling (PCC) when a large load is suddenly applied to the grid and will cause a sudden voltage drop or sag. In this paper 22% voltage drop is caused by a sudden load change.

The application of STATCOM will lead to mitigate voltage drop within the standard limits. It will compensate voltage at the Point of common coupling (PCC) as shown in Fig. 2 by injecting reactive power to the grid. In this case the STATCOM works in capacitive operation mode to mitigate voltage sag. The application of STATCOM to mitigate voltage sag is depicted in Fig.6 which will decrease the voltage drop to less than 8% which is acceptable and less than 10% limit which is defined by many associations like Information Technology Industry Council (ITI).

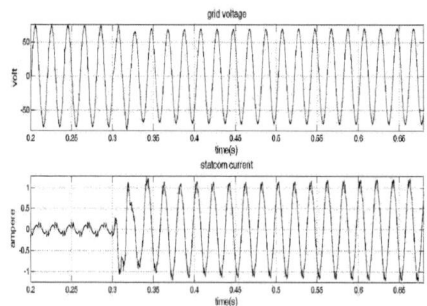

Fig.6 STATCOM current and grid voltage during sag

978-1-4577-0007-1/11 $26.00 © 2011 IEEE

Normalized voltage amplitude generated by Goertzel block is depicted in Fig. 7. According to this figure the controller is fast enough to adjust the voltage at the PCC in less than three cycles. It is obvious from Fig. 7 that 2% settling time is smaller than three voltage cycles (60ms) as in typical schemes which is comparatively good [7]. It is also comparable with complex and computational methods such as Fuzzy-PI-Based controller and Self-Tuning PI Controller Using Particle Swarm Optimization which have been recently proposed and have about the same response time [8, 9].

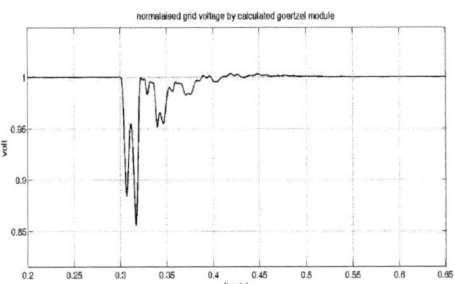

Fig.7 Normalized voltage generated by Goertzel block in STATCOM

The frequency spectrum of the STATCOM current during the step load change (voltage sag) is depicted in Fig. 8. According to the figure the THD is 4.5% which complies to IEEE 519-92 standard.

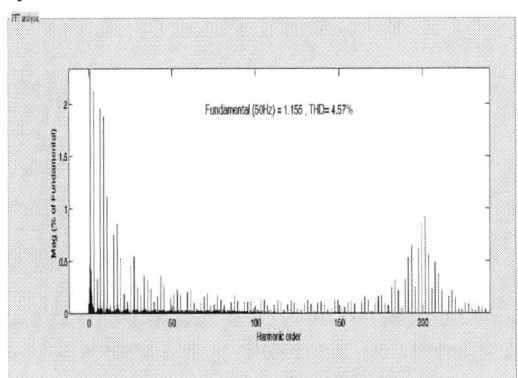

Fig 8 frequency spectrum of STATCOM current

IV. CONCLUSION

A new STATCOM controller based on Goertzel sag detection and a new current control mode is proposed and simulated. The sag detection method is suitable for DSP implementation. It does not require extensive calculations such as in FFT nor is it prone to noise interference.

The proposed current control method has also great superiority over conventional voltage control methods in terms of simplicity and harmonic spectrum. It is a fixed frequency approach and does not have the filtering and EMI problems as in the hysteresis method.

The control method is also applied on a cascade multilevel inverter to make a multilevel STATCOM.

The controller is a single phase controler and can handle all single phase voltage sag disturbances. The simulations show that STATCOM can effectively work to mitigate voltage sag and maintain voltage in safe limits.

ACKNOWLEDGMENT

The authors would like to thank Universiti Teknologi Malaysia (UTM) and ministry of higher education Malaysia for their financial support of the project.

REFERENCES

[1] "IEEE Recommended Practice for Monitoring Electric Power Quality", IEEE Std 1159-1995
[2] Naidoo, R.; Pillay, P.; "A New Method of Voltage Sag and Swell Detection" , IEEE Transactions on Power Delivery ,Volume 22, Issue 2, (April 2007) Page(s):1056 – 1063.
[3] Giroux, P.; Sybille, G.; Le-Huy, H.;" Modeling and simulation of a distribution STATCOM using Simulink's Power System Blockset", The 27th Annual Conference of the IEEE Industrial Electronics Society, 2001. IECON '01.Volume 2, (29 Nov.-2 Dec. 2001) Page(s):990 - 994 vol.2
[4] rulphchasseing,"digital signal processing and applicationswith the C6713 and C6416 DSK ", john wiley& sons publications,(2005).
[5] E. Najafi and A. H. M. Yatim, "A novel current mode controller for a static compensator utilizing Goertzel algorithm to mitigate voltage sags," Energy Conversion and Management, vol. 52, pp. 1999-2008, 2011.
[6] Najafi, E.; Yatim, A.H.M; , "A D-STATCOM based on Goertzel algorithm for sag detection and a novel current mode controller," Industrial Electronics and Applications (ICIEA), 2010 the 5th IEEE Conference on , vol., no., pp.1006-1011, 15-17 June 2010
[7] Ben-Sheng Chen; Yuan-Yih Hsu; , "A Minimal Harmonic Controller for a STATCOM," IEEE Transactions on Industrial Electronics, vol.55, no.2, pp.655-664, Feb. 2008
[8] An Luo; Ci Tang; Zhikang Shuai; Jie Tang; Xian Yong Xu; Dong Chen; , "Fuzzy-PI-Based Direct-Output-Voltage Control Strategy for the STATCOM Used in Utility Distribution Systems," IEEE Transactions on Industrial Electronics, , vol.56, no.7, pp.2401-2411, July 2009
[9] Chien-Hung Liu; Yuan-Yih Hsu; , "Design of a Self-Tuning PI Controller for a STATCOM Using Particle Swarm Optimization," IEEE Transactions on Industrial Electronics, vol.57, no.2, pp.702-715, Feb. 2010

Design of a Current Mode PI Controller for a Single-phase PWM Inverter

S. M. Cherati[1], N. A. Azli[2], S. M. Ayob and A. Mortezaei

Power Electronics and Drive Research Group (PEDG), Energy Research Alliance
Universiti Teknologi Malaysia, 81310 UTM Johor Bahru, Malaysia
sam.mcherati@fkegraduate.utm.my[1]
naziha@ieee.org[2]

Abstract- **This paper presents the design of current mode PI controller for single-phase PWM inverter. The controller is comprised of inductor current as the inner loop and output voltage as the outer feedback loop. The control design is carried out using *Sisotool*, which is provided in Matlab. By using this tool, users can examine the effects of changing the gain control values to system's transient response and stability, simultaneously. Hence, simplify and speed up the control design process. To evaluate the performance of the designed controller, the inverter is simulated under several types of load disturbances. From the result, it is shown that the designed controller exhibits a good transient response i.e. fast rising and settling time with small overshoot when subjected to step load disturbance.**

I. INTRODUCTION

Power inverter is an important part of many DC to AC conversion equipments such as uninterruptible power supply (UPS), induction motor drive and automatic voltage regulator (AVR) systems. In these systems, it is the major requirement for the power inverter to be capable of producing and maintaining a stable and clean sinusoidal output voltage waveform regardless of the type of load connected to it. The main key to successfully maintain this ability is to have a feedback controller [1].

Currently, there are various control methods that have been proposed for inverter. Among the prevalent methods are the Voltage Mode Control (VMC) and Current Mode Control (CMC). In VMC method, the control parameters design is easy. The implementation can be considered as inexpensive since it requires only one voltage sensor. However the main drawback of VMC is that the system is prone to instability and very sensitive to large input variations. Moreover, VCM also does not control the current. Thus, power electronic switches are vulnerable to over-current damage or overloads. To achieve better performance, robustness and also immune to input disturbance and current protection, current mode (CMC) is more preferable. Unlike VMC, CMC has an additional inner loop. Usually, the inductor current is used as the inner loop. The inductor current is sensed and used to control the duty cycle.

In this paper, accurate design of PI parameter is discussed for voltage and current loop and their proper bandwidth and phase margin are calculated. Both the transfer function and the state-space models of the inverter are provided.

II. DYNAMIC MODEL OF SINGLE-PHASE INVERTER

Fig. 1, shows the equivalent circuit of a single-phase full bridge inverter with connected load. In this study, control based on the linear strategy theory is presented. Solid-state switches and connected rectifier are nonlinearity source of a system, so for proper control design, the system must be linearized around its operating point. If the designed controller is robust enough, the system can work around its operating point with high performance which means a wide range of bandwidth is required [2]. In this work, a nominal resistive load of $R = 10\,\Omega$ is set as the operating point for linearization.

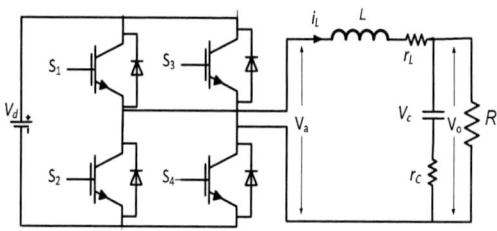

Fig. 1. Full bridge single-phase inverter

A. State-Space Model

The resistive load and its related filter operate as a continuous time second-order system. The state variables of the system are capacitor voltage (V_c) and inductor current (i_L). The dynamic model of Fig. 1, can be represented by the following equations.

$$\begin{pmatrix} i_L' \\ V_c' \end{pmatrix} = \begin{pmatrix} -\dfrac{r_L r_c + R(r_L + r_c)}{L(R + r_c)} & -\dfrac{R}{L(R + r_c)} \\ \dfrac{1}{C(1 + r_c/R)} & \dfrac{1}{RC(1 + r_c/R)} \end{pmatrix} \begin{pmatrix} i_L \\ V_c \end{pmatrix} + \begin{pmatrix} \dfrac{1}{L} \\ 0 \end{pmatrix} V_a \quad (1)$$

$$V_o = \begin{pmatrix} \dfrac{r_c}{r_c + R} & \dfrac{R}{R + r_c} \end{pmatrix} \begin{pmatrix} i_L \\ V_c \end{pmatrix} \quad (2)$$

One of the important factors of load effect on the inverter is output impedance which in this case as rectifier is a load with nonlinear nature; the output voltage will be highly distorted. For determining the output impedance, circuit laws can be applied as shown in Fig. 2.

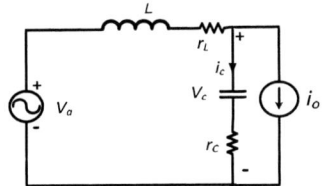

Fig. 2. Equivalent circuit of a single-phase inverter

$$V_o(s) = \frac{r_c C s + 1}{LCS^2 + C(r_L + r_c)s + 1} V_a - \frac{r_c LCS^2 + (L + r_L r_c C)s + r_L}{LCS^2 + C(r_L + r_c)s + 1} I_o \quad (3)$$

r_L and r_c are ignored in many modeling since they are generally very small but in this paper only r_c is disregarded ($r_c = 0$). By using $s = j\omega$ and determining Z as output impedance in $V_o = ZI_o$,

$$Z = -\frac{jL\omega + r_L}{-LC\omega^2 + jCr_L\omega + 1} \quad (4)$$

It can be clearly seen in (4) that, by increasing the switching frequency, the output impedance will be reduced. This means that lower component values of the LC filter are required [3].

By ignoring the internal resistance of the filter capacitor ($r_c = 0$), (1) and (2) can be simplified as (5) and (6).

$$\begin{pmatrix} i_L' \\ V_c' \end{pmatrix} = \begin{pmatrix} -\frac{r_L}{L} & -\frac{1}{L} \\ \frac{1}{C} & \frac{1}{RC} \end{pmatrix} \begin{pmatrix} i_L \\ V_c \end{pmatrix} + \begin{pmatrix} \frac{1}{L} \\ 0 \end{pmatrix} V_a \quad (5)$$

$$V_o = \begin{pmatrix} 0 & 1 \end{pmatrix} \begin{pmatrix} i_L \\ V_c \end{pmatrix} \quad (6)$$

B. Analysis of multiple feedback loop control

A DC motor has some similarity to an inverter in terms of using cascade control (multi-loop control) [4]. In a DC motor system, the armature current and stator voltage are used as current and voltage feedback loop for achieving sufficient steady-state and good transient performance. A single-phase inverter has the same scenario. The inductor current of the filter acts as an inner loop parameter while the output voltage is the outer loop parameter. Fig. 3 illustrates the voltage and current loop [5].

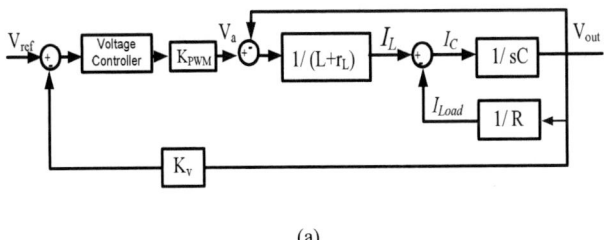

(a)

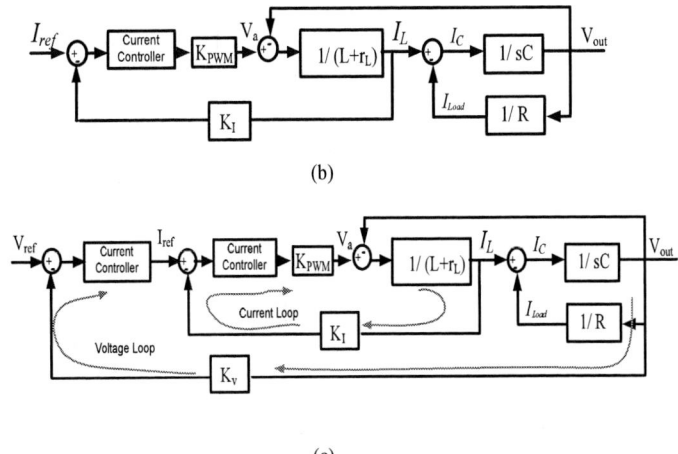

(b)

(c)

Fig. 3. Control schemes for a single-phase inverter. (a) Voltage feedback loop control. (b) Current feedback loop control. (c) Outer voltage and inner current feedback loop control

K_v and K_I of Fig. 3, represent voltage and current sensors gain respectively. Fig. 3(c), shows the overall current mode control structure. The variable K_{PWM} is the PWM gain, which is defined as $\frac{V_{dc}}{V_{Tri}}$. The controller for the loops can be either PI, Sliding mode, Fuzzy, Deatbeat etc. [6]-[9]. For this work, a conventional PI controller is used.

III. CONTROL SYSTEM DESIGN

The design of PI controller can be done using several methods. It can be designed using Ziegler-Nicholas, Pole-placement method and Frequency response [10]. However, the latter method will be used to design the controller. For simplicity, *Sisotool* from MATLAB/SIMULINK is used to tune the controller and evaluate the suitability of bandwidth and stability. As mentioned earlier, the control system is designed for a linear load (R = 10 Ω) and tested on a nonlinear load such as a rectifier. The simulation testing parameters are as given in Table 1.

TABLE 1. SIMULATION TESTING PARAMETERS

V_{dc}	215 V
V_{out}	200 V_{p-p}
r_L	0.3 Ω
Filter Inductor (L)	1.2 mH
Filter Capacitor (C)	13.2 μF
Linear Load (R)	10 Ω
Switching Frequency	40 kHz
Current sensor gain (K_i)	0.06
Voltage sensor gain (K_v)	0.01
Nonlinear Load (Diode Rectifier)	R = 100 Ω C = 500 μF
Voltage Reference	1 V

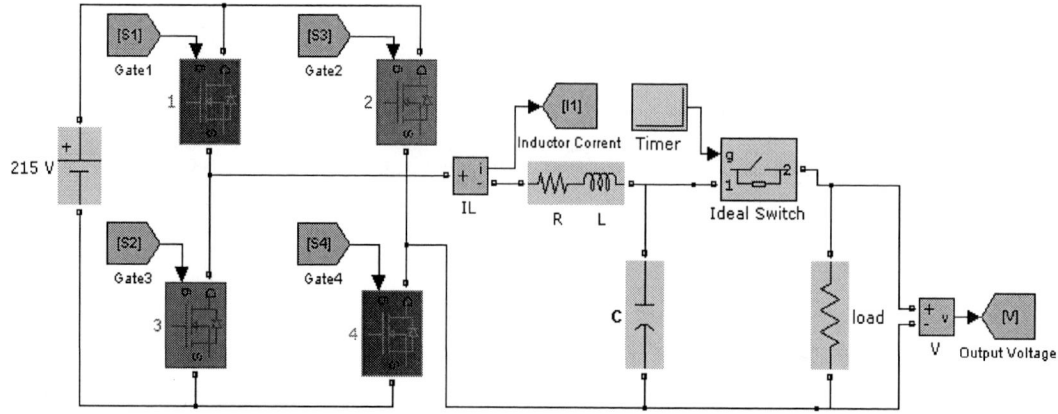

Fig. 4. The single-phase circuit with its control system

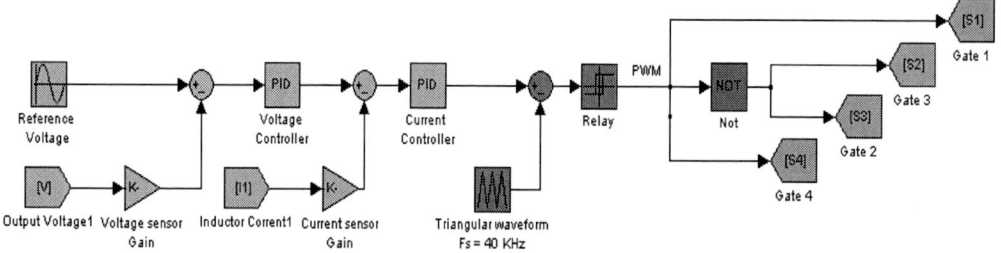

Fig. 5. The single-phase inverter simulation model

PI controllers are used as the voltage and current controllers. The PI controller transfer function is represented as:

$$K_p + \frac{K_i}{s} = K_p \frac{(s+K_i/K_p)}{s} \qquad (7)$$

For obtaining accurate value of K_p and K_i, the *Sisotool* in MATLAB/SIMULINK can be used. Since the voltage loop and current loop are decoupling, each PI controller can be designed separately. In this work, the inner loop will be firstly designed. The control bandwidth for the inner loop should be larger than the outer loop, since the inductor current has faster response compared to the output voltage. Since the switching frequency is 40 kHz, the control bandwidth for the current loop should be within 4 kHz. The bandwith for the voltage loop is set to be within 400 Hz. Both loops should be designed with phase margins of within 65°.

The PI controller for the current loop can be designed by using the transfer function between the inductor current and the current reference. Based on Fig. 3(b), the transfer function of $\frac{I_L}{I_{ref}}$ can be obtained as follows:

$$\frac{I_L}{I_{ref}} = K_{pwm} \frac{(RCs+1)}{RLCs^2+(CRr_L+L)s+(R+r_L)} \qquad (8)$$

$$K_{pwm} = \frac{V_{dc}}{V_{tri}} \qquad (9)$$

The complete transfer function by incorporating the PI controller can be expressed as in (10).

$$\frac{I_L}{I_{ref}} = K_p \frac{(s+K_i/K_p)}{s} \times \frac{V_{dc}}{V_{tri}} \times \frac{(RCs+1)}{RLCs^2+(CRr_L+L)s+(R+r_L)} \qquad (10)$$

From (10), it can be shown that the controller adds a zero at the left side of s-plan and zero at the origin to compensate the second-order transfer function. Fig. 6 shows the resulting bode-plot of the closed-loop current mode.

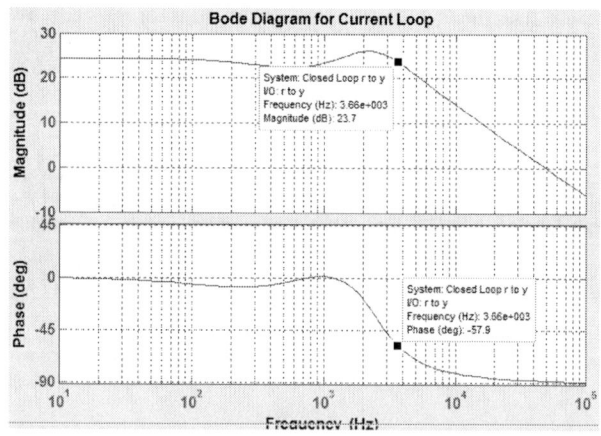

Fig.6. Bode plot of current loop system

From Fig. 6, the bandwidth of the current loop is $3.62 \times 10^3 Hz$ while the phase margin is 65°, which are acceptable values. To design the outer loop controller, it is assumed that the inner loop is unity as illustrated in Fig. 7. Equation (11) expresses the complete transfer function of Fig. 7 with PI controller.

$$\frac{V_o}{Vref} = K_p \frac{(s + K_i/K_p)}{s} \times \frac{R}{RCs+1} \qquad (11)$$

It should be noted that the outer loop should have a bandwith of 400 Hz. Fig. 8 shows the resulting bode plot and root locus of the voltage loop.

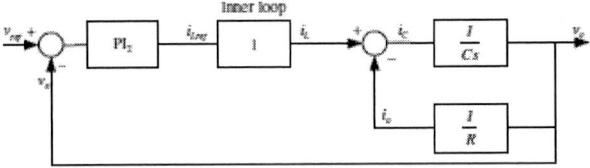

Fig. 7. Voltage loop block diagram with the designed current loop.

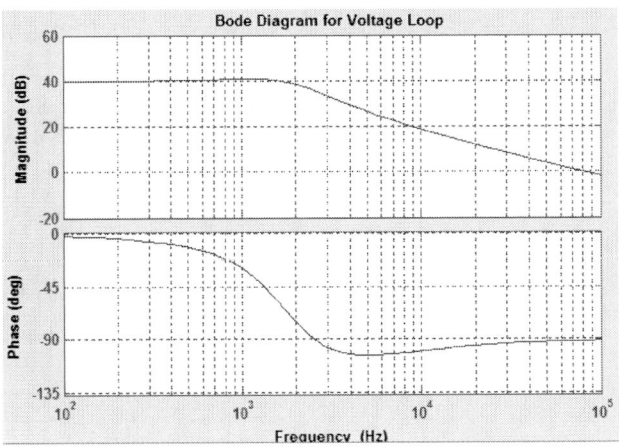

Fig.8. Bode plot and root locus of the voltage loop

It can be clearly seen that, the bandwidth is lower than the inner loop (942 Hz) while its phase margin is 63°.

IV. SIMULATION RESULTS

The control system has been designed for a linear load (R = 10 Ω). The following figures (Fig. 9 and Fig. 10) illustrate the voltage and current waveforms of the load when the system becomes no-load at t = 0.004.

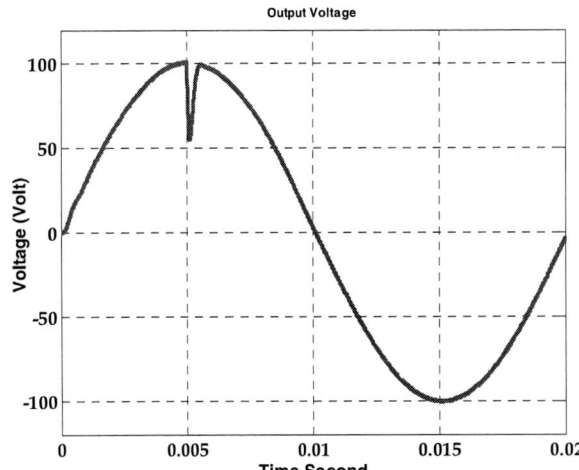

Fig. 9. Output voltage

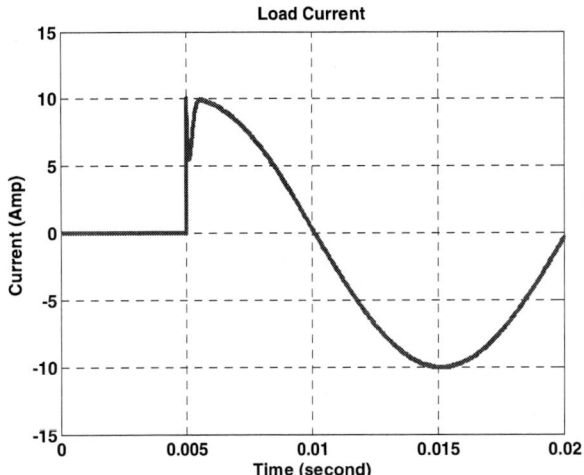

Fig. 10. Load current

V. CONCLUSION

In this paper a current mode PI controller for a single-phase PWM inverter has been designed. The control structure is comprised of two loops and has been arranged in a cascaded fashion. Two systems variables namely the inductor current and the output voltage are sensed as the feedback variables. The *Sisotool* from MATLAB/SIMULINK has been used to tune and design the PI control parameters for both loops. The performance is verified by subjecting the inverter system with different types of load. The simulation results have shown that the controller is capable of producing good output voltage regulation.

978-1-4577-0007-1/11 $26.00 © 2011 IEEE

REFERENCES

[1] Bass, R.M.; Krein, P.T." State-plane animation of power electronic systems: a tool for understanding feedback control and stability "Applied Power Electronics Conference and Exposition, 1990.

[2] Hong Yi, Jiyang Dai and Jiaju Wu; "Research on Modeling and Control of The Single-Phase Inverter System with A Nonlinear Load" Proceedings of the 7th World Congress on Intelligent Control and Automation June 25 - 27, 2008, Chongqing, China

[3] H. Deng, R.Oruganti and D. Srinivasan, "Modelling and Control of Single-Phase UPS Inverter: A Survey", in Proc. PEDS-Kuala Lumpur, Kuala Lumpur, pp. 848-853, 2005.

[4] W. Leonhard, "Control of electrical drives", Springer-Verlag, 1984.

[5] Ying-Yu Tzou, Shih-Liang Jung;" Full control of a PWM DC-AC converter for AC voltage regulation", IEEE Transactions on Aerospace and Electronic Systems.vol.34,no.4, pp: 1218 – 1226, 1998.

[6] Naser M.Abdel-Rahim,etc,"Analysis and design of a multiple feedback loop control strategy for single-phase voltage-source UPS inverter",IEEE trans.power electrp.vol.11,July 1996, pp.532-541.

[7] Zhang Kai,etc,"Deadbeat control of PWM inverter with repetitive disturbance'.1999 APEC.Record.1999, pp.1026-1031.

[8] S.M. Ayob, N.A. Azli, Z. Salam, "Single input PI-fuzzy controller for single phase inverter system", International Review of Electrical Engineering, vol. 3, no.3, pp. 418 – 425, June 2008.

[9] Michael J.Ryan, ets, "Control Topology options for single-phase UPS inverter",IEEE trans.Indus.Appli.,Vol.33,March/April 1997, pp.493-501.

[10] N. Mohan, T. M. Undeland, W. P. Robbins, " Power Electronics, Converters, Applications, and Design", 3th edition, John Wiley & Sons, New York, 2003.

978-1-4577-0007-1/11 $26.00 © 2011 IEEE

Comparison of Adaptive and Fixed-Band Hysteresis Current Control Considering High Frequency Harmonics

Hani Vahedi

Islamic Azad University, Sari Branch Sari, Iran

hvahedi@ieee.org,

Yasser Rahmati Kukandeh

Shahed University of Technology Tehran, Iran

y.rahmati.k11@gmail.com

Mahsa Ghapandar Kashani

Mapna Group Tehran, Iran

Mahsa_ka6214@yahoo.com

Aliakbar Dankoob

Islamic Azad University, Aliabad Katoul Branch Aliabad, Iran

aliakbardankoob@yahoo..com

Abdolreza Sheikholeslami

Babol University of Technology Babol, Iran

asheikh@nit.ac.ir

Abstract - Shunt active power filters (APF) are widely used in power systems to eliminate the current harmonics and to compensate reactive power due to their accurate and fast operation. In this paper the instantaneous power theory is used to extract the harmonic components of system current. Then fixed-band hysteresis current control is explained. Because of fixed-band variable frequency disadvantages, the adaptive hysteresis current control (AHCC) is introduced that leads to fix the switching frequency and reduce the high frequency components in source current waveform. Due to these advantages of AHCC, the switching frequency and switching losses will be diminished appropriately. Some simulations are done in Matlab/Simulink. The Fourier Transform and THD results of source and load currents and the instantaneous switching frequency diagram are discussed to prove the efficiency of this method. The Fourier Transform and THD results of source and load currents are discussed to prove the validity of this method.

Index Terms - Active Power Filter, High Frequency Harmonics, Hysteresis Current Control, Instantaneous Power Theory, Instantaneous Switching Frequency.

I. INTRODUCTION

The Problem of reactive power burden has been a concern for power engineers for many years. This problem is aggravated in harmonic environment caused by power electronic converters. This nonlinear current having a high amount of harmonics distorts the ac voltage at the point of common coupling (PCC) and therefore affects the other neighboring loads connected to the same system. Excessive reactive power demand increases feeder losses and reduces the active power flow capability of distribution system, whereas unbalancing affects the operation of transformers and generators. The advent of custom power device technology has proved to be a boon for the electric distribution system facing these problems (reactive power burden, harmonics, and unbalanced loading). The custom power device such as shunt active filter (APF) is found to be quite suitable to cater to the aforesaid problems. The major issue related to the effective operation of APF is its controllability to compensate reactive power, harmonics, and unbalanced loading [1-3].

Fig. 1 shows a 3-phase power system includes voltage source, nonlinear load and a 3-phase APF which has been connected to it through interface inductors. APFs are based on voltage source inverters (VSI). The control of APF depends on two major factors: the first is extraction of reference currents, and second is the method of generation of pulse width

modulation (PWM) signals using these extracted reference currents. The techniques reported in the literature for reference current extraction is instantaneous reactive power theory [4]

Many switching methods are used to produce switching pulse leads to generate reference current. Hysteresis current control (HCC) has been noticed more than other current control techniques, due to simplicity and quicker dynamic response [5-9].

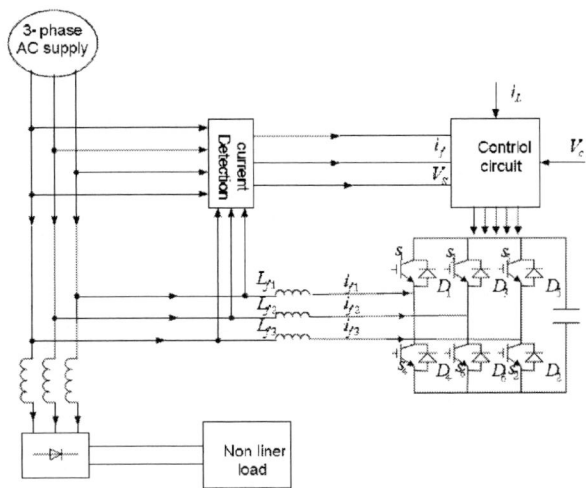

Fig. 1 Power System Diagram with APF.

The main problem of HCC is its variable switching frequency which leads to variable high frequency components in source current waveform, audio noises and increase switching losses. One of performed methods that can solve this problem is the AHCC method that builds variable band for current tracking, hence the switching speed becomes smooth and the frequency switching will be fixed considerably.

Furthermore, different frequency components in current waveform will be appeared due to different switching frequencies that make it difficult to design appropriate filters to eliminate these components and make noises affects measuring devices. To overcome this problem, an adaptive hysteresis current control (AHCC) has been introduced. Using this method, variable hysteresis bandwidth is calculated instantaneously, leads to reduce the switching frequency variation, thus the fixed-band HCC issues will be amended.

978-1-4577-0007-1/11 $26.00 © 2011 IEEE

In this paper, in section II, the instantaneous power theory has been explained to extract the harmonics components of current waveform. Then in section III, the fixed-band HCC and AHCC have been clarified. Finally, some simulations have been done with Matlab/Simulink and the results consist of switching frequency diagrams and current THD in high frequency range have been discussed in section IV. By comparing the results of simulations, the advantages of AHCC in fixing switching frequency and modifying above mentioned problems, especially reducing the high frequency components in source current waveform and switching losses have been proved.

II. INSTANTANEOUS POWER THEORY

One of the popular compensation reference current extraction methods is the instantaneous reactive power theory (p-q theory) presented by Akagi. Although there are some problems with this theory, it is well-established and simple in implementation. The p-q theory could be briefly reviewed as follow [4]:

Assume a three-phase load with the instantaneous voltages as $v(t)=[v_a(t)\ v_b(t)\ v_c(t)]^t$ and the instantaneous currents as $i_l(t)=[i_{la}(t)\ i_{lb}(t)\ i_{lc}(t)]^t$ (Fig. 1). Using (1), $v(t)$ and $i_l(t)$ can be converted to o-α-β coordination where C is the matrix (2):

$$\left[v_{0\alpha\beta}(t)\right]^t=C\left[v_{abc}(t)\right]^t,\left[i_{0\alpha\beta}(t)\right]^t=C\left[i_{abc}(t)\right]^t \quad (1)$$

$$C=\sqrt{\frac{2}{3}}\begin{bmatrix}\frac{1}{\sqrt{2}} & \frac{1}{\sqrt{2}} & \frac{1}{\sqrt{2}} \\ 1 & \frac{-1}{2} & \frac{-1}{2} \\ 0 & \frac{\sqrt{3}}{2} & \frac{-\sqrt{3}}{2}\end{bmatrix} \quad (2)$$

Let's assume that the zero sequence current ($i_{l0}(t)$) is null. Thus, the instantaneous active ($p(t)$) and reactive ($q(t)$) powers can be calculated as:

$$\begin{bmatrix}p(t)\\q(t)\end{bmatrix}=\begin{bmatrix}v_\alpha(t) & v_\beta(t)\\-v_\beta(t) & v_\alpha(t)\end{bmatrix}\begin{bmatrix}i_{l\alpha}(t)\\i_{l\beta}(t)\end{bmatrix} \quad (3)$$

$p(t)$ and $q(t)$ can be decomposed to the average parts ($\bar{p}(t),\bar{q}(t)$) and the oscillating parts ($\tilde{p}(t),\tilde{q}(t)$). It is notable that $\bar{p}(t)$ is produced by the fundamental harmonic of the positive sequence component of the load current. Therefore, in order to compensate the harmonics and the instantaneous reactive power, compensation reference currents can be extracted as follow:

$$\begin{bmatrix}i_{f\alpha}^*(t)\\i_{f\beta}^*(t)\end{bmatrix}=\begin{bmatrix}v_\alpha(t) & v_\beta(t)\\-v_\beta(t) & v_\alpha(t)\end{bmatrix}^{-1}\begin{bmatrix}-\tilde{p}(t)\\-q(t)\end{bmatrix} \quad (4)$$

$$\begin{bmatrix}i_{fa}^*(t) & i_{fb}^*(t) & i_{fc}^*(t)\end{bmatrix}^t=C^{-1}\begin{bmatrix}0 & i_{f\alpha}^*(t) & i_{f\beta}^*(t)\end{bmatrix}^t \quad (5)$$

III. HYSTERESIS CURRENT CONTROL

Hysteresis current control is used for generating the switching pulses. Among the various current control techniques, HCC is the most extensively used technique because of the noncomplex implementation, outstanding stability, absence of any tracking error, very fast transient response, inherent limited maximum current, and intrinsic robustness to load parameters variations. As indicated in [6-8]

a review of used current control techniques for PWM converters reveals that HCC shows certain superiority for active power filter applications. HCC provides a better low-order harmonic suppression than PWM control, which is the main target of the active power filter. It is easier to realize with high accuracy and fast response. However, as a disadvantage its switching frequency might fluctuate.

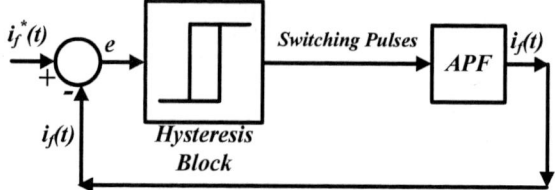

Fig. 2 fixed-band hysteresis current control loop.

In the HCC technique the error function is centered in a preset hysteresis band. When the error exceeds the upper or lower hysteresis limit the hysteretic controller makes an appropriate switching decision to control the error within the preset band and send these pulses to VSI to produce the reference current as shown in Fig. 2.

The outputs of the hysteresis blocks are directly fed as the firing pulse of VSI switches.

A. Fixed-band Hysteresis Current Control

In fixed-band HCC, the hysteresis bandwidth (HB) has been taken as a small portion related to system current and in many researches it has been taken as 5% of main current which will be HB=0.9A, here.

B. Adaptive Hysteresis Current Control (AHCC)

As above-mentioned, the crucial concern with the fixed band hysteresis current control is producing a varying modulation frequency of the power converter which, in turn, results in increasing the switching losses. To avoid this situation, adaptive hysteresis current controller methods with the variable hysteresis band have been recommended in literature [6-8]. Hence, a variable hysteresis band is defined for each phase so that the switching frequency remains almost constant.

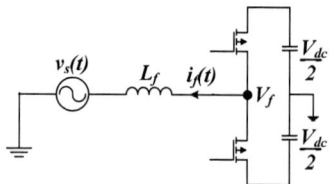

Fig. 3 Single- phase diagram of a power system with APF.

The variable hysteresis band (HB) formula can be calculated based on Fig. 1. The following KVL equation can be easily achieved:

$$\frac{di_f(t)}{dt}=\frac{1}{L_f}(V_f-v_s(t)) \quad (6)$$

Where V_f is the inverter-side voltage and can be elaborated as below:

978-1-4577-0007-1/11 $26.00 © 2011 IEEE

$$V_f = \begin{cases} \dfrac{V_{dc}}{2} & \text{the upper switch is ON} \\[2mm] \dfrac{-V_{dc}}{2} & \text{the lower switch is ON} \end{cases} \qquad (7)$$

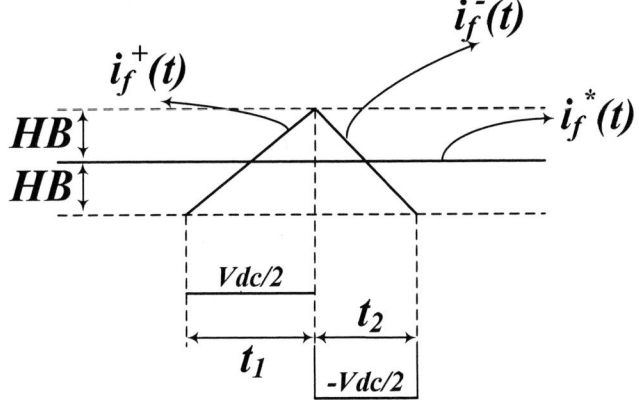

Fig. 3 The upper and lower bands of the reference compensation current.

Having paid attention to Fig. 3, the below relations can be obtained:

$$\frac{di_f^+(t)}{dt} = \frac{1}{L_f}(V_f - v_s(t)) \qquad (8)$$

$$\frac{di_f^-(t)}{dt} = \frac{-1}{L_f}(V_f + v_s(t)) \qquad (9)$$

Where $i_f^+(t)$ and $i_f^-(t)$ are the rising current and the falling current, respectively. Furthermore, the following relations can be extracted:

$$\begin{cases} \dfrac{di_f^+(t)}{dt} \times t_1 - \dfrac{di_f^*(t)}{dt} \times t_1 = 2HB \\[2mm] \dfrac{di_f^-(t)}{dt} \times t_2 - \dfrac{di_f^*(t)}{dt} \times t_2 = -2HB \end{cases} \qquad (10)$$

$$f = \frac{1}{t_1 + t_2} \qquad (11)$$

Where t_1 and t_2 are switching intervals and f is the switching frequency.

By substituting (8), (9) and (11) in (10), the hysteresis band (HB) can be achieved as follow:

$$HB = \frac{V_{dc}}{8fL_f} - \frac{L_f}{2fV_{dc}}\left(\frac{v_s(t)}{L_f} + \frac{di_f^*(t)}{dt}\right)^2 \qquad (12)$$

The adaptive HB should be derived instantaneously during each sample time to keep the switching frequency constant.

IV. SIMULATION RESULTS

To verify validity of the proposed method some simulations are done using MATLAB/Simulink. The nonlinear load consists of a three-phase diode rectifier with a DC-side resistive load. It should be mentioned that the nonlinear load is connected to the grid via inductances (L_f=4 mH). Besides, the load voltages have an rms value of 155V-50Hz leads to a rms value of 18 A for system current.

The source voltage has been remained sinusoidal and does not contain any harmonics. Fig. 4 shows comparative diagrams of load current, filter current and source current

respectively for Fixed-band HCC and AHCC methods simulation. These diagrams show a good filtering leads to eliminate the source current harmonics, so the source current contains just the main component.

TABLE 1
APF SIMULATION PARAMETERS

Supply Phase Voltage	155 V
Grid Frequency	50 Hz
Load Resistanc R_l	10 Ω
Inverter Side Inductance L_f	4 mH
Rectifier Side Inductance L_l	3 mH
APF dc-link Voltage V_{dc}	500 V
Fixed Hysteresis Bandwidth	0.9 A

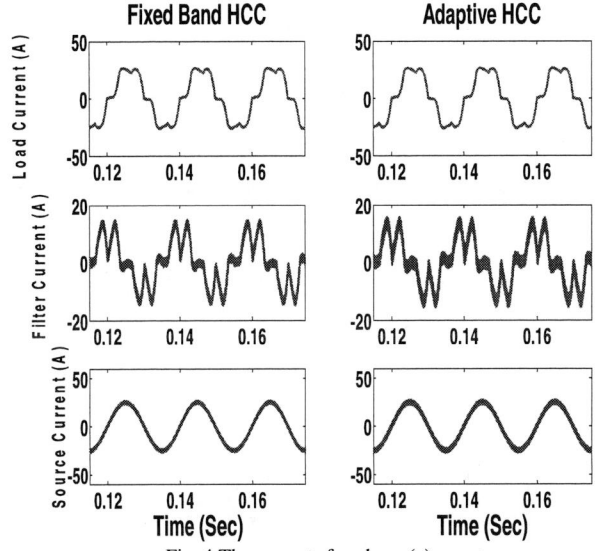

Fig. 4 The currents for phase (a).

TABLE 2
THD% VALUE OF LOAD AND SOURCE CURRENTS

	Source Current THD%			Load Current THD%		
	Phase a	Phase b	Phase c	Phase a	Phase b	Phase c
Fixed-band	2.15	2.59	2.83	24.42	24.31	24.55
AHCC	3.73	3.81	3.53	24.42	24.21	24.55

The THD results in Table 2 show that AHCC method works properly to track the reference current and there was a good filtering process. But the following figures show the difference between fixed-band and AHCC. The AHCC distinction will be proved by figures 5 and 6.

Figure 5 shows the instantaneous switching frequency for the fixed-band HCC and AHCC. It is obvious that the switching frequency in fixed-band HCC varies in vast range (in this case it changed from 15 KHz to 25 KHz) and cause audio noises and inject high frequency components in source current that makes it difficult to design appropriate filters for eliminating them. In AHCC method, the instantaneous

978-1-4577-0007-1/11 $26.00 © 2011 IEEE   187

switching frequency remains constant with little deviation contrary to conventional fixed band hysteretic current control method. In practical application, it is necessary to kept switching frequency to a certain limits, in order to determine switching device and decrease its switching losses [9].

Fig. 6 proves that many high frequency components have been injected to the source current due to variable switching frequency, but in Fig. 6, the AHCC results prove the fact that this method has worked properly results in fixing switching frequency (12 KHz to 15 KHz). This result influences the source current THD especially in high frequency range. Since the variation range of switching frequency has been limited to small domain, the high frequency components of source current have been reduced to a narrow range which is apparent in Fig. 6.

As the variable switching frequency cause audio noises, the AHCC fixes this problem by constant switching frequency, too.

The vast range of high frequency components of current harmonic is just a source for audio noises as well as producing switching losses due to the switch resistance. Each harmonic order should be multiplied by the square of the switch resistor to obtain the power losses so in fixed-band method this value is higher than AHCC.

Besides, calculating the number of switching On-Off pulses proves the fact that fixing switching frequency decrease the switching number and the switching number has a direct relation with the switching losses. In this simulation for 0.2Sec, the switching number has been changed from approximately 11000 to 6000 respectively from fixed-band to AHCC method. So the switching losses is reduced about 50%.

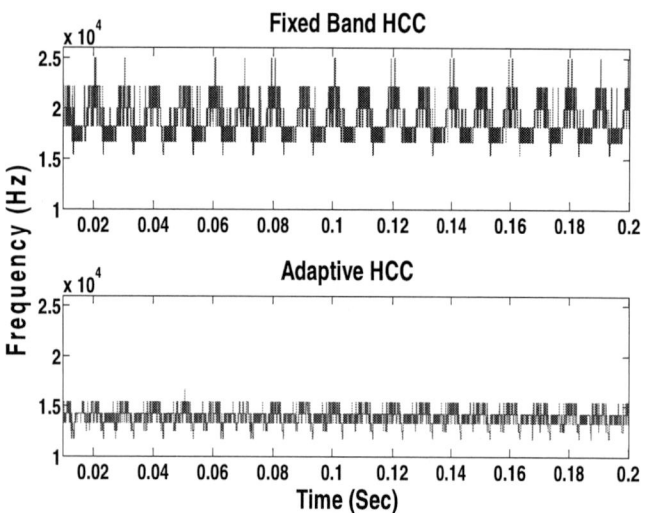

Fig. 5 Instantaneous switching frequency

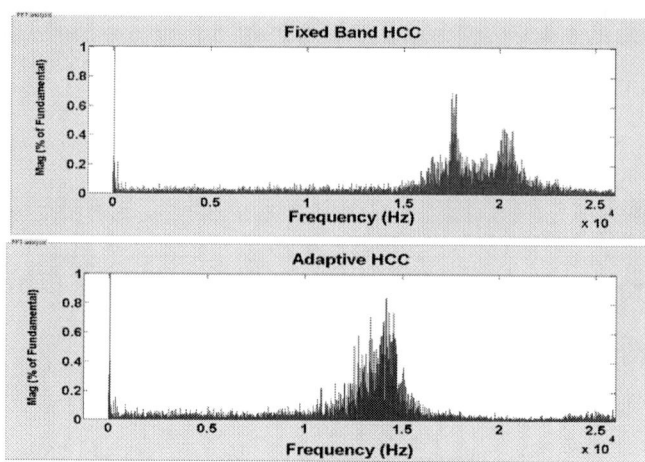

Fig. 6 FFT analysis of the source current in high frequency range

IV. CONCLUSION

Shunt active power filters are the most suitable devices in power networks eliminate the current harmonics and compensate the reactive power. Instantaneous power theory is one of the effective methods which has been explained in this paper. Afterwards, the hysteresis current control has been clarified with two modes: fixed-band and AHCC. The simulation results proved that AHCC technique made the fixed switching frequency that results in reducing the high frequency components of source current and switching losses.

REFERENCES

[1] Bhim Singh, Jitendra Solanki, "An Implementation of an Adaptive Control Algorithm for a Three-Phase Shunt Active Filter", IEEE Trans, Industrial Electronics, vol. 56(8), 2009.

[2] M. Tavakoli Bina, E. Pashajavid, "An efficient procedure to design passive LCL-filters for active power filters", Elsevier Journal Electric Power Systems Research, 2009, Vol. 79, Iss. 4, pp. 606-614.

[3] Ali Emadi, Abdolhossein Nasiri, Stoyan B. Bekarov, "Uninterruptable Power Supplies and Active Filters", Illinoise Institute of Technology, 2005

[4] Wenjin Dai, Yongtao DaiTingjian Zhong, "A New Method for Harmonic and Reactive Power Compensation", IEEE Int. Conf. on Industrial Technology, ICIT, 2008, pp.1-5

[5] M. P. Kazmierkowski and L. Malesani, "Current control techniques for three-phase voltage source PWM converters: a survey", IEEE Trans. Industrial Electronics, vol. 45(5), pp. 691–703, 1998.

[6] B.K.Bose, "An Adaptive Hysteresis Band Current Control Technique of a Voltage Feed PWM Inverter for Machine Drive System", IEEE Trans. On Ind. Elec., 1990, pp. 402-406

[7] Hani Vahedi, Abdolreza Sheikholeslami, "Variable hysteresis current control applied in a shunt active filter with constant switching frequency", IEEE Conf. on Power Quality, PQC, 2010, pp.1-5

[8] Hani Vahedi, Abdolreza Sheikholeslami, "The Source-Side Inductance Based Adaptive Hysteresis Band Current Control to be Employed in Active Power Filters", International Review on Modeling and Simulation Journal, Vol. 3, No. 5, October 2010, pp. 840-845.

[9] G. Vazquez, P. Rodriguez, R. Ordonez, T. Kerekes,; R. Teodorescu, "Adaptive Hysteresis Band Current Control for Transformerless Single-Phase PV Inverters", IEEE Conf. on Industrial Electronics, IECON, 2009, pp. 173-177.

Pulse Density Modulated Soft-Switching Single-Phase Cycloconverter

Taufik Taufik and Jesse Adamson
Electrical Engineering Department
California Polytechnic State University
San Luis Obispo, California, USA

Anton Satria Prabuwono
Faculty of Information Science and Technology
Universiti Kebangsaan Malaysia
UKM Bangi, Selangor D.E., Malaysia

Abstract-**Single stage cycloconverters generally incorporate hard-switching at turn on and soft-switching at turn off which limit limits their use in lower frequency applications. This paper presents a proposed solution to this problem using a pulse density modulated soft-switching cycloconverter or abbreviated PDMSS cycloconverter. Unlike standard cycloconverters, the controller in PDMSS cycloconverter lets only complete half cycles of the input waveform through to the output. This requires a much greater frequency step down from the input to the output. The analysis and design of the PDMSS cycloconverter will be described using a simulink model whose results demonstrate the ability of PDMSS to produce the sinusoidal waveforms.**

I. INTRODUCTION

Many electrical loads use power electronics. These include cell phones, computer, stereos, televisions and various home appliances. It is common for these devices to use power electronics extensively. For example, different circuitry in various parts of a computer requires different voltage and current levels that must be regulated by advanced power electronics. Cars use power electronics as well in their computer systems, auxiliary equipment and alternators. Beyond cars, expanding in size, power electronics are seen in airplanes, trains and ships. Stationary applications include power plants, sub stations, wind turbines and other centralized and distributed generation.

Power electronics creates and adjusts voltage level in the traditional way using AC generated magnetic fields. The advancement is in techniques to adjust voltage and current levels as well as convert between DC and AC of various frequencies. With power electronics, solid state switches are used to generate AC signals at high frequencies. The higher frequencies are often needed to reduce magnetic component size and improve efficiency. Cycloconverter is one commonly used AC-AC conversion technique that lowers the frequency and may also be used to change the number of phases of electrical power as it is being transferred from one power bus to another.

In this paper, a new modulation technique is applied to conventional single-phase cycloconverter. The technique attempts to improve converter's efficiency by implementing soft-switching. This provides several potential benefits as well as tradeoffs. The design is simulated to explore its strengths, weaknesses and potential for further development.

II. BACKGROUND

The simplest cycloconverter is the single-phase to single-phase topology. This is implemented with four bi-directional switches in an H-bridge configuration as shown in Figure 1.

In this configuration, the switches are operated in pairs. Switching on only S1 & S4 causes the current through the load to flow in forward polarity with AC-IN, while switching on only S2 & S3 causes the current to flow in reverse polarity with AC-IN. This operation allows current to flow either direction through the load, for any given polarity of the AC-IN source. The switches can, therefore, be activated in such a way that the average or filtered output waveform is the desired wave shape (sinusoidal) and frequency.

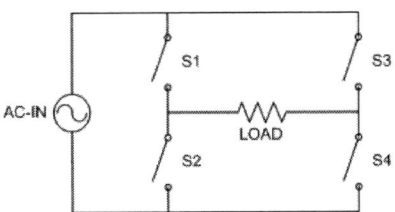

Fig. 1. Single Phase Cycloconverter with Bidirectional Switches

Standard power Field Effect Transistors (FET's), and Insulated Gate Bipolar Junction Transistors (IGBT's) are unipolar; that is, they only block and allow current flow in one direction. For this reason, they are not viable realizations of S1 through S4. Triacs are standard bidirectional switches sometimes used for small scale cycloconverters. They are not currently available for large current applications.

Standard transistors can be arranged to form a bidirectional switch as seen in Figure 2 [1]. This can work for medium applications like electric vehicles and industrial motor drives. Their advantage is the high switching speeds of standard FET's or IGBT's. Their disadvantage is the extra voltage drop created by the diode in series with the switch. This diode is essential for two reasons: first to create a path for current to flow around which ever transistor happens to reverse biased, and second to prevent the transistor from being reverse biased with significant voltage.

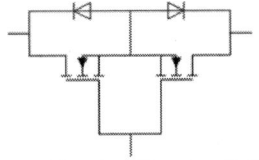

Fig. 2. Bidirectional Transistor Switch

FET's have an inherent diode in the reverse direction making them incapable of reverse blocking. This inherent diode is poor quality, however, so an additional fast diode is usually designed into the silicon substrate as well. If this is

978-1-4577-0007-1/11 $26.00 © 2011 IEEE

the case, then no additional discrete diodes are required in the design [2].

IGBT's, unlike FET's, are capable of reverse blocking, however, their reverse blocking capability is usually a small fraction of their forward blocking capability. For this reason, most power IGBT's are produced with an additional diode as well. The current standard switch for cycloconverters is the thyristor. This is a three terminal device, like standard transistors. Unlike standard transistors, however, thyristors have very high reverse voltage blocking capabilities. Also, unlike transistors, thyristors cannot be turned off. They are semi-controllable switches that are turned on by a pulse of current injected into their gate and pulled out of their cathode. This semi-controllability fits suitably in cycloconverters because the AC input automatically reverse biases the thyristors for half of each cycle.

Like transistors, thyristors are unidirectional. For this reason, thyristor cycloconverters require twice as many switches as a bidirectional switch cycloconverter as depicted in Figure 3.

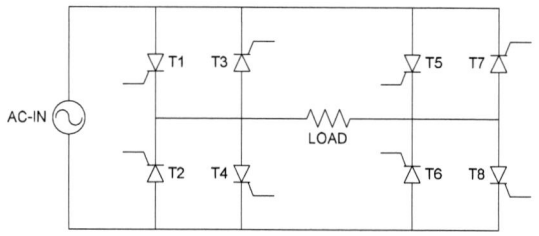

Fig. 3. Single Phase Cycloconverter with Unidirectional Switches

Though the standard single stage cycloconverter has been around for many years and remains in mainstream use in many applications today, there are many possibilities for improvement on this topology. These range from new types of switches to new switching control algorithms. A potential improvement from a control standpoint is soft-switching. Soft-switching is turning a switch on or off with zero voltage or current across the switch. There are four elements of soft-switching, some combination of which must be utilized for a converter to qualify as soft switching. They are zero voltage switching (ZVS) at turn on or turn off and zero current switching (ZCS) at turn on or turn off. In actuality, it is not possible for the current or voltage to be exactly zero for the entire turn-on and turn-off transitions. For this reason, soft-switching cannot completely eliminate, but can only reduce, switching losses in real world applications.

To facilitate soft-switching, the voltage across the switch and/or current through the switch must propagate in such a way as to automatically cross the zero axes so the switch can be turned off softly. In cycloconverters, the AC input provides these zero crossing transitions automatically. In addition, soft turn-off is inevitable in a standard cycloconverter using thyristors. This is because the thyristors only turn off when reverse biased, which happens automatically just after the zero crossing when the current through and voltage across them is very low. Turn on in a standard cycloconverter, however, is controlled by injection of gate current. This is performed by a controller using a reference waveform.

The controller's job is to make the output waveform track the reference by commanding the switches on and off at

proportional intervals. To do this softly requires turning the switch on at the beginning of the forward half cycle and letting the entire pulse through. This is shown in Figure 5. The drawback of letting only full pulses through is the requirement of a much higher input to output frequency ratio and/or more output voltage filtering. This increased input frequency, however, has a much lighter impact on efficiency due to the soft switching. This is especially advantageous considering the slow switching speed of thyristors.

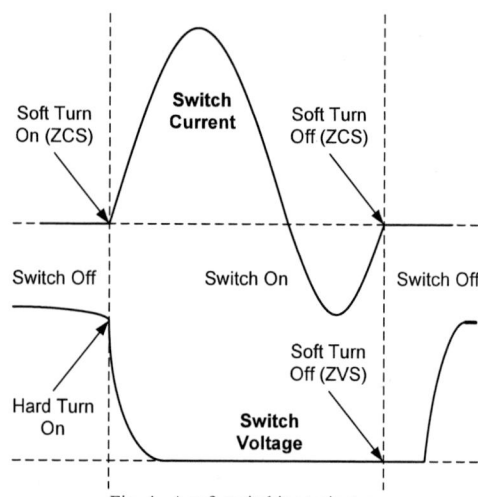

Fig. 4. A soft-switching trajectory

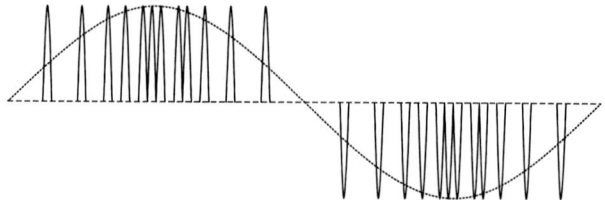

Fig. 5. Monopolar Pulse Density Modulation

This idea of letting pulses through in discrete time to represent an output signal was developed under the formal name of Pulse Density Modulation (PDM). The origins are in analog to digital converters, namely delta sigma converters. In PDM, a digital pulse train is passed through an analog filter to produce a continuous waveform. This method is in contrast to popular modulation techniques in power electronics like Pulse Width Modulation (PWM) and Pulse Frequency Modulation (PFM). These three are compared and contrasted in Table I.

TABLE I
COMPARISON OF MODULATION TECHNIQUES

	PWM	PFM	PDM
On Time	Continuous	Continuous/Discrete	Discrete
Off Time	Continuous	Discrete/Continuous	Discrete
Frequency	Discrete	Continuous	Discrete

The simplest realization of PDM is the single bit delta sigma converter control loop shown in Figure 6 [3]. In the delta sigma, a reference waveform is added to a negative feedback signal and fed into an integrator. The output of the integrator is compared to a threshold. After the reference waveform causes the integrator to rise above the threshold, the converter outputs one discrete pulse. This pulse is negatively fed back to the initial summing junction, reducing the integrator. This regulation produces an output pulse train whose integrated area is linearly proportional to the reference wave's integrated area. This output pulse train is then run through a low pass filter to reconstruct the reference wave. This is an example of bipolar switching. To make this monopolar switching would require two thresholds: one high and one low. It would also require the ability to output two discrete pulses, one positive and one negative, as well a feedback signal from each to the integrators input summing junction.

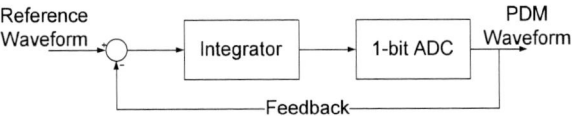

Fig. 6. Delta Sigma A/D converter

The idea of a cycloconverter utilizing soft-switching based on pulse density modulation has been around for several years. For example, PDM for a three-phase cycloconverter was analyzed, modeled, and simulated in [4]. Others such as [5], [6] and [7] made revisions to the three-phase PDM cycloconverter, such as using various resonant converters as the input to the cycloconverter. Additional variant on the soft cycloconverter was reported in [1] which uses an isolated input to the cycloconverter. This topology employs a center tapped transformer and two bidirectional switches, allowing half of the transformer output voltage for the high half cycle and half for the low half cycle. A similar converter using bidirectional switches is described in [8] which targets Solid Oxide Fuel Cells (SOFC's). The use of soft-switching cycloconverter has expanded to applications such as electric vehicle [9], induction heaters [10], and magnetrons [11].

Despite the previous research done into this niche of soft switching cycloconverters, there is certainly more research needed to advance their practical applications. Several applications have been identified where high-frequency input to low-frequency output cycloconverters with high efficiency could thrive. In this paper, in an attempt to further improve cycloconverter technology and improve soft-switching cycloconverter credibility, a full bridge, soft switching, thyristor based, grid frequency, PDM controlled cycloconverter is proposed, simulated and investigated.

III. DESIGN

The soft-switching cycloconverter design can be broken down into several blocks: the load, the filter, the thyristors, the drivers, the controller logic block and the controller analog block as illustrated in Figure 7. The main input power flows through the thyristors. They are triggered by the drivers via control lines from the logic block. The logic block decides which thyristors to turn on by comparing signals in the controller analog block.

Though the power and signals flow from left to right, the design actually starts with the single phase cycloconverter topology (Figure 8) embodied in the Thyristor block. From here the design moves backwards to the drivers needed to fire the thyristors. The controller is then designed to operate the drivers. The controller is broken into analog and digital subsections because the design of these blocks is significantly different. The simulation load will be purely resistive.

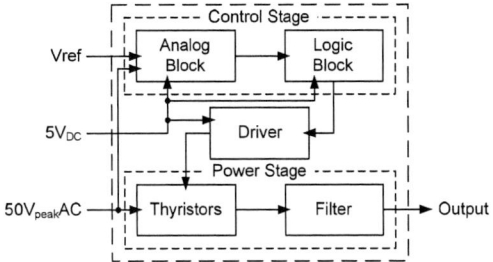

Fig 7. Soft-switching cycloconverter block diagram

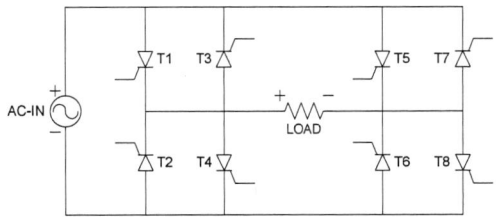

Fig. 8. Single-phase cycloconverter

The thyristors are activated by the driver circuitry in pairs and therefore arranged into four switch groups. Each switch group allows current to flow in one direction through the load from one half cycle of the AC source. Each switch group is activated by a single control signal.

The control signals are named based on the thyristors they drive, referenced from Figure 8. These are shown in the first column of Table II. Each control signal is designed to let one pulse (one half cycle of the input waveform) through to the output. Each signal can only map a signal input polarity to a single output polarity, thereby requiring four control signals to control the four input polarity to output polarity combinations.

TABLE II
CONTROL SIGNALS FROM LOGIC BLOCK TO THYRISTOR DRIVERS

Control Signal	Thyristors	Pulse Mapping
to18	T1 & T8	Positive Input to Positive Output
to27	T2 & T7	Negative Input to Positive Output
to36	T3 & T6	Negative Input to Negative Output
to45	T4 & T5	Positive Input to Negative Output

Input signals include chip power, a reference sine wave and the main high frequency AC power source. These inputs will be compared along with the PDM integrator and transformed into discrete signals that the logic block will use to decide when to fire which thyristors. In order to decide when to fire the thyristors, the Analog Block starts by comparing several values and encoding them in digital signal lines. First the voltage polarity of the high frequency input is needed. This signal is named "resp". It will be high when the high frequency source is in its positive half cycle, or resonating positively. This signal is used to make sure thyristors are reverse biased when their drivers are activated and deactivated so they are ready to turn on as soon as they start to become forward biased. This ensures soft turn on. The drivers can switch off when the thyristors are forward biased because this does not turn off the thyristors. They continue to conduct until reverse biased.

Likewise the reference polarity will be encoded in the "refp" signal. This signal will be positive when the reference is positive. It will be used to ensure that output pulse polarity always matches reference polarity. This is necessary for the converter switching to be classified as monopolar. Additionally, a "CLK" signal will be needed to determine exactly when to turn the thyristors on and off in their respective half cycles. The clock should transition somewhere in the middle of each half cycle. This signal, along with "resp" and "refp" hold all the required information to line up the output pulses.

The only needed information remaining is when to emit a pulse. This is where the PDM integrator comes in and requires two signals because the switching scheme is monopolar. The "intGreater" signal is triggered high when the integrators output voltage rises above the high threshold, therefore requiring a positive output pulse that is negatively fed back to the integrator to bring it back down. Likewise, a high "intLess" signal signifies the integrator output voltage dropped below the low threshold requiring a negative output pulse that is negatively fed back to bring it back up. These signals are summarized in Table III.

TABLE III
SIGNALS FROM ANALOG BLOCK TO LOGIC BLOCK

resp	Positive for main input voltages positive half cycle
refp	Positive for reference voltages positive half cycle
CLK	Rising edge around the middle of each main input half cycle
intGreater	Positive when a positive output pulse is needed
intLess	Positive when a negative output pulse is needed

The Logic Block receives these five signals and outputs six signals. Four of the outputs are the driver control signals listed in Table II and two signals are feedbacks to the Analog Block. The two feedback signals are shown in Table IV as OR gated combinations of the control signals.

TABLE IV
FEEDBACK SIGNALS FROM LOGIC BLOCK TO ANALOG BLOCK

lastPulseP	to18 OR to27
lastPulseN	to36 OR to45

The logic block can now be determined. Starting with the first row of Table II it is seen that signal to18 maps a positive input pulse to a positive output pulse. This should happen when the input is in its negative half cycle (resp is low), the reference is in its positive half cycle (refp is high) and a positive pulse is required (intLess is high). The other signals are found similarly:

$$to18 = \overline{resp} \cap refp \cap \text{int} Greater \qquad (1)$$

$$to27 = resp \cap refp \cap \text{int} Greater \qquad (2)$$

$$to36 = resp \cap \overline{refp} \cap \text{int} Less \qquad (3)$$

$$to45 = \overline{resp} \cap \overline{refp} \cap \text{int} Less \qquad (4)$$

The standard output filter for inverters is and LC low pass. This is configured with the inductor in series with the load, blocking high frequencies, and the capacitor parallel to the load, shorting high frequencies to ground. This presents a problem with the cycloconverter. However, because when all the switches are off, there is no path for the inductor to discharge. For this reason, the L and C must be swapped to form a CL filter. This allows inductor to pull current from the capacitor when the thyristors are off.

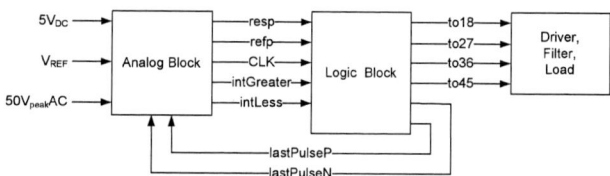

Fig. 9. Soft-switching cycloconverter block and signal diagram

The converter is now ready for simulation. There are several details that will come into play at different stages of simulation. The basic implementation has been split into concise blocks with connecting signals. The complete layout is organized and fits together as shown in Figure 9.

IV. SIMULATION

From control standpoint, the proposed soft-switching cycloconverter can be broken down into three blocks as previously shown in Figure 9. The first two blocks are the analog and logic control sections, while the last block combines the rest of the converter. In order to get a better understanding of the control system and to verify the control techniques described in earlier, a Simulink model was developed as shown in Figure 10. The simulation starts with the "Vref" block in the center left hand side, where the 60 Hz reference sine wave is generated. This goes through an amplifier to scale the wave, thereby adjusting the output average pulse density and filtered peak voltage.

The reference is then split and fed into a comparator to determine the "refp" signal, as well as into the integrator summing junction. The integrator summing junction also adds in the feedback signal "lastPulseN" and subtracts the feedback signal "lastPulseP" to form the input to the "discreteIntegrator". Using these signals the integrator output tracks the build-up of discrepancy between the average output and the reference input. This output is monitored by two comparators, one for the high threshold that triggers a high output pulse and one for the low threshold that triggers a low output pulse.

At the top left the input source "AC V Src" is measured and compared by "COMP2" to generate the "resp" signal. The same signal is also full bridge rectified using an absolute value block "FBRect" and compared to the value 45. This generates the clock signal and concludes the analog block from the block diagram in Figure 9.

The analog block signals are measured by the virtual oscilloscope "OS1". The signals now enter the logic block. In this block they are combined using the "NOT" and

"AND" blocks as described by equations 1 through 4. After this the signals are fed into D flip flops and outputted to the drivers. This concludes the logic block. The driver signals "to18" and "to27" represent the positive half cycle of the

output waveform. Adding (logical OR) these two waveforms together produces "lastPulseP", and likewise "to36" and "to54" added together produce "lastPulseN".

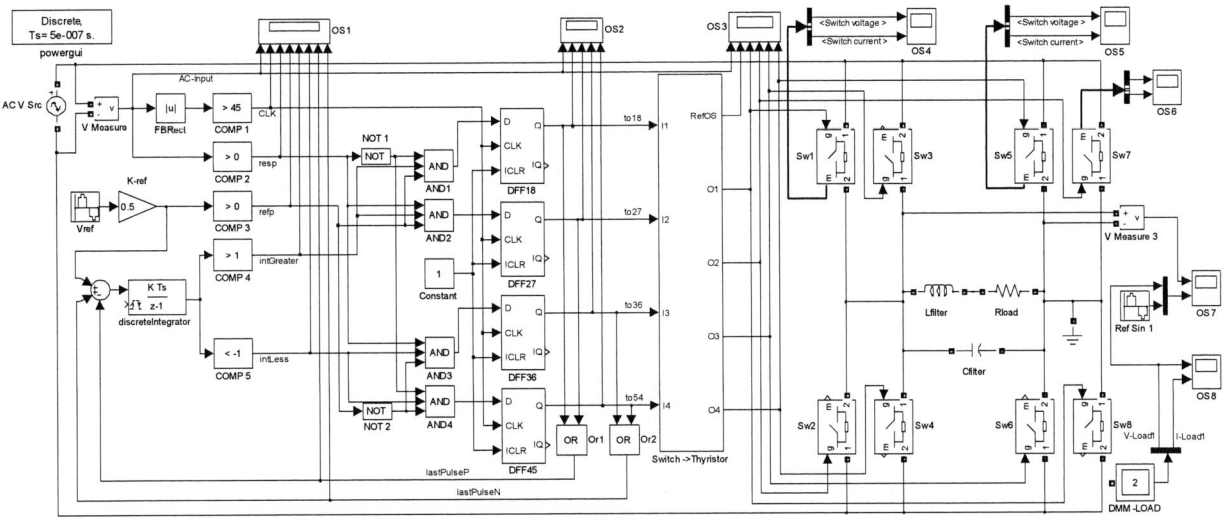

Fig 10. Simulink model

The next block in the Simulink model is named "Switch->Thyristor". This is a subsystem consisting of four D flip flops and a clock signal which, for proper operation, must be set to twice the frequency of the "AC V Src". This block makes up for a deficiency in Simulink switch models. This subsystem and its output lines are not present in the physical design. The issue is that Simulink switch models don't behave like PSpice switch models or actual switches. Their gates are not modeled from a circuit standpoint, with voltages and currents, but rather from a digital control standpoint with either a high or low (on or off) value.

There are several switches in the Simulink Power Library ("powerlib"); however, they all seem to simply be a resistor if their gate is high and an open circuit if their gate is low. The only difference between these switches is the addition of parallel diodes and snubber circuitry. They are all essentially bidirectional fully controllable switches. To account for this, the "Switch->Thyristor" block adjusts the gate signals so the thyristor is only on when forward biased.

The thyristor driver block is not incorporated into the Simulink model because, as previously mentioned, the switch used is controlled by Simulink signals and not gated at the circuit level using voltage and current as in traditional circuit analysis software. The eight thyristors are labeled in the simulink model as "Sw1" through "Sw8". The switch voltage drop and current are observed by scopes "OS4", "OS5" and "OS6", however, these don't contain clearly decipherable trajectories and so were not included in this paper.

The unfiltered output from scope "OS7" is seen in Figure 11. This waveform shows the monopolar pulse density modulation of 50V-peak pulses. It is noted that these pulses, or half cycles of the high frequency input, are staggered so their density varies sinusoidally. To convert this to a clean sinusoid, however, the high frequency components must be filtered out. The input and output voltage zero crossing

transition is shown in Fig. 12 for both converter input and referenced output. Notice how some positive pulses in "V-Load" come from positive pulses in "AC-Input", while other positive pulses in "V-Load" come from negative pulses in "AC-Input" and likewise with negative pulses in "V-Load".

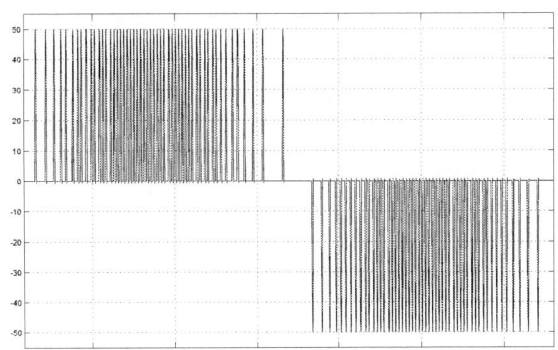

Fig. 11. Simulink Unfiltered Output "OS7" with R = 3.5 Ω

One filtered output waveform is shown in Figure 13. This is distorted due to the capacitor being too small and unable to supply the inductor with enough current when the thyristors are off. To compensate for this, the capacitor was increased by a factor of 3, while the inductor was decreased by a factor of 3. The effect, as seen in Figure 14, is less distortion, but more ripple due to a smaller inductor. The waveform is also slightly higher in amplitude.

Both filtered waveforms above are fairly rough looking. Smoothing them out requires increasing both the inductor and capacitor, which will in turn lower the cutoff frequency of this low pass filter. Increasing both values results in the output waveform in Figure 15. This larger filter produces a significantly smoother waveform; however, it cuts into the fundamental amplitude, reducing it by almost one third.

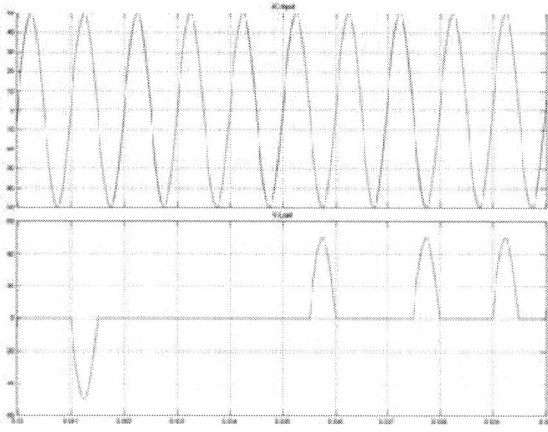

Fig. 12. AC input and output voltage zero crossing transitions

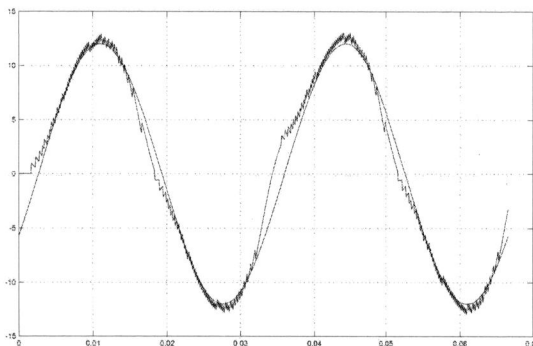

Fig. 13. Simulink Filtered V_{OUT} with 11.1 mH, 400 µF and 3.5 Ω

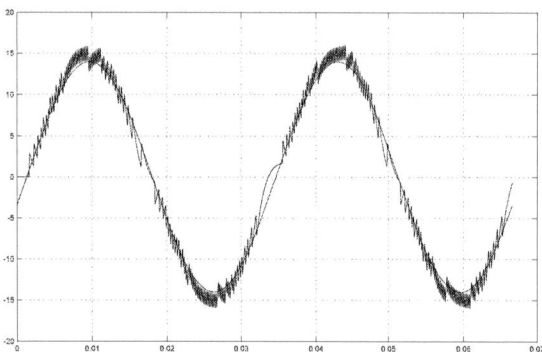

Fig. 14. Simulink Filtered V_{OUT} with 3.7 mH, 1200 µF and 3.5 Ω

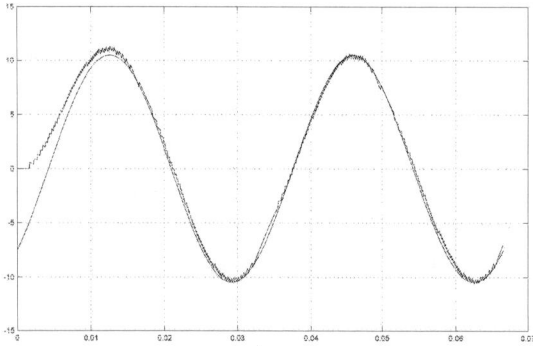

Fig. 15. Simulink Filtered V_{OUT} with 20 mH, 2000 µF and 3.5 Ω

V. CONCLUSION

A new pulse density modulation technique with the provision of soft-switching transition for single-phase cycloconverter has been proposed and described in this paper. The design and simulink simulation have been performed whose results demonstrates that the cycloconverter indeed possesses the soft-switching property. Simulation results also produced the expected sinusoidal waveform which demonstrates the functionality of the cycloconverter. They further prove that the soft-switching technique could be performed without compromising the quality of sinusoidal output. Currently, hardware prototype is being developed and later tested. Results from the hardware measurements will be compared to those obtained from simulations and will be reported in future conferences.

ACKNOWLEDGMENT

The authors would like to thanks Faculty of Information Science and Technology, Universiti Kebangsaan Malaysia for financial support.

REFERENCES

[1] S. H. Hosseini, M. Sabahi, A. Y. Goharrizi, "Multi-function zero-voltage and zero-current switching phase shift modulation converter using a cycloconverter with bi-directional switches", *Power Electronics IET* , vol. 1, issue 2, pp. 275-286, 2008.

[2] W.K. Chen, *Linear Networks and Systems*, Brooks/Cole Engineering Division, Belmont, CA

[3] R. C. Jaeger and T. N. Blalock, *Microelectronic Circuit Design*, 2nd Edition, McGraw-Hill Press, 2004.

[4] M. Xianmin and Y. Qian, "Study and Analysis of Three-phase High-frequency AC Cycloconverter Based on Pulse Density Modulation", *Proceedings of Electrical Machines and Systems Sixth International Conference*, vol.1, pp. 415-418, 2003.

[5] H. Yonemori and M. Nakaoka, "Advanced Soft-Switching Sinewave PWM High-Frequency Inverter-Link Cycloconverter Incorporating Voltage-Clamped Quasi-Resonant and Capacitive Snubber Techniques", *Proceedings of Industry Applications Society Annual Meeting*, pp. 795-802, 1991.

[6] H. Yonemori, K. Muneto, M. Nakaoka, and T. Maruhashi, "Modern High-Frequency Inverter Linked Cycloconverter Type Sinewave Power Processing System Using Multi-Resonant Soft-Switching PWM Strategy", *Proceedings of Telecommunications Energy Conference*, pp. 566-573, 1991.

[7] H. Yonemori, A. Chibani, and M. Nakaoka, "New Soft-Switching Phase-Shifted PWM High-Frequency Inverter-Linked Cycloconverter Incorporating Voltage-Clamped Quasi-Resonant Technique", *Proceedings of Power Electronics Specialists Conference*, pp. 283-290, 1991

[8] R. K. Burra, R. Huang, and S. K. Mazumder, "A Low-Cost Fuel-Cell (FC) Power Electronic System (PES) for Residential Loads", *Proceedings of Telecommunications Energy Conference*, pp. 468-473, 2004.

[9] F. Gustin, and A. Berthon, "Simulation of a multi motor soft switching converter for electric vehicle applications", *Proceedings of Power Electronics and Motion Control Conference*, vol. 2, pp. 660-664, 2000.

[10] H. Sugimura, S. Mun, S. Kwon, T. Mishima, and M. Nakaoka "Direct AC-AC Resonant Converter using One-Chip Reverse Blocking IGBT-Based Bidirectional Switches for HF Induction Heaters", *Proceedings of Industrial Electronics*, pp. 406-412, 2008.

[11] H. Sugimura, B. Saha, S. Mun, E. Hiraki, H. Omori, and M. Nakaoka, "Direct High Frequency Soft Switching PWM Cyclo-Converter-Fed AC-DC Converter without DC Link for Consumer Magnetron Drive", *Proceedings of International Conference on Power Electronics*, pp. 1185-1190, 2007.

978-1-4577-0007-1/11 $26.00 © 2011 IEEE

A Novel Soft Switching Bidirectional Coupled Inductor Buck-Boost Converter for Battery Discharging-Charging

A. Mirzaei[*], A. Jusoh[*], Z. Salam[*], E. Adib[**], H. Farzanehfard[**]

[*]Dept. of Electrical Eng., Universiti Teknologi Malaysia, Johor Bahru, Malaysia
[**]Dept of Electrical and computer Eng., Isfahan University of Technology, Isfahan, Iran

Abstract- **This paper proposes a novel zero voltage transition (ZVT) auxiliary circuit applied to a bidirectional converter for interface circuit between ultracapacitors or batteries with DC bus in electric vehicle (EV), fuel cell electric vehicle (FCEV) and hybrid electric vehicle (HEV). This auxiliary circuit provides almost soft switching condition for all switching elements while the control circuit remains PWM. So, the energy conversion through the converter is highly efficient. The proposed converter acts as a ZVT Buck to charge ultracapacitor or battery. On the other hand, it acts as a ZVT Boost to discharge ultracapacitor or battery. Buck operation of the proposed bidirectional converter is analyzed and its operating modes are discussed. The simulation has been done using PSPICE software and its results verify aforementioned capabilities of this converter.**

*Keywords—*Battery; DC-DC converter; Soft Switching; Ultracapacitor (UC)

I. INTRODUCTION

Bidirectional DC–DC converters allow the transfer of power between two DC sources in either direction. They are increasingly used in applications such as DC uninterruptible power supplies [1-2], battery chargers [3], multiplexed-battery systems [4], computer systems, aerospace systems [5-6], DC motor drives circuits and electric vehicles [7-8].

In electric vehicle the fuel cell (FC) is the main power source. However, it suffers from low power density, reduced efficiency at light load demand, low power transfer rate in transitory situations and high cost per Watt. It is clear that some other source should be used to allow faster power delivery to the traction system. Consequently, it is not necessary sizing the FC to the peak power demand, but only to the average power. In addition, this procedure increases the FC lifetime.

For batteries and fuel cells, the chemical to electrical energy conversion depends on a slow electrochemical process. Thus, a usual solution employs also an ultracapacitor (UC) bank to improve the time response of the supply system under sudden load disturbances.

Energy storage devices (battery and UC) are used together also to save more energy during regenerative braking phases [9-10].

The major challenge in designing EV converter is converting the electrical output from the storage system to DC bus and vice versa. In addition, the DC-DC converter must be implemented cost effectively with suitable weight and volume. Meanwhile, the high efficiency, reliability and simple control techniques are three key items which strongly important in EV converters. Because of the limited efficiency of hard-switching converters, soft-switching techniques are gaining popularity. Many soft-switched converter topologies are described elsewhere [11-14].

Zero-voltage transition (ZVT) and zero-current transition (ZCT) converters are new family of soft switched converter. In these converters, an auxiliary circuit containing resonant elements and an auxiliary switch is used that provide soft switching at switching instances and is usually incapable of transferring energy from an input source to output [15-20]

In this paper, a high efficiency bidirectional soft switching converter is developed that can be used as an interface circuit between UC or battery and DC bus in EV. The proposed converter acts as a ZVT Buck to charge UC or battery. On the other hand, it acts as a ZVT Boost to discharge UC or battery. Meanwhile, the control circuit is simple and remains PWM. Moreover, it has fast dynamic response.

The proposed auxiliary circuit consists of two unidirectional switches, one diode, three inductors which are coupled and two resonant capacitors.

This paper is organized as follow. In the second section, description and operation of the circuit in Buck operation is discussed. In section three simulation results and efficiency curve are shown. Conclusions drawn from this paper are presented in the last section.

II. DESCRIPTION AND OPERATING PRINCIPLE OF THE CIRCUIT

The proposed interface circuit is shown in Fig. 1. This converter operates in two modes of operation: (1) it acts as a ZVT Buck to charge UC or battery (2) it acts as a ZVT Boost to discharge UC or battery. During Buck operation mode, S_1 and D_2 are on while During the Boost mode, S_2 and D_1 are on. In addition, the converter is comprised by auxiliary switches (S_{a1}, S_{a2}), coupled inductors (L_{r1}/ L_{r2}/L_s), one diode (D_{a3}) and two resonant capacitors (C_{r1} and C_{r2}).

The auxiliary circuit provides the soft switching condition for S_1 and S_2. In order to simplify the theoretical analysis, it is assumed that all components are ideal. Meanwhile, inductor L_f is large enough to assume its current is constant in a switching

978-1-4577-0007-1/11 $26.00 © 2011 IEEE

cycle. Also, the leakage inductance of coupled inductors is very small and can be ignored. The converter operation is analyzed in buck mode. Since soft switching is achieved similarly in both buck and boost modes, the detail description of boost operating intervals is omitted.

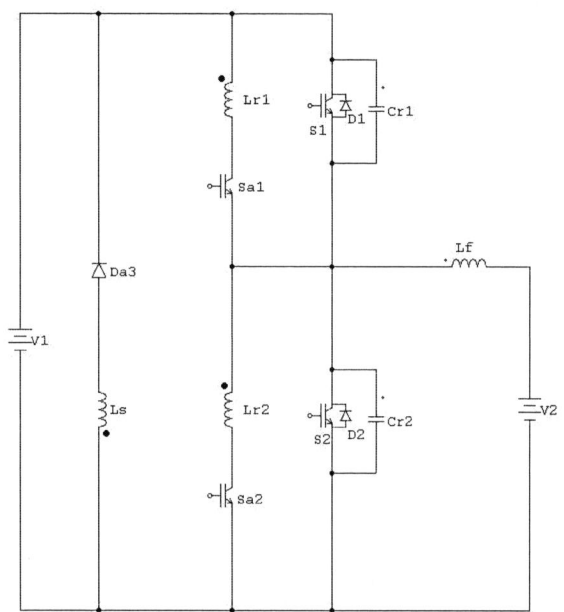

Fig.1. The proposed bidirectional converter

The proposed converter in Buck operation is shown in Fig. 2. The converter theoretical waveforms are shown in Fig. 3. During one switching cycle, the proposed converter has seven operating intervals (or modes). Equivalent circuit for each operating interval is shown in Fig. 4

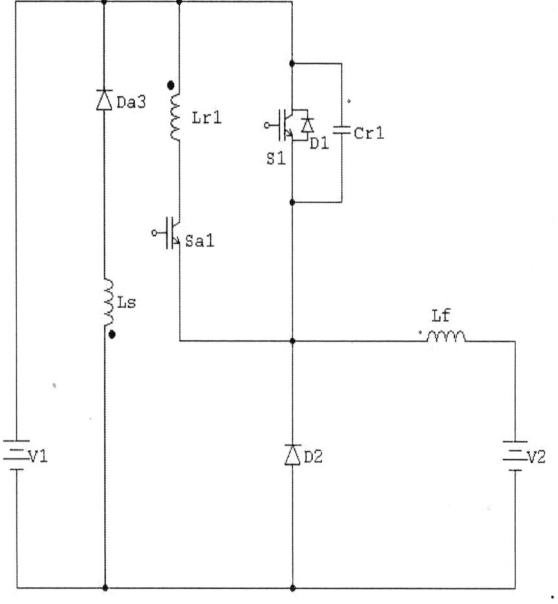

Fig.2. The proposed converter in Buck operation

Interval 1 [t_0-t_1]: Before t_0, D_2 is on and output current flows through it. Furthermore at t_0, S_{a1} is softly turned on under ZCS due to series resonant inductor L_{r1}. The input voltage (V_1) is place across L_{r1} and its current increases linearly. This interval ends when L_{r1} current reaches I_o and D_2 turns off under ZC condition. Based on the equivalent circuit in Fig. 4a, the duration of interval 1 can be written as:

$$t_1 - t_0 = \frac{L_{r_1} I_o}{V_1} \qquad (1)$$

The L_{r1} current during this interval can be described as:

$$I_{Lr_1} = \frac{V_1}{L_{r_1}} t \qquad (2)$$

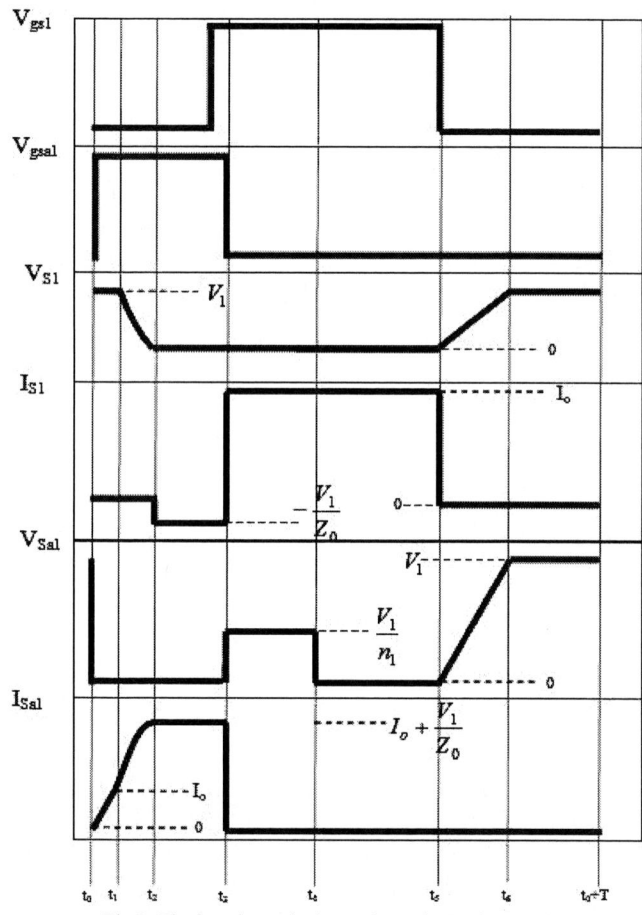

Fig.3. The key theoretical waveforms in Buck operation

Interval 2 [t_1-t_2]: At t_1, a resonance starts between C_{r1} and L_{r1}. Thus C_{r1} discharges and L_{r1} current increases in a resonance manner. At t_2, L_{r1} current reaches to maximum and C_{r1} is fully discharged. Based on the equivalent circuit in Fig. 4b, the L_{r1} current and C_{r1} voltage can be respectively described as:

978-1-4577-0007-1/11 $26.00 © 2011 IEEE   196

$$I_{Lr_1} = I_o + \frac{V_1}{Z_0} \sin(\omega_0(t - t_1)) \tag{3}$$

$$V_{cr_1}(t) = V_1 \cos(\omega_0(t - t_1)) \tag{4}$$

$$Z_0 = \sqrt{\frac{L_{r_1}}{C_{r_1}}} \tag{5}$$

$$\omega_0 = \frac{1}{\sqrt{L_{r_1} C_{r_1}}} \tag{6}$$

Interval 3 [t_2-t_3]: If the resonance continues, C_{r1} voltage will become negative. Thus, D_1 conducts and clamps S_1 voltage to zero and S_1 can be turned on under ZVS condition. Since the voltage across L_{r1} during this interval is zero its current remains constant. Based on the equivalent circuit in Fig. 4c, the L_{r1} and S_1 currents can be respectively described as:

$$I_{Lr_1} = I_o + \frac{V_1}{Z_0} \tag{7}$$

$$I_{S_1} = I_o - I_{Lr_1} = -\frac{V_1}{Z_0} \tag{8}$$

$$Z_0 = \sqrt{\frac{L_{r_1}}{C_{r_1}}} \tag{9}$$

Interval 4 [t_3-t_4]: At t_3, S_{a1} almost turns off ZVS. In this condition, L_{r1} current drops gradually to zero and current flows through L_s and D_{a3}. Then, the secondary voltage of the Flyback transformer is clamped to V_1. The secondary voltage is reflected to the primary and the primary voltage equals V_1/n_1, where V_1 is the input voltage and n_1 is the turn ratio of the Flyback transformer. When S_{a1} turns off, the voltage across this switch is V_1/n_1. By selecting large values for n_1 (i.e., $n_1 > 5$), the voltage across S_{a1} is very small and the switch turns off almost ZVS. At t_3, the energy entirely transfers from L_{r1} to L_s and L_s current rises to maximum. Based on the equivalent circuit in Fig. 4d, the L_s current can be described as:

$$I_{Ls} = \frac{I_o}{n_1} + \frac{V_1}{n_1 Z_0} - \frac{V_1}{L_{S_1}}(t - t_3) \tag{10}$$

This interval ends when L_s current reaches zero. Also in this interval energy is transferred from UC or battery to DC bus.

Interval 5 [t_4-t_5]: In this interval, the converter operates like a regular Buck converter. So, the energy is transferred from battery or UC to DC bus through S_1 and all other semiconductor devices are off, Fig. 4e.

Interval 6 [t_5-t_6]: in this interval, S_1 is turned off and L_f current charges C_{r1}. Since C_{r1} voltage changes linearly from

zero to V_1, S_1 is turned off under ZV condition. Based on the equivalent circuit in Fig. 4f, the duration of interval 6 and C_{r1} voltage can be written as:

$$t_6 - t_5 = \frac{C_{r_1} V_1}{I_o} \tag{11}$$

$$V_{C_{r1}} = \frac{I_o}{C_{r_1}}(t - t_6) \tag{12}$$

Interval 7 [t_6-t_0+T]: when C_{r1} voltage reaches V_1, D_2 starts to conduct and the energy flows from L_f to DC bus, Fig. 4g.

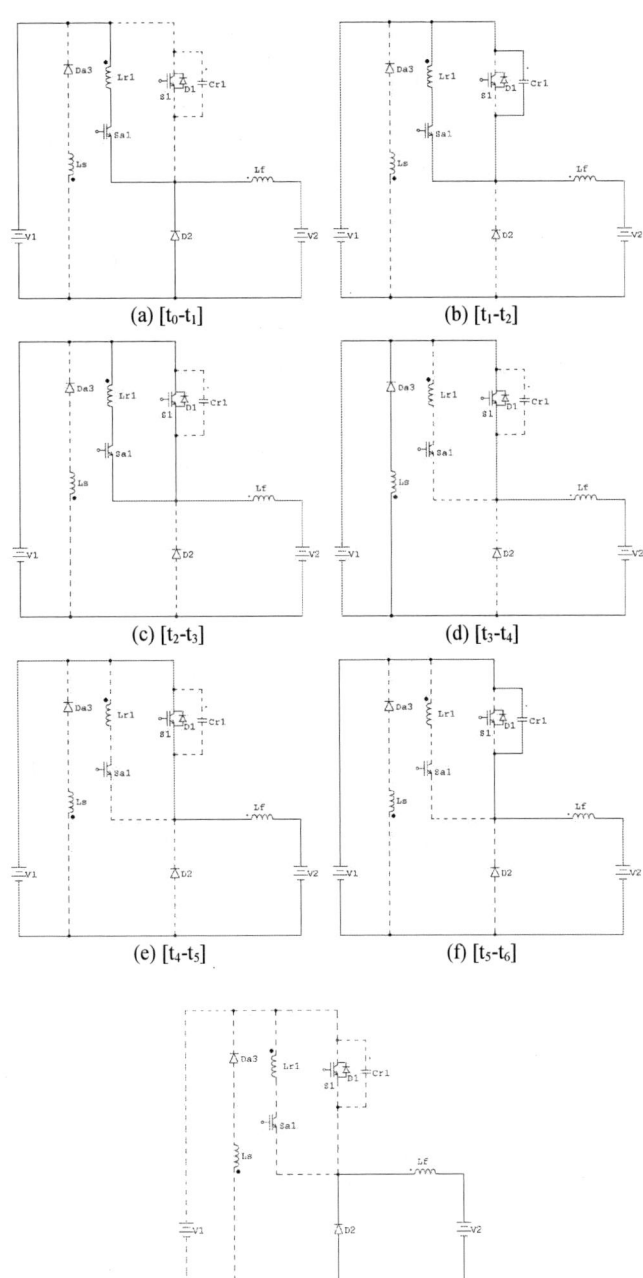

Fig.4. Equivalent circuits for each stage during Buck operation

III. SIMULATION RESULTS

To verify the above analysis, the proposed ZVT Buck converter is simulated using PSPICE software. The parameters are shown in table I. The model is shown in Fig.5.

Table I. The parameters of components

component	parameter
Input voltage	120V
Output voltage	48V
Switching frequency	100kHz
power	240W
Filter inductor	316.8µH
Filter capacitor	3.125µF
Switches	IRF540
Resonant inductors (L_{r1}/L_s) / Coupling (k)	2.4µH/86.4µH/0.995
Resonant capacitor (C_{r1})	2nF

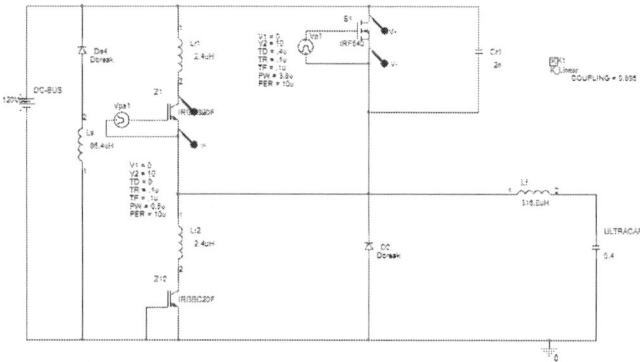

Fig.5. Simulation model of the ZVT Buck converter

The simulation results are shown in Fig.6.

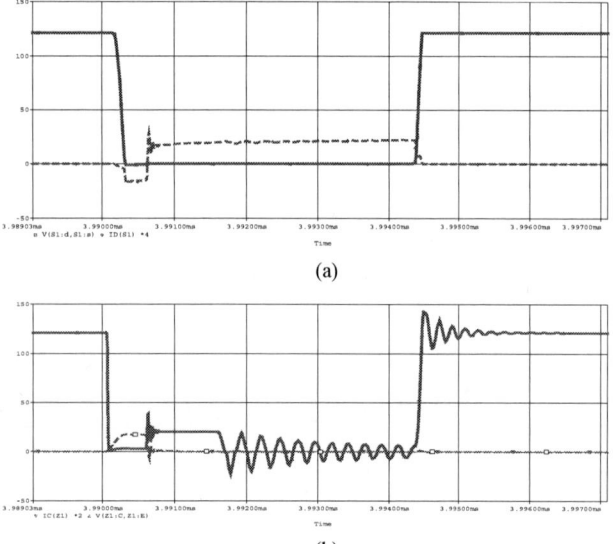

(a)

(b)

Fig.6. Simulation results for the proposed ZVT Buck converter. (a) The voltage and current of main switch (S1); (b) The voltage and current of auxiliary switch (Sa1)

Fig.6 (a) shows the voltage across S_1 and the current through it in Buck operation. The ZVS condition for the main switch is obviously illustrated. Also, Fig.6 (b) shows the voltage across S_{a1} and the current through it in Buck operation. The achieved ZC condition for the auxiliary switch turn-on instant and almost ZV condition for its turn-off instant are also apparent in this figure.

The efficiency curves for proposed ZVT Buck and conventional Buck converter in nominal power and lower power are shown in Fig.7.

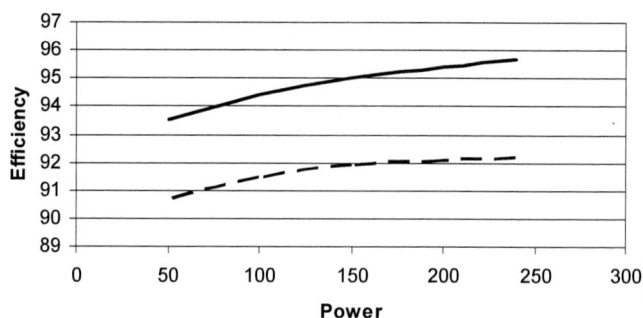

Fig.7. Converter efficiency curve (continuous line) versus hard switching counterpart (broken line) in Buck operation

The nominal power of the proposed ZVT Buck converter is 240 Watt. So; in this power the efficiency is maximized. But, when the power is less than nominal power (because loss of the auxiliary circuit is increased than main circuit) the efficiency is decreased as same as all soft switching converters. Fig.7 shows that efficiency of the proposed ZVT Buck converter at nominal power is higher than 95%.

IV. COCLUSION

This paper proposes a novel zero voltage transition (ZVT) auxiliary circuit applied to a bidirectional converter for interface circuit between ultracapacitors or batteries with DC bus in electric vehicle (EV), fuel cell electric vehicle (FCEV) and hybrid electric vehicle (HEV). This auxiliary circuit provides almost soft switching condition for all switching elements while the control circuit remains PWM. So, the energy conversion through the converter is highly efficient. The proposed converter acts as a ZVT Buck to charge ultracapacitor or battery. On the other hand, it acts as a ZVT Boost to discharge ultracapacitor or battery. The bidirectional converter has less switching losses with soft switching technique and has high efficiency. The measured efficiency of the proposed converter in Buck mode is more than 95% at full load. In case of light load, the efficiency decreases due to the fact that the circulating energy involved in the resonant process is constant and independent of load. The comparison of the proposed converter for ultracapacitor or battery interface circuit with other interface circuits exhibits its superior performance. The converter is analyzed in buck mode and its different operating modes are presented. The simulation results justify the theoretical analysis.

ACKNOWLEDGEMENT

The authors wish to acknowledge the Universiti Teknologi Malaysia for financial support under Fundamental Research Grant Scheme (FRGS) (Vot 78588).

REFERENCES

[1] K. Venkatesan, "Current mode controlled bidirectional flyback converter," in *Proc. IEEE PESC'89*, 1989, pp. 835–842.

[2] M. Jain, P. K. Jain, and M. Daniele, "Analysis of a bi-directional DC-DC converter topology for low power application," in *Proc. CCECE'97*, 1997, pp. 548–551.

[3] D. M. Sable, F. C. Lee, and B. H. Cho, "A zero-voltage- switching bidirectional battery charger/discharger for the NASA EOS satellite," in *Proc. IEEE APEC'92*, 1992, pp. 614–620.

[4] R. K. Williams and W. Grabowski, "Single package 30-V battery disconnect switch facilitates battery multiplexing in notebook computers," in *Proc. IEEE APEC'98*, 1998, pp. 691–699.

[5] B. Ray, "Single-cycle resonant bidirectional DC/DC power conversion," in *Proc. IEEE APEC'93*, 1993, pp. 44–50.

[6] Z. R. Martinez and B. Ray, "Bidirectional DC/DC power conversion using constant frequency multi-resonant topology," in *Proc. IEEE APEC'94*, 1994, pp. 991–997.

[7] F. Caricchi, F. Crescimbini, F. G. Capponi, and L. Solero, "Study of bi-directional buck-boost converter topologies for application in electrical vehicle motor drives," in *Proc. IEEE APEC'98*, 1998, pp. 287–293.

[8] F. Caricchi, F. Crescimbini, and A. Di Napoli, "20 kW water-cooled prototype of a buck-boost bidirectional DC-DC converter topology for electrical vehicle

[9] P. M. Hunter and A. H. Anbuky, "VRLA battery rapid charging under stress management," *IEEE Trans. Ind. Electron.*, vol. 50, no. 6, pp. 1229–1237, Dec. 2003.

[10] I. Buchmann, "Learning the basics about batteries," Battery University. com, 2003 [Online]. Available: www.batteryuniversity.com

[11] N. Liyong, J. Jiuchun and Z. Weige, "Study on Optimum Design Procedure of Charger Based on Full-Bridge Phase-Shifted ZVZCS Converter," *IEEE Vehicle Power and Propulsion Conference (VPPC)*, Harbin, China, pp. 1–4, September 2008.

[12] H. Tao, J. L. Duarte and M. A. M. Hendrix, "Three-Port Triple-Half-Bridge Bidirectional Converter With Zero-Voltage Switching," *IEEE Transaction on Power Electronic*, vol. 23, pp. 782–792, March 2008.

[13] H. Cha, J. Choi, W. Kim and V. Blasko, "A New Bi-directional Three-phase Interleaved Isolated Converter with Active Clamp ," *Applied Power Electronics Conference and Exposition, APEC 2009*, pp. 1766–1772, February 2009.

[14] G. Calderon-Lopez, A. J. ForsythCha, "High-Power Dual-Interleaved ZVS Boost Converter with Interphase Transformer for Electric Vehicles ," *Applied Power Electronics Conference and Exposition, APEC 2009*, pp. 1078–1083, February 2009.

[15] M. L. Martins, J. L. Russi, J. R. Pinheiro, H. A. Grundling and H. L. Hey, "Unified Design for ZVT PWM Converters with Resonant Auxiliary Circuit," *Proc. IEE-Electr. Power Appl*, vol. 151, no. 3, pp. 303-312, March 2004.

[16] P. J. M. Menegaz, J. L. F. Vieira, and D. S. L. Simonetti, "A ZVT DC-DC Self Resonant Boost Converter with Improved Features," *in Proc. IEEE Power Electronics Specialists Conf. (PESC), Aachen, Germany*, pp. 1631-1654, 2004.

[17] N. Jain, P. K. Jain, and G. Joos, "A Zero Voltage Transition Boost Converter Employing A Soft Switching Auxiliary Circuit with Reduced Conduction Losses," *IEEE Trans. Power Electron*, vol. 19, no. 1, pp. 130-139, January 2004.

[18] E. Adib and H. Farzanehfard, "Family of zero-current transition PWM converters," *IEEE Trans. Ind. Electron.*, 2008, vol. 55, no. 8, pp. 3055–3063.

[19] E. Adib and H. Farzanehfard, "Family of zero current zero voltage transition PWM converters," *Inst. Eng. Technol. Power Electron.*, 2008, vol. 1, no. 2, pp. 214 –223.

[20] A.K. Panda, H. N. Pratihari, B. P. Panigrahi and L. Moharana, "A Zero Voltage Transition Synchronous Buck Converter with an Active Auxiliary Circuit," *DSP Journal*, 2009, vol. 9, Issue 1, pp.41-49.

AUTHOR INDEX

Adamson, J. 189
Adib, E. 195
Akmeliawati, R. 92
Al-Mashhadany, Y. 110
Anwari, M. 16, 57, 69
Awan, A. 28
Ayob, S. 180
Aziz, J. 104
Azli, N. 46, 116, 180
Baharom, R. 34
Bahrainian, F. 1
Baiju, M. 51
Bakar, M. 86
Behrouzian, E. 1
Bina, M. 155
Cherati, S. 138, 143, 180
Dankoob, A. 185
Das, B. 40
Daud, M. 122
Duerbaum, T. 98
Farzanehfard, H. 195
Fujita, G. 170
Funabashi, T. 170
Gholamian, S. 138
Gobbi, R. 132
Hamide, M. 16
Hamzah, M. 34
Hamzah, N. 34
Hassan, M. 165
Ibrahim, Z. 75
Idris, A. 34
Idris, N. 69, 116, 122
Ishaque, K. 5, 10
Jacob, B. 51
Jamali, M. 138
Jamaludin, I. 75, 159
Jidin, A. 122
Jusoh, A. 16, 195
Kashani, M. 185
Kaykhosravi, A. 46
Khamis, A. 159
Khosravi, F. 46
Kuebrich, D. 98
Kukandeh, Y. 185
Lada, M. 75, 159
Lazi, J. 75, 159
Lesan, S. 143
Liu, Y. 40
Mahmoodi, S. 116
Majid, M. 165
Malik, M. 28
McCarty, M. 57
Mirzaei, A. 195
Mirzaie, M. 138
Mohammed, B. 63

Mohindo, O. 159
Mortezaei, A...116, 180
Muhammad, N. 81
Nagarajan, R. 86
Najafi, E. 176
Nandan, J. 132
Nejad, M. 128
Ngan, M. 22
Nomura, M. 170
Nordin, N. 116
Prabuwono, A. 57, 189
Rahman, H. 165
Rahmani, R. 165
Rao, K. 63
Rizqiawan, A. 170
Saad, A. 86
Sadati, S. 149
Sahid, M. 81
Salam, Z. 5, 10, 195
Samadaei, E. 143, 149
Shamsudin, A. 5, 10
Sheikholeslami, A. 155, 185
Stahl, J. 98
Sulaiman, M. 75
Sutikno, T. 122
Tabesh, A. 1
Taghipour, M. 128
Taheri, H. 5, 10
Tan, C. 22
Taufik, T. 57, 189
Tayyebi, M. 165
Vahedi, H. 155, 185
Watson, N. 40
Wong, J. 69
Yakub, M. 92
Yao, L. 104
Yatim, A. 81, 176
Yazdani-Asramil, M. 149
Zamani, A. 1

9781457700071